FOODBORNE DISEASES

Second Edition

Food Science and Technology
International Series

A complete list of books in this series appears at the end of this volume.

Foodborne Diseases

Second Edition

Edited by

Dean O. Cliver

School of Veterinary Medicine
Department of Population Health and Reproduction
University of California
Davis, California

Hans P. Riemann

Haven Laboratory
Banning, California

ACADEMIC PRESS

An imprint of Elsevier Science

Amsterdam · Boston · London · New York · Oxford
Paris · San Diego · San Francisco · Singapore · Sydney · Tokyo

Copyright © 2002 by ELSEVIER SCIENCE LTD
except Chapter 3 "*Salmonella*" by Gray and Fedorka-Cray, Chapter 4 "*Shigella*" by Lampel and Maurelli, and
Chapter 6 "*Campylobacter jejuni* and Related Organisms" by Altekruse and Swerdlow.

First edition published 1990.

Academic Press
An imprint of Elsevier Science
84 Theobald's Road, London WC1X 8RR, UK
http://www.academicpress.com

Academic Press
An imprint of Elsevier Science
525 B Street, Suite 1900, San Diego, California 92101-4495, USA
http://www.academicpress.com

ISBN 0-12-176559-8

Library of Congress Catalog Number: 2002103914

A catalogue record for this book is available from the British Library

Cover image: *Salmonella enteridis* and *Escherichia coli* 0157:H7. The blue are the *Salmonella* and
the green are the *E. coli* due to the insertion of fluorescent protein plasmids into the bacteria.

Typeset by Charon Tec Pvt. Ltd, Chennai, India
Printed and bound in Spain by Grafos S.A. Arte Sobre Papel, Barcelona

02 03 04 05 06 07 GF 9 8 7 6 5 4 3 2 1

Contents

Contributors

Sean F. Altekruse (Ch. 6), Division of Cancer Epidemiology and Genetics, National Cancer Institute, MSC#7234, Rockville, MD 20852, USA

Rhona S. Applebaum (Ch. 23), National Food Processors Association, 1350 I Street, NW, Suite 300, Washington, DC 20005-3305, USA

Merlin S. Bergdoll [deceased] (Ch. 16), Food Research Institute, University of Wisconsin – Madison, 1925 Willow Drive, Madison, WI 53706-1187, USA

Dane T. Bernard (Ch. 23), Keystone Foods, 5 Tower Bridge, 300 Bar Harbor Drive, Suite 600, W. Conshohocken, PA 19428, USA

Robert L. Buchanan (Ch. 5), US Food and Drug Administration, Center for Food Safety and Applied Nutrition, Harvey W. Wiley Federal Building, 5100 Paint Branch Parkway, College Park, MD 20740-3835, USA

Jean C. Buzby (Ch. 2), Economic Research Service, US Department of Agriculture, 1800 M Street NW, Room N4165, Washington, DC 20036-5831, USA

Fun S. Chu (Ch. 19), Food Research Institute and Department of Food Microbiology and Toxicology, University of Wisconsin – Madison, 1925 Willow Drive, Madison, WI 53706-1187, USA

Dean O. Cliver (Chs 11, 12, and 22), Department of Population Health and Reproduction, School of Veterinary Medicine, University of California, Davis, CA 95616-8743, USA

John H. Cross (Ch. 13), Uniformed Services University of the Health Sciences, Department of Defense, 4301 Jones Bridge Road, Bethesda, Maryland 20814-4799, USA

J. P. Dubey (Ch. 13), US Department of Agriculture, Agricultural Research Service, Animal and Natural Resources Institute, Parasite Biology, and Systematics Epidemiology Laboratory, Beltsville, Maryland 20705-2350, USA

Paula J. Fedorka-Cray (Ch. 3), Food Safety Pathogen Reduction, Richard Russell Research Center, US Department of Agriculture, 950 College Station Road, Athens, GA 30605-2720, USA

Pina M. Fratamico (Ch. 5), US Department of Agriculture, Agricultural Research Service, Eastern Regional Research Center, 600 East Mermaid Lane, Wyndmoor, PA 19038, USA

Elisabeth Garcia (Ch. 21), Department of Food Science and Technology, University of California, Davis, CA 95616, USA

Jeffrey T. Gray (Ch. 3), Food Safety Pathogen Reduction, Richard Russell Research Center, US Department of Agriculture, 950 College Station Road, Athens, GA 30605-2720, USA

Mansel W. Griffiths (Ch. 18), Department of Food Science, Ontario Agricultural College, University of Guelph, Guelph, Ontario, Canada N1G 2W1

Maha N. Hajmeer (Ch. 22), Department of Population Health and Reproduction, School of Veterinary Medicine, University of California, Davis, CA 95616-8743, USA

Linda J. Harris (Ch. 10), Department of Food Science and Technology, One Shields Avenue, University of California, Davis, CA 95616, USA

Susan L. Hefle (Ch. 14), Department of Food Science and Technology, University of Nebraska, 356 Food Industry Complex, Lincoln, NE 68583, USA

Keith Ito (Ch. 17), National Food Processors Association, 6363 Clark Avenue, Dublin, CA 94568, USA

Eric A. Johnson (Ch. 15), Food Research Institute, University of Wisconsin – Madison, 1925 Willow Drive, Madison, WI 53706-1187, USA

Vijay K. Juneja (Ch. 8), US Department of Agriculture, Agricultural Research Service, Eastern Regional Research Center, 600 East Mermaid Lane, Wyndmoor, PA 19038-8598, USA

Georg Kapperud (Ch. 7), Norwegian Institute of Public Health, PO Box 4404 Nydalen, N-0403, Oslo, Norway

Ronald G. Labbe (Ch. 8), University of Massachusetts, Department of Food Science, Amherst, MA 01003, USA

Keith A. Lampel (Ch. 4), Food and Drug Administration, HFS-237, 200 C Street SW, Washington, DC 20204, USA

Suzanne M. Matsui (Ch. 12), Division of Gastroenterology P304, Stanford University, School of Medicine, Stanford, CA 94305-5487, USA

Zdeněk Matyáš (Foreword), Centrum Hygieny Potravinových Řetězců, Institutu Hygieny a Epidemiologie, 612 42 Brno, Palackého 1-3, Czech Republic

Anthony T. Maurelli (Ch. 4) Department of Microbiology and Immunology, Uniformed Services University of the Health Sciences, F. Hébert School of Medicine, 4301 Jones Bridge Road, Bethesda, MD 20814-4799, USA

K. Darwin Murrell (Ch. 13), US Department of Agriculture, Agricultural Research Service, 10300 Baltimore Avenue, Building 005, Beltsville, Maryland 20705-2350, USA

Nina Gritzai Parkinson (Ch. 17), National Food Processors Association, 6363 Clark Avenue, Dublin, CA 94568, USA

Hans P. Riemann (Co-editor), Haven Laboratory, 195 East Lincoln, Banning, CA 92220, USA

Tanya Roberts (Ch. 2), Economic Research Service, US Department of Agriculture, 1800 M Street NW, Room N4081, Washington, DC 20036–5831, USA

Riichi Sakazaki [deceased] (Ch. 9), c/o Asuka Jun-Yaku, 2-4 Kanda-Sudacho, Chiyoda-ku, Tokyo 101, Japan

Edward J. Schantz (Ch. 15), Food Research Institute, University of Wisconsin – Madison, 1925 Willow Drive, Madison, WI 53706-1187, USA

Heidi Schraft (Ch. 18), Department of Biology, Lakehead University, Thunder Bay, Ontario, Canada P7B 5E1

Virginia N. Scott (Ch. 23), National Food Processors Association, 1350 I Street NW, Suite 300, Washington, DC 20005, USA

James L. Smith (Ch. 5), US Department of Agriculture, Agricultural Research Service, Eastern Regional Research Center, 600 East Mermaid Lane, Wyndmoor, PA 19038, USA

Jill A. Snowdon (Ch. 2), SGA Associates, 9541 Warfield Road, Gaithersburg, MD 20882, USA

David L. Swerdlow (Ch. 6), Surveillance Branch, Division of HIV/AIDS Prevention, National Center for HIV, STD and TB Prevention, Centers for Disease Control and Prevention, Atlanta, GA, USA

Steve L. Taylor (Chs 1, 14, and 20), University of Nebraska, Department of Food Science and Technology, Filley Hall, East Campus, Lincoln, NE 68583-0919, USA

Carl K. Winter (Ch. 21), Department of Food Science and Technology, University of California, Davis, CA 95616, USA

Amy C. Lee Wong (Ch. 16), Food Research Institute, University of Wisconsin – Madison, 1925 Willow Drive, Madison, WI 53706-1187, USA

Foreword

Incidents of foodborne disease harm people in all countries of the world. It is true that some of the diseases have been controlled at least in some areas, but others are emerging or reemerging. In spite of extensive knowledge on diseases such as salmonellosis, cholera and botulism, these diseases have not diminished and in fact they are causing increasing problems in many countries and are significantly affecting human health, as well as productivity and the international trade in foods. A number of factors are responsible for occurrence of foodborne diseases. They originate from environmental, social and economic conditions prevailing in the regions concerned. Foodborne disease and its epidemiology involve the entire chain of production, processing and distribution of food. The level of community sanitation is important, and the role of food habits and culture is increasingly being recognized in both developed and developing countries. Rapid urbanization, technological advances, international movements of people, animals and food products, centralization of food processing, long chains of food distribution and changing habits have all modified the conventional approaches to the epidemiology of foodborne disease.

Properly planned, organized, managed and staffed national and international food hygiene programs are able to contribute to prevention of foodborne diseases. In addition to food safety, they are able to improve public health in several other ways, such as quality and quantity of food, reduction of fraud, adulteration, prevention of dumping and others. The successful outcome of such programs is largely dependent on public participation, and, indeed, public education is the basis of effective and long-lasting improvements in the level of food hygiene.

A multiplicity of persons need to be educated and trained about hazardous situations to which foods are exposed and about food hygiene measures. This includes cooks, housewives, food producers, manufacturers, distributors, managers, and any other food workers, and of course food inspectors, food hygienists, technologists, sanitarians and laboratory personnel.

Education and training must take into consideration the number of etiological agents (infectious or toxic) that can cause foodborne diseases. The number is considerable and includes bacteria, viruses, prions, mycotoxins, parasites, chemicals and radionuclides.

Moreover, behavioral factors related to food are also numerous and may have a cultural or psychophysiological basis, or may result from popular beliefs about health-promoting or invigorating properties of certain types of food.

Food is of universal interest, and outbreaks of foodborne diseases can cause considerable emotional reaction in the community. Sometimes these community reactions get translated into political problems, especially if badly or over-reported in the mass media.

This book, written by very able authors, presents an in-depth description of our current understanding of the infectious and toxic pathogens associated with food.

The book will be certainly well accepted by all readers throughout the world. I wish it has great success.

Prof. MVDr. Zdeňek Matyáš
Professor emeritus, Veterinary and Pharmaceutical University,
Brno-Czech Republic
Formerly Chief, veterinary public health,
World Health Organization, Geneva

Preface

The first edition of *Foodborne Diseases* appeared in 1990. It was based on a course taught at the University of Wisconsin – Madison that was coordinated by Cliver, so most of the authors were faculty who had participated in the course. The book was apparently used fairly widely, partly because the editorial style allowed people, whose skills in technical English were limited, to grasp the concepts presented.

With Cliver's move to the University of California, Davis, in 1995, he joined forces with Riemann, and they agreed to undertake a second edition of *Foodborne Diseases*, updating the content while attempting to serve the same readership as the first edition. They cast a much wider net in recruiting authors, which may have contributed to some of the delays experienced in completing the book. All the same, this has been done, and they thank the contributors for the excellent quality of their work. The topics covered are largely the same as in the first edition, with some changes in organization.

One lamentable loss occurred when Dr Alfred E. Harper declined to update his chapter on Diet and Chronic Diseases and Disorders from the first edition. It was this chapter that de-bunked the *red wine phenomenon* before the term was coined, by showing that French and American men die at the same rate, despite the apparent lack of deaths from heart disease among the French. This kind of healthy skepticism is badly needed in all aspects of food safety – we commend the reader to the first edition to learn about this and other important information contributed by Dr Harper.

Food safety is a dynamic field. Problems recur long after they were 'solved', and new problems arise that even long-term practitioners never imagined. This is a time when people are so anxious about the safety of their food that they forget that many in the world, and even in the US, do not have enough to eat. This anxiety is leading to the destruction of large quantities of allegedly unsafe food and to the imposition of costly 'safety measures' that in some instances have no basis in science. We have tried to present in this book the information needed to make scientifically valid decisions in food safety, in reasonably easy-to-understand terms. Books like this are inevitably out-dated before they emerge from the bindery, but we believe that the readers for whom this was written and edited will find it useful.

Dedication

I dedicate this edition to my family, my mentors, and my students, who together have helped me to sustain my sense of wonder through a rather long scientific career.

D.O.C.

Principles

Disease Processes in Foodborne Illness

Steve L. Taylor

I. Classification of Disease Processes in Foodborne Illness

Foodborne diseases can be classified into five categories: infections, intoxications, metabolic food disorders, allergies, and idiosyncratic illnesses. Infections and intoxications can affect everyone. Metabolic food disorders, allergies, and idiosyncratic illnesses are sometimes referred to collectively as individualistic adverse reactions to foods because they affect only certain individuals in the population.

II. Structure and Function of the Digestive Tract

A. Anatomy and Digestive Functions

The digestive or gastrointestinal (GI) tract, especially the small intestine, is a primary site of action for foodborne infectious agents because they are ingested with food. All foodborne diseases start here. Let us, therefore, begin with a brief review of the anatomy and physiology of the GI tract.

Foodborne Diseases 2nd Edn
ISBN: 0-12-176559-8

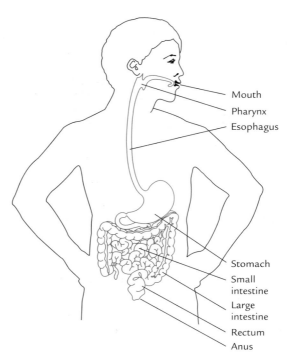

Figure 1.1 The organs of the gastrointestinal tract.

The GI tract or alimentary canal is a hollow tube beginning at the mouth and terminating at the anus (Fig. 1.1). The folded tube is several times longer than the height of the individual it serves. Food moves through the open area of the tube, which is called the lumen. The lumen is technically outside the body and is separated from the interior of the body by a lining known as the mucosa. The GI tract has many additional layers, some of which comprise smooth muscles that move intestinal contents through the lumen by rhythmic contraction (peristalsis). Some regions of the GI tract are separated by sphincters, circular muscles that control the passage of food.

The primary physiological functions of the GI tract are digestion and absorption. Digestion is the process of breaking down food into components that can be absorbed or taken into the body.

Food enters via the mouth, where it is chewed by the teeth and mixed with saliva, which moistens the food, lubricates the upper portions of the GI tract, and provides amylase, a digestive enzyme that degrades starch into absorbable sugar subunits. The esophagus is the conduit from the mouth to the stomach. Food passes through the esophagus by rhythmic contractions and gravity. A sphincter at the bottom of the esophagus allows food to enter the stomach but also prevents digestive juices and food from being moved back into the esophagus.

The stomach functions partly as a reservoir until the lower portions of the GI tract are ready to receive and process the food; the stomach of an adult can hold 1–2 liters of food and fluid. The stomach also produces HCl and pepsin, a proteolytic enzyme, and mixes these with the food. Digestion begins in earnest in the stomach, but very

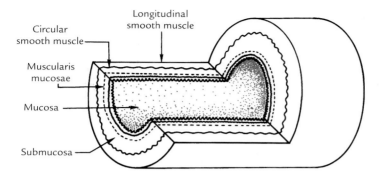

Figure 1.2 Diagram of the layers of the small intestine. Reprinted with permission from P. L. Smith (1986). Gastrointestinal physiology. *In* 'Gastrointestinal Toxicology' (K. Rozman and O. Hanninen, eds), pp. 1–28. Elsevier, New York.

little absorption occurs, with the exception of small amounts of water, alcohol, and certain drugs. The contents empty through the pylorus (another sphincter) into the small intestine over a period of hours.

The small intestine, the major site for digestion of food and absorption of its nutrients, is divided into three sections: duodenum, jejunum, and ileum. Digestion occurs primarily in the duodenum or first portion of the small intestine. The duodenum receives secretions from the pancreas with enzymes that digest fats, carbohydrates, and proteins. The duodenum also receives bile, which is secreted from the liver and stored in the gall bladder. Bile emulsifies fats for easier digestion and absorption.

Absorption of the nutrients occurs throughout the length of the small intestine. A cross-section of the jejunum is depicted in Fig. 1.2. The small intestine has a large surface area for absorption due to numerous fingerlike projections (villi) that extend into the intestinal lumen. To envision the millions of villi lining the surface of the small intestine, imagine a tube lined with terrycloth. Figure 1.3 shows an enlarged version of one villus with an adjoining crypt. Projecting from the villi are even smaller strands called microvilli, which contain many of the enzymes and receptors involved in nutrient absorption. This convoluted surface multiplies the absorptive area approximately 600-fold in comparison to a flat, smooth surface. If flattened, the absorptive surface of the small intestine would measure approximately 200 square meters, the size of a tennis court.

Absorption takes place via the columnar epithelial cells (Fig. 1.3) near the tip of the villus – a term that describes their general shape and location as part of the intestine's epithelial layer. The mature absorptive cells of the villus differentiate from precursor cells in the crypts. The villous tip cells have a short half-life; they are constantly being sloughed into the intestinal lumen and replaced by cells migrating from the crypts. The epithelial layer covering the villus also contains various specialized cells. Among the most common of these is the goblet cell (Fig. 1.3), which secretes mucin that adheres to the outer surface of the villus. Endocrine cells are also present in the epithelial layer and secrete gastrointestinal hormones. The lamina propria is the area just below the surface epithelial cells. The lamina propria contains phagocytic cells, such as macrophages, lymphocytes, and plasma cells, which defend against invasion by foreign microorganisms or proteins.

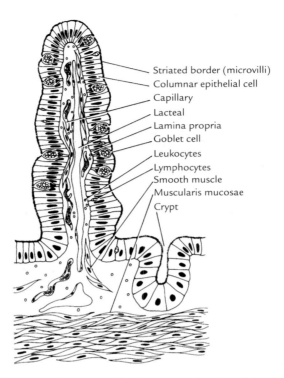

Striated border (microvilli)
Columnar epithelial cell
Capillary
Lacteal
Lamina propria
Goblet cell
Leukocytes
Lymphocytes
Smooth muscle
Muscularis mucosae
Crypt

Figure 1.3 A villus from the small intestine with adjacent crypt showing cellular diversity.

Each villus contains a network of capillaries, which carry water-soluble nutrients to the blood, and a lacteal (a terminal portion of the lymphatic system), which carries fat-soluble products of digestion. The lymph eventually empties into the bloodstream. The small intestine normally absorbs 90% or more of the energy value of the consumed food. Micronutrients such as minerals and vitamins are absorbed somewhat less efficiently.

The large intestine (or colon or bowel) is the last major section. It is a collecting chamber for solid waste; contents usually spend 24 hours or more in this section. Bacteria flourish in the colon due, in part, to the slow movement of the contents. The bacteria feed on remaining food macronutrients. Some of the products of bacterial action are absorbed from the colon; up to 10% of the available energy from foods is absorbed from the large intestine. The bacteria may synthesize important quantities of certain micronutrients, including vitamin K and some of the B vitamins. Since very little, if any, oxygen is available in the large intestine, the bacteria in the lumen of the large intestine are obligate or facultative anaerobes.

B. Fluid Balance

The word *diarrhea* is derived from the Greek term meaning to flow through. Diarrhea is a change in the frequency, fluidity (consistency), or volume of the feces. Some

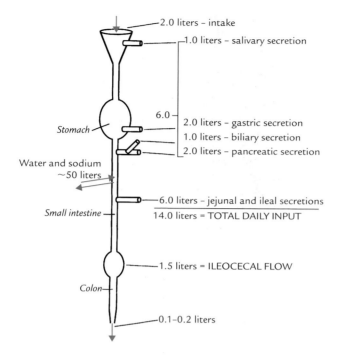

Figure 1.4   Normal fluid balance in the gastrointestinal tract.

disturbance of the mechanisms for secretion and absorption of water in the GI tract results in an increased loss of water with the feces.

Before addressing diarrhea as a disturbance in fluid balance, we will look at normal fluid fluxes in the GI tract. The normal fluid balance in the adult GI tract is depicted in Fig. 1.4. Besides the variable but quantifiable total input of 14 liters of fluid per day, there is a constant flux of water and sodium across the mucosa of the small intestine. In normal, healthy adults, this may total as much as 50 liters/day in each direction. The flux of fluid in the distal small intestine favors absorption – by the time the contents enter the colon, the 14-liter volume has been reduced to 1.5 liters. The contents are concentrated even further in the cecum and ascending colon. The volume of water in the normal stool is only 0.1–0.2 liters/day. The colon has the capacity to absorb about three times as much water as it does under normal conditions. Thus, in some cases of diarrhea, the volume of water entering the colon must be substantial to exceed this reserve capacity. Even with diarrhea, the water excreted in the feces is only a small percentage of the total fluid entering the intestinal tract.

Diarrhea can be defined as an increase in fecal water content. At least four mechanisms can be involved (Fig. 1.5): (1) defective absorption of solutes, (2) increased secretion of solutes, (3) structural abnormalities in the intestine, and (4) altered intestinal motility.

Decreases in the absorption of solutes can occur as the result of overeating, enzyme or bile acid deficiencies, or ingestion of poorly absorbed ions, such as those that occur

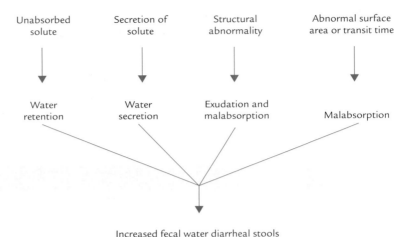

Figure 1.5 Diarrhea mechanisms.

in certain laxatives. This 'osmotic' diarrhea can occur in the small or large intestines or both.

Large quantities of water and various ions, especially sodium, normally are transported in both directions across the intestinal mucosa. The duodenum is rather 'leaky' and allows passive movement of water and sodium, principally from blood to lumen. The jejunum absorbs sodium, chloride, and bicarbonate against their epithelial electrochemical gradients. In the ileum, sodium and chloride are actively absorbed, but bicarbonate is secreted, all against their respective electrochemical gradients. Disturbing these processes can result in an abnormal secretion of electrolytes and water or 'secretory' diarrhea. Bacterial enterotoxins frequently interfere with intestinal electrolyte transport, stimulating the secretion of electrolytes and water. The colon plays a lesser role in secretory diarrhea.

Structural abnormalities in the intestinal tract can lead to increased stool volume by interfering with absorption of nutrients or by exudative processes. Damage to the absorptive epithelial cells of the small intestine can compromise their effectiveness in the absorption of nutrients. This leads to an increase of unabsorbed solutes in the lumen and to osmotic diarrhea. Severe mucosal damage can lead to the release of cells and other structural materials into the lumen and increases the hyperosmolarity of the luminal contents. In both cases, the active absorption of electrolytes against the electrochemical gradient can be lost, since the carriers are located in the epithelial cells that have been damaged.

While increased motility is not solely responsible for most cases of diarrhea, it may play a role in each of the other types. Normally, an increased flow of luminal contents is compensated by an available excess area of mucosal absorptive surface. However, if the increased flow occurs in combination with a decrease in mucosal absorptive surface area due to cellular damage, diarrhea can ensue. Similarly, the increased volume of fluid encountered in osmotic or secretory diarrhea stimulates intestinal motility.

Table 1.1 Pathogenesis of diarrhea

Organ	Malfunction
Small intestine	Decreased absorption of fluid and electrolytes
	Incomplete absorption of nutrients
	Increased secretion of fluid and electrolytes
Large intestine	Decreased absorption of fluid and electrolytes

Table 1.2 Pathophysiological mechanisms of foodborne infectious diarrheal agents

Abnormal electrolyte and water transport with normal mucosal permeability
  Secretion stimulated by bacterial enterotoxins and endotoxins
Abnormal mucosal permeability
  Increased permeability to fluid and electrolytes
    Invasive bacteria
    Enterotoxins
  Abnormal permeability to plasma proteins or protein-losing enteropathy
    Mucosal structural damage caused by viral, bacterial, or parasitic agents

Malabsorption problems more frequently involve the small intestine, while inflammatory problems are often localized in the large intestine. The pathogenesis of diarrhea is often described in part by its site (Table 1.1).

The pathophysiology of diarrhea caused by foodborne infectious agents is summarized in Table 1.2. Of the many potential mechanisms in diarrhea, foodborne infectious agents are implicated in only a few. The characteristics of the diarrheal stools will vary with the mechanism. If electrolyte balances are compromised, the stools will be watery; severe cases are sometimes referred to as rice-water stools because of the large ratio of water to solids. If severe mucosal injury has occurred, blood loss into the feces is often observed; the stools are black or red from the presence of blood. Foul-smelling stools are frequently a sign that the fat absorption has been poor.

C. Immune Functions

The cells active in the gut-associated lymphatic tissue (GALT) are distributed throughout the GI tract and are divided among three general sites: the Peyer's patches, the mucosa itself, and the mesenteric lymph nodes. The Peyer's patches are located principally in the ileum but are also scattered throughout the small intestine. The Peyer's patches contain macrophages and lymphoid cells, primarily B cells (50–70%) and T cells (11–40%). The outer layer of the Peyer's patches consists of columnar epithelial cells and M cells. M cells play a key role in the uptake (by endocytosis) and transport of antigens to the lymphocytes and macrophages of the Peyer's patches. M cells have no lysosomes and do not modify the antigen, but simply transport it from the lumen to the other cells. The mechanism of action of the M cells of the Peyer's

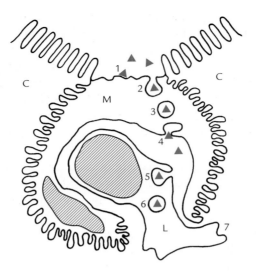

Figure 1.6 Schematic representation of uptake and transport of an antigen by an intestinal M cell including: (1) adherence of antigen to the M cell surface; (2) endocytosis; (3) phagosome formation; (4) exocytosis into the intercellular space; (5) uptake by a lymphocyte; (6) phagosome formation; and (7) lymphocyte migration. C, epithelial cells; L, lymphocyte. Reprinted with permission from A. Naukkarinen and K. J. Syrjanen (1986). Immunoresponse in the gastrointestinal tract. *In* 'Gastrointestinal Toxicology' (K. Rozman and O. Hanninen, eds), pp. 213–245. Elsevier, New York.

patches is depicted in Fig. 1.6. The appendix in humans also contains structures that resemble Peyer's patches.

The intestinal mucosa is another important site of GALT. The lamina propria contains macrophages and lymphocytes (B and T cells, including mature plasma cells, and null cells, including K or killer cells and NK or natural killer cells). The lymphocytes of the lamina propria interact with the overlying epithelium. There are also lymphocytes between the epithelial cells of the mucosal surface. The origin and nature of these lymphocytes are not yet certain. Lymphocytes from the blood or lymph are occasionally found in the gut lumen.

The mesenteric lymph nodes are yet another major component of the GALT. They serve as collection points for the lymph that drains from the intestinal villi and Peyer's patches. The mesenteric lymph nodes experience constant antigenic stimulation and contain many large germinal centers. The medulla of the mesenteric lymph nodes contains many mature plasma cells, while the cortex contains numerous T cells.

Many cell types are involved in the gut-associated immune reactions. Macrophages participate in the processing and transport of antigens. B lymphocytes are mediators of the humoral (antibody-producing) immune response, while T cells are responsible for cell-mediated reactions. Plasma cells are the progeny of B cells, and synthesize and secrete an antibody molecule with specificity for one site on one particular antigen. T lymphocytes can regulate (help or suppress) the activity of the B cells. T cells can also be sensitized to antigens and participate directly in cell-mediated immunity. K cells are very important in defense against bacteria and parasites. They recognize antigens that have been coupled with antibodies and destroy these antigen–antibody complexes. NK cells exert spontaneous killing activity without prior sensitization to the antigen and independent of the presence of an antigen–antibody complex. Mononuclear phagocytes serve important accessory roles in humoral or cellular immunity, as antigen-presenting cells or as cytotoxic cells. These phagocytes can also secrete many biologically active substances, including prostaglandins and interferons. Intestinal mast cells can secrete a variety of pharmacologically active substances that are responsible for localized anaphylactic reactions. Mast cells can also engulf bacteria. However, the major role of the mast cells is in allergic disease, which will be discussed later.

Two types of immune responses can occur in the GI tract upon exposure to an antigen: antibody-mediated (humoral) and cell-mediated. The B cells, as plasma cells, produce several different classes of antibodies: IgG, IgM, IgA, IgD, and IgE. IgG antibodies are the most prevalent and important in serum, but IgA plays a pre-eminent role in intestinal immunity. Most of the plasma cells in the lamina propria produce IgA; IgA-producing cells outnumber IgG-producing cells by 20- to 30-fold. Plasma cells in the salivary glands also produce predominantly IgA. IgA and, to a lesser extent, IgM are secreted into the lumen and are affixed to epithelial surfaces, providing antibody-mediated immunity to specific antigens. True food allergies involve the abnormal production of allergen-specific IgE antibodies in the gut (see Section VI).

Cell-mediated immunity in the GI tract has received less scientific study. Cytotoxic T cells specific for the triggering antigen are involved in cell-mediated responses. Cell-mediated responses may be quite important in the control of certain viral and bacterial infections of the GI tract. Celiac disease, also known as gluten-sensitive enteropathy, likely involves abnormal cell-mediated immune responses (see Section VI, D).

III. Infections

Infections are diseases caused by the presence of viable, usually multiplying, microorganisms at the site of inflammation. In the case of foodborne infections, the viable microorganisms have been ingested with food. The dose required to produce an infection varies with the type of microorganism, even though the microorganism will usually multiply in the GI tract or some other organ of the body to produce the infectious disease.

A. Types of Agents Involved

Microorganisms involved in foodborne infections include bacteria, viruses, rickettsia, protozoa, and parasites. Common bacteria involved in foodborne infections are *Salmonella*, *Campylobacter*, and *Clostridium perfringens*. Bacteria implicated less frequently in foodborne infections include *Listeria monocytogenes*, *Escherichia coli*, including *E. coli* 0157:H7, *Yersinia enterocolitica*, *Shigella*, *Vibrio cholerae*, *Vibrio parahaemolyticus*, *Bacillus cereus*, and *Streptococcus* group A. Viruses are identified as causative agents of foodborne disease less often than bacteria. These types of microorganisms are more difficult to recover from foods, which may be partially responsible for the less frequent implication of these microorganisms. Examples of viruses that can be acquired via foods are the Norwalk gastroenteritis virus and hepatitis A virus.

Protozoa and parasites also cause foodborne infections in the US, but are more frequently encountered in other countries. Examples of protozoal agents that can be transmitted via foods include *Giardia lamblia* and *Entamoeba histolytica*. Parasites implicated in foodborne infections historically include roundworms (*Trichinella spiralis*), tapeworms (*Taenia saginata*), and flukes (*Fasciolopsis buski*). More recently, *Toxoplasma gondii*, *Cyclospora*, and *Cryptosporidium* have been widely implicated in foodborne and waterborne illnesses in the US.

B. Pathogenesis

1. Sites of action
Since the infectious microorganisms are ingested with food, they must at least pass through the GI tract and may cause symptoms there before reaching other tissues. Some foodborne disease agents can invade intestinal tissue, from which they can reach the blood and other body tissues. Such organisms may or may not elicit GI symptoms. Depending upon the organism, the dose, and other factors, the organism may attack virtually any tissue or organ in the body.

2. Symptoms
The symptoms of foodborne infections depend upon the nature and dose of the infecting organism, and the organs or tissues affected. Since the organisms often must multiply in the intestinal tract or other infected tissues, the onset of symptoms is usually delayed by 8 hours to several days, depending on the initial dose and the growth characteristics of the infecting organism. The delayed onset of symptoms is a characteristic feature of foodborne infections and allows infections to be differentiated from intoxications. Since the GI tract is frequently the site of action for foodborne infections, GI symptoms such as nausea, vomiting, and diarrhea are among the most common symptoms encountered in infectious episodes.

a. Gastrointestinal symptoms
Diarrhea has already been discussed with regard to fluid balance. The other principal GI symptom is vomiting. The driving force for vomition, contraction of the abdominal muscles and diaphragm, propels the contents out of the stomach and upper small intestine through the esophagus and mouth. Vomiting, or emesis, is controlled both by the autonomic nervous system, acting on smooth muscle, and the central nervous system, acting on somatic muscles. A center in the brain controls the vomiting reflexes.

Vomiting occurs with many diseases, not just foodborne infections. Intoxications of certain types can also induce vomiting. GI infections commonly produce vomiting because the GI tract contains numerous receptors that signal and stimulate the vomiting center in the brain. The majority of vomiting receptors are in the GI tract, especially the duodenum, which has been called the organ of nausea. These receptors can be stimulated by rapid emptying of the stomach, acute distention of the small intestine, and inflammation or irritation of the mucosal lining of the small intestine. Toxins produced by bacteria can stimulate these abdominal receptors, interact directly with the neurons signaling the vomiting center, or directly stimulate the chemoreceptor trigger zone, an area located at the base of the brain's fourth ventricle near the opening to the spinal canal. For example, the enterotoxins produced by certain strains of *Staphylococcus aureus* growing in foods interact with abdominal receptors for the autonomic nervous system, and elicit nausea and vomiting within 6 hours of ingestion of the offending food. The chemoreceptor trigger zone reacts directly to chemicals in the blood and is more important in vomiting associated with toxic chemicals than in vomiting associated with infectious diseases. Once the vomiting center is stimulated, vomiting occurs via nerve impulses sent from the brain to the diaphragm and abdomen.

Vomiting should be differentiated from regurgitation, which also involves the ejection of previously swallowed food from the mouth. However, regurgitation does not require a neural reflex. It often occurs because the esophageal sphincter is incompetent or because abdominal pressure from exercise or the changing of position causes some release of material through the sphincter.

b. Other symptoms

The other symptoms associated with foodborne infectious agents are too numerous to discuss in much detail. Some of the infectious agents in foods cause only gastrointestinal symptoms. *Vibrio cholerae*, which causes cholera, a profuse watery diarrhea, does not invade intestinal or other tissues. Some cause extraintestinal symptoms only on rare occasions or among individuals compromised in their defensive abilities. *Listeria monocytogenes*, for example, causes septicemia, stillbirths, and other extraintestinal problems, but attacks only individuals who are pregnant, elderly, or immunocompromised. *Yersinia enterocolitica* usually causes only intestinal symptoms such as diarrhea and abdominal pain. Occasionally, this organism can cause extraintestinal symptoms such as arthritis. However, some organisms cause extraintestinal infections routinely. An example is the hepatitis A virus, which elicits its symptoms in the liver.

3. General characteristics of infectious organisms

Infectious microorganisms must possess certain characteristics if they are to be involved in foodborne disease. First, they must be viable *in vivo*; the microorganisms must survive the acidic conditions of the stomach and the action of the various digestive enzymes to reach their site of action in a viable condition. Many microorganisms would be unable to survive in the acidic conditions of the stomach but, when they are ingested with food, the buffering capacity of the food reduces the acidity of the stomach and increases their likelihood of survival. If the site of action is extraintestinal, the microorganisms must be able to penetrate the mucosal barrier and reach the site of action in a viable state. Since infections usually require the multiplication of the microorganisms *in vivo*, the microorganism must be able to establish itself within the GI lumen or within some other tissue site in the body. In the intestines, colonization of the lumen requires that the microorganism be able to compete effectively with the resident microflora. Competition is especially important in the colon, where large numbers of bacteria normally reside. The microorganism must also be able to withstand any bodily defense mechanisms. Some infectious bacteria produce enterotoxins that mediate their effects on the host. Infectious microorganisms may be categorized as either invasive or noninvasive.

4. Invasive infections

Invasive infections are caused by microorganisms that penetrate tissues and perhaps multiply intracellularly. Among invasive bacteria associated with foodborne infections, the classic example is *Shigella*. Others include *Salmonella*, *Yersinia enterocolitica*, *Campylobacter*, some enteropathogenic *Escherichia coli* (especially the O157:H7 serotype), and *Listeria monocytogenes*. Viruses are invasive, as are some of the protozoa and many parasites. Among protozoa, *Entamoeba histolytica* invades tissues but not cells, while *Toxoplasma gondii* invades cells and multiplies intracellularly. *Trichinella spiralis* is a good example of an invasive parasite. Although these microorganisms can invade extraintestinal tissues and cause damage and symptoms at sites beyond the intestine, they must first invade intestinal tissues and often cause intestinal symptoms.

Invasive microorganisms cause diarrhea by damaging the intestinal mucosa, causing a loss of structural integrity and a disruption of electrolyte and fluid balance. If the invasion occurs in the small intestine, the result is usually watery diarrhea. For example, *Salmonella* typically invades the ileum. If the invasion occurs in the colon, the result can sometimes be bloody diarrhea. *Shigella* typically invades the colon, and symptoms of shigellosis range from mild, watery diarrhea from a mild acute inflammation to severe dysentery with significant blood loss from severe, diffuse ulcerative lesions.

Bacterial invasion of intestinal cells can result in symptoms other than diarrhea. The inflammatory response to the invading bacteria can elicit fever, chills, and tenesmus. Damage to intestinal cells can result in altered nutrient absorption and immunological abnormalities. Abdominal pain and cramping, nausea, and vomiting are also frequently associated with invasive infections of the intestine.

If more disseminated infections occur as the result of the invasion, then many other symptoms will result from the invasion of other body tissues. Septicemia, in which the bacteria live and proliferate in the blood, can be a life-threatening complication with certain types of invasive bacteria such as *Salmonella typhi*.

Invasive infections usually occur in two stages: penetration and multiplication. Once invasive bacteria have penetrated the intestinal mucosa, they will either begin to multiply intracellularly or be transported to extraintestinal tissues. *Shigella* penetrates colonic epithelial cells and multiplies to produce inflammatory lesions. In severe cases, the lamina propria may also be invaded. *Yersinia enterocolitica* penetrates the Peyer's patches and multiplies. *Shigella* and *Y. enterocolitica* rarely disseminate to other tissues, but can in severe cases. The invasive bacterium must be able to survive and prosper in the microenvironment of the cells that it invades for an infectious disease to occur.

With invasive bacteria, enterotoxins and cytotoxins may sometimes play a secondary role in the pathogenesis of the infection. Enterotoxins are substances that induce intestinal fluid loss without altering the morphology of the intestine, while cytotoxins are substances that elicit fluid loss and also severely alter intestinal morphology. The cytotoxins in particular may participate in the damage to the intestinal mucosa that occurs during invasive infections.

5. Noninvasive infections

Some foodborne infectious bacteria are incapable of penetrating cells and multiplying intracellularly. Some foodborne protozoa (e.g. *Giardia lamblia*) and parasites also cause illness without actually penetrating intestinal tissues.

Noninvasive infectious bacteria must be able to survive and proliferate in the intestinal lumen. These bacteria often colonize the small intestine because the number of competing bacteria is far lower than in the colon. Adherence to the intestinal surface allows the bacteria to divide and produce enterotoxins without being removed along with other luminal contents by peristalsis. The adherence is quite specific and involves flexible filaments on the bacterial surface interacting with sugar residues on the surface of the microvilli.

These noninvasive bacteria, such as *Vibrio cholerae* and enterotoxigenic *Escherichia coli*, produce enterotoxins that mediate the disease symptoms. The symptoms (principally diarrhea) are usually confined to the intestinal tract, since these bacteria cannot reach other tissues. The enterotoxin produced by *V. cholerae* is a protein with a molecular weight of 84 000 Da. The enterotoxin increases the activity of an enzyme, adenylate cyclase, which causes active secretion of chloride ion from crypt epithelial cells and prevents the absorption of sodium and chloride ions in the villous tip epithelial cells. The increased osmolarity of the lumen is balanced by the secretion of water. The volume of diarrheal stools in cholera can be quite profound, and death can ensue from dehydration. Enterotoxigenic *E. coli* produces a heat-labile toxin that is virtually identical to the cholera enterotoxin. Some strains of enterotoxigenic *E. coli* produce a heat-stable enterotoxin that is much smaller and activates guanylate cyclase rather than adenylate cyclase.

C. Defense Mechanisms

Infections do not always produce disease. Pathogenic bacteria, viruses, protozoa, or parasites can be present, especially in small numbers, in the GI tracts of asymptomatic individuals. Humans tolerate microorganisms in their GI tracts because numerous defense mechanisms prevent disease. Only when the defense mechanisms are overwhelmed will infections cause illness.

1. Intestinal factors

The intestinal mucosa serves as a physical barrier – one cell thick – to the entry of potentially infectious bacteria into the body. Individuals with pre-existing intestinal damage are more susceptible to intestinal infections than individuals with an intact intestinal epithelium.

The microflora of the intestinal tract provides additional protection. The existing bacteria tend to be fairly stable and compete very effectively for nutrients. Some release inhibitory substances into the lumen. For example, certain colonic bacteria produce volatile fatty acids that are inhibitory to other bacterial species.

Goblet cells, spaced intermittently along the intestinal epithelium, produce mucus which discourages bacterial colonization of the small intestine. Infectious micro-organisms must penetrate this mucous barrier to reach the epithelial surface that possesses the receptors for adherence. Some infectious bacteria elaborate enzymes that hydrolyze the mucus, allowing passage of the bacterial cell through this barrier. Other infectious bacteria penetrate the mucous barrier by other means. Bacterial motility is associated with virulence for some species of infectious bacteria. These motile bacteria may simply bore through the mucous layer.

Bile acids, degradation products of cholesterol, are inhibitory to some bacteria. The bile acids are secreted into the duodenum through the bile duct from the liver. Their primary function is to aid in the digestion and absorption of fats and fat-soluble vitamins. Some infectious bacteria are quite resistant to the effects of the bile acids. When bacterial overgrowth occurs in the small intestine, some bacteria will hydrolyze the bile acids, leading to poor absorption of fats and steatorrhea, a foul-smelling fatty type of diarrhea.

Intestinal motility is also a protective mechanism. The normal transit time from the ingestion of food until the elimination of waste in a healthy adult is 24–72 hours. This constant movement of material tends to clear the lumen of potentially infectious microorganisms.

The diet may exert some influence through factors that either promote or inhibit bacterial growth. A high-fiber diet may simply entrap bacterial cells and prevent their access to mucosal surfaces. Also, the diet provides the bulk for elimination of waste and stimulates peristalsis.

2. Phagocytosis

Phagocytosis, a process by which certain cells (phagocytes) engulf and destroy infecting microorganisms and their toxins, is an extremely important defense mechanism. Several categories of phagocytes exist. Leukocytes, including neutrophils, monocytes, and eosinophils, circulate with the blood. Other phagocytes, called macrophages, are localized in tissues. Macrophages are located all along the GI tract but are particularly prevalent in the ileal Peyer's patches. Leukocytes can also invade the lamina propria.

The process of phagocytosis is depicted in Fig. 1.7. Bacterial cells or macromolecules that adhere to cellular membranes are engulfed by a process known as endocytosis. The membrane of the cell actually surrounds the particle or macromolecule and pinches off, forming a vacuole containing the foreign material. The macromolecule-laden vacuole, known as a phagosome, then fuses with a cellular organelle called a lysosome. The lysosome contains a variety of hydrolytic enzymes, including proteases, lipases, and carbohydrases, which are capable of destroying certain infectious microorganisms and their toxins. Lysozyme, one of the constituent enzymes of macrophage lysosomes, may be particularly important in killing bacterial cells, but

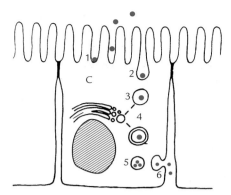

Figure 1.7 Schematic representation of phagocytosis in an intestinal epithelial cell (C) with: (1) adherence of a macromolecule to the microvillus epithelial surface; (2) endocytosis; (3) phagosome formation; (4) phagolysosome formation; (5) residual vacuole containing undegraded material; and (6) exocytosis into the extracellular space. Reprinted with permission from A. Naukkarinen and K. J. Syrjanen (1986). Immunoresponse in the gastrointestinal tract. *In* 'Gastrointestinal Toxicology' (K. Rozman and O. Hanninen, eds), pp. 213–245. Elsevier, New York.

the most important killing mechanisms of macrophages and leukocytes are oxidative. These mechanisms involve either hydrogen peroxide or free radicals, such as super-oxide anion, singlet oxygen, or hydroxyl radical, produced by the macrophages and leukocytes. Microorganisms killed in this manner can then be digested by lysosomal action. Some microorganisms are resistant and are thus able to persist inside macrophages.

3. Immunological mechanisms

Immunological mechanisms are extremely important in defense against infections. The GI tract is one of the major lymphatic organs of the body. Because the GI tract is exposed directly to environmental antigens, it is often the initial organ to mount an immunological defense against pathogenic microorganisms. The gut is the major site for sensitization of immunocompetent cells that are subsequently recruited to other tissues and organs in the body. Immunological responses are somewhat delayed on the initial exposure to an antigen and often occur only after some symptoms have been encountered. Once immunity to an infecting microorganism is attained, immunological defense mechanisms assume paramount importance in the control of subsequent infections by the same microorganism.

Among the most important components of the GALT are the antibodies attached to intestinal epithelial surfaces. These antibodies neutralize viruses, control bacterial proliferation, and prevent access of enterotoxins to their sites of action. Secretory IgA bound to the surface of intestinal epithelial cells prevents bacterial adherence, which limits the opportunities for colonization and penetration. IgA can also neutralize the enterotoxins produced by bacteria such as *Vibrio cholerae*. Both humoral and cell-mediated immune responses are involved in the defense against viral infections. In the case of viral agents that replicate in the mucosa, secretory IgA antibodies can prevent adherence and replication. With invasive viruses, IgG in the serum can be important.

Parasitic infections usually induce the synthesis of antiparasitic IgE antibodies that combine with receptors on mast cell membranes, interact with the parasitic antigens, and induce localized anaphylactic reactions. Parasitic infections of the GI tract tend to be persistent, demonstrating that immunological defense mechanisms alone are inadequate to control such infections.

IV. Intoxications

Intoxications are disease states caused by a hazardous dose of a toxic chemical that are neither mediated immunologically (allergies) nor primarily the result of a genetic deficiency (metabolic disorders). The agents are nonviable, in contrast to infections, in which viable organisms are involved. All chemicals are toxic at some dose, but only those which produce noticeable adverse reactions under the dose and circumstances of exposure are hazardous to our health. Therefore, with chemical intoxications, it is very important to provide information on the dose and circumstances of exposure.

A. Types of Agents Involved

Foodborne toxicants can be of either natural or synthetic origin. Some natural toxicants are normal unavoidable constituents of various animal, plant, or microbial species. Other natural toxicants are more properly classified as natural contaminants of food because they originate from sources such as bacteria or molds, which contaminate foods on occasion. The natural toxicants in foods are the subject of Chapter 14.

Synthetic toxicants in foods are man-made chemicals that can enter foods either intentionally or unintentionally. Agricultural chemicals (insecticides, herbicides, etc.) and food additives would be examples of chemicals intentionally added to foods. These synthetic chemicals are not typically hazardous at the doses that occur in foods following the proper use of these substances. However, these chemicals can be hazardous at higher doses that may result from improper usage. Industrial chemicals, such as polychlorinated biphenyls (PCBs), are examples of synthetic chemicals that enter foods unintentionally.

Foodborne toxicants can be further classified according to their mode of action. For example, solanine, a naturally occurring constituent of potatoes, is a neurotoxin because it has the ability to inhibit acetylcholinesterase, an enzyme critical to nerve transmission. Parathion and malathion are synthetic insecticides which share this mode of action. Another example would be the aflatoxins, natural contaminants of foods arising from the growth of certain molds, which are carcinogens.

B. Pathogenesis

1. Sites of action

Chemical toxicants have many different sites of action. Some affect several tissues and organs while others are quite tissue specific. Table 1.3 provides a list of major organs

Table 1.3 Sites of action for foodborne toxicants	
Tissue site	**Toxin affecting the site**
Blood components	Nitrite
Brain	Domoic acid
Gastrointestinal tract	Staphylococcal enterotoxins
	Trichothecene mycotoxins
Heart	Erucic acid (rapeseed oil)
Kidney	Ochratoxin
Liver	Aflatoxins
	Ethanol
Lung	Paraquat
Nervous system	Saxitoxin
	Botulinum toxins
Skeletal system	Lead (Pb)
Skin	Histamine
	Trichothecene mycotoxins

and tissues and examples of foodborne toxicants that affect these sites. Intoxications are more likely than infections to affect parts of the body remote from the GI tract.

2. Symptoms

Many different symptoms can be produced by foodborne intoxications, depending on the nature and dose of the toxicant, its mechanism of action, and the target organ(s) affected. Symptoms can also vary with age, sex, nutritional status, the existence of other disease states, and from one individual to another.

Exposure to a foodborne toxicant can be either acute or chronic. Acute exposure involves a one-time ingestion of the toxicant or, occasionally, multiple exposures over a fairly short time. For example, illness would occur within a few minutes after ingestion of poisonous mushrooms or a few hours after the ingestion of staphylococcal enterotoxins. Chronic exposure involves ingestion of the toxicant with foods over a long period of time, sometimes a lifetime, though not typically on an everyday basis. An example of chronic exposure is the prolonged ingestion of a carcinogenic contaminant such as the aflatoxins.

Effects can also be acute or chronic. Acute symptoms occur a few minutes to a few hours after exposure to the toxicant, while chronic symptoms require long periods of time to develop. The vomiting associated with ingestion of the staphylococcal enterotoxins is an example of an acute effect, while the liver cancer occurring from exposure to aflatoxin is a chronic effect. The acute effects of exposure to foodborne toxicants occur only after acute exposure. Chronic effects can result from either acute or chronic exposure. Aflatoxin produces chronic effects from chronic exposure. However, chronic effects can also result from acute exposure; long-term neurological problems are experienced by individuals who have eaten foods contaminated with mercury.

An important distinction should be made between onset times in foodborne intoxications and infections. Symptoms of acute intoxications typically begin a few minutes to a few hours after eating the contaminated food. Foodborne infections have a

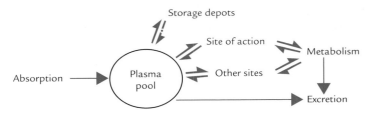

Figure 1.8 Schematic representation of toxin distribution.

delayed onset time of 8 hours to several days. This difference is often critical in distinguishing among the many potential causes of foodborne disease outbreaks.

3. Characteristics of toxins

Foodborne toxins must retain their toxicity through storage, preservation, and preparation of the food. Food processing and preparation are often important in the destruction or removal of toxins. Many foodborne toxins are heat stable.

Foodborne toxins must retain their toxicity until they reach their site of action. First, they must withstand digestive processes. If the toxin acts at some site other than the intestinal tract, it must be able to cross the intestinal barrier and find its way to the target organ and site of action (Fig. 1.8). Lipophilic toxins, such as the aflatoxins, can easily cross the lipid membranes of cells. Water-soluble toxins will usually cross only in their un-ionized forms, which are dependent on the pH of the GI tract. Some toxins cross by using carrier systems that exist for nutrients. For example, lead (Pb) reaches the circulation by using the carrier system that exists for calcium (Ca). Some toxins are deposited in body storage depots. The bone serves as a storage depot for Pb, while the adipose tissue is a storage site for many lipophilic toxins. The toxin can also be excreted through the kidney (urine), bile, or feces. Urinary excretion is a common route of removal for many toxins, but the lipophilic toxins must be metabolized into more hydrophilic forms so that they are compatible with urine, which is an aqueous medium. Metabolism does not invariably result in detoxification. Some foodborne toxins exist as pro-toxins and are metabolized (bioactivated) to more toxic forms. The aflatoxins, for example, are bioactivated in the liver.

Toxins interfere with crucial biochemical processes, often at fairly low concentrations. The dose of the toxin must exceed the capacity of the body to detoxify the chemical.

C. Defense Mechanisms

Adverse reactions occur only when the dose of the chemical exceeds the capacity of the body to detoxify or eliminate the toxin. A variety of defense mechanisms are available to prevent foodborne intoxications.

1. Intestinal factors

Fiber can bind and entrap toxicants and lead to their elimination with the feces. Vomiting is another means for clearing a noxious substance from the GI tract.

Intestinal bacteria metabolize and detoxify or bioactivate certain foodborne toxicants. Intestinal motility limits the time a potentially toxic chemical spends in the intestinal lumen. Pre-existing intestinal damage promotes the absorption of foodborne toxins, so the integrity of the gut wall offers some protection against foodborne intoxications.

2. Phagocytosis

Phagocytosis is much less important in protection against intoxicants than against infections. Also, some toxins are able to impair the function of phagocytic cells and can enhance the chances of acquiring a secondary infection.

3. Immunological mechanisms

Toxins must be capable of acting as antigens if they are to provoke an immunological response. Many of the low-molecular-weight toxins are incapable of acting as antigens. Some can act as haptens by binding to proteins and provoking an immunological response. Proteinaceous toxins can act directly as antigens. Humoral immune responses can play a significant role in protection against proteinaceous toxins.

4. Enzymatic defense mechanisms

The primary defense mechanism against chemical intoxications is enzymatic detoxification. Most chemicals can be detoxified enzymatically, although this defense mechanism can be overwhelmed at high doses. The particular enzymes involved in detoxification will vary depending upon the chemical, mode of administration, and dose and frequency of exposure.

a. General types of reactions

The enzymatic reactions involved in the detoxification of foodborne chemicals can be divided into two distinct 'phases': Phase I reactions, which include oxidation, reduction, and hydrolysis; and Phase II reactions, which are predominantly conjugation reactions involving the products of the Phase I reactions. In general, the products of enzymatic metabolism are more water-soluble and therefore more easily excreted than their precursors.

The most important organs for enzymatic detoxification of foodborne chemicals are the liver, followed by the kidney and the intestine. Some of the most important enzymes are localized intracellularly in the smooth endoplasmic reticulum.

b. Factors affecting metabolism

Many factors can affect enzymatic detoxification (Table 1.4). With foodborne intoxications, we are usually interested in the human species, but variations in detoxification assume some importance when the toxic potential of a foodborne chemical is evaluated in some other species. Enzyme induction can play a particularly significant role. The cytochrome P-450-containing monooxygenase system can be induced by exposure to many of its substrates. Subsequently, other substrates are more efficiently metabolized because of the higher level of this critical enzyme.

Table 1.4 Factors affecting enzymatic detoxification mechanisms

Species	Disease states	Enzyme inhibitors
Strain	Nutritional status	Altitude
Age	Route of administration	Gravity
Sex	Diet	Dose size
Time of day	Pregnancy	Dose vehicle
Season	Enzyme inducers	

V. Metabolic Food Disorders

A metabolic food disorder is a disease state caused by exposure to a chemical in foods that is toxic to certain individuals because they display some genetic deficiency in their ability to metabolize the chemical or because the chemical exerts some unusual effect upon their metabolic processes. Two examples follow.

A. Lactose Intolerance

Lactose intolerance results from a deficiency of the enzyme lactase, or β-galactosidase, in the intestinal mucosa. Lactose, the principal sugar in milk, cannot be absorbed unless it is first hydrolyzed into galactose and glucose. As a result of the enzyme deficiency, the undigested lactose passes into the colon and is metabolized by colonic bacteria. Abdominal cramping, flatulence, and frothy diarrhea occur within a few hours after consumption of dairy products and are usually self-limited.

Lactose intolerance is a genetically acquired trait. Symptomatic lactose intolerance can appear in early childhood, but may also appear later in life. The prevalence of lactose intolerance among Caucasian Americans is 6–12%, but may be as high as 60–90% in ethnic groups such as Greeks, Arabs, Jews, Black Americans, Japanese, and other Asians.

The usual treatment for lactose intolerance is avoidance of dairy products. However, only the most severely affected patients need to practice total avoidance of dairy products. Many can tolerate some lactose in their diets, and thus can comfortably consume small quantities of milk. Lactose-intolerant individuals can often tolerate yogurt and acidophilus milk because these products contain lactose-digesting bacteria. Lactose-hydrolyzed milk is also available in the marketplace for these individuals.

B. Favism

Favism is an acute hemolytic anemia experienced by some individuals after consumption of broad beans (fava beans) or inhalation of the pollen of the *Vicia faba* plant. The major symptoms are typical of hemolytic anemia: pallor, fatigue, dyspnea (shortness of breath), nausea, abdominal and/or back pain, fever, and chills. Rarely, more serious symptoms are noted, including hemoglobinuria, jaundice, and renal failure. The onset time ranges from 5 to 24 hours. Prompt and spontaneous recovery is

Table 1.5 Common allergenic foods		
Cows' milk	Fish	Tree nuts (almonds, walnuts, etc.)
Crustacea (shrimp, crab, lobster)	Peanuts	Wheat
Eggs	Soybeans	

usual after ingestion of the beans or inhalation of the pollen ceases. Favism is most prevalent when the *V. faba* plant is blooming and the pollen is in the air, and when the edible broad beans are available in the market.

Favism affects individuals with a deficiency of the enzyme glucose-6-phosphate dehydrogenase (G6PDH) in their red blood cells. Fava beans contain several oxidants, including vicine and convicine, which damage the erythrocytes of sensitive individuals. This is a metabolic disorder in which the foodborne chemical has an abnormal effect on the host's metabolism. G6PDH deficiency is among the most common genetic deficiencies in human populations worldwide. It is particularly prevalent among Oriental Jewish communities in Israel, Sardinians, Cypriot Greeks, Black Americans, and certain African populations, but is virtually absent in northern European nations, North American Indians, and Eskimos. However, the disease occurs only among susceptible individuals in the areas of the world where fava beans are eaten, primarily in the Mediterranean region, the Middle East, China, and Bulgaria.

VI. Allergy

Food allergy is a disease state caused by exposure to a particular (often proteinaceous) chemical to which certain individuals have a heightened sensitivity (hypersensitivity) that has an immunological basis. Like the metabolic food disorders, food allergies affect only certain individuals in the population; the prevalence in the US ranges from less than 1% of adults up to 4–8% in infants.

A. Types of Agents Involved

The most common allergenic foods are listed in Table 1.5. These foods are responsible for 90% or more of all food allergies on a worldwide basis. However, any food that contains protein has the potential to elicit allergic reactions. Over 160 different allergenic foods have been described in the medical literature. Cows' milk is the most common allergenic food among infants, perhaps because it is their most common food. Eggs and peanuts are also common allergenic foods among infants in the US. The most common allergenic foods among adults in the US are peanuts and crustacea. Many infants outgrow their food allergies, often during their first few years. Cows' milk and egg allergies are commonly outgrown, while peanut allergy persists. The mechanisms involved in the development of tolerance are not completely understood. The prevalence of a particular food allergy depends on the inherent immunogenicity of the food and on the frequency with which that food is consumed. Thus, the prevalence of specific food allergies can vary from one country or one culture to another based upon eating patterns. For example, soybeans and seafoods are common allergenic foods in Japan and hazelnut allergy is quite prevalent in European countries.

The allergens in these foods are usually proteins, although low-molecular-weight substances can occasionally act as allergens. For example, penicillin, an antibiotic that is allergenic to some individuals, may occur in certain foods if it has been used to treat some animal disease. While the majority of food allergens are proteins, only a few of the many different proteins in foods are allergens. The identities of only a few of the food allergens have been established. For example, in milk, the most common allergens are casein, β-lactoglobulin, and α-lactalbumin. However, cows' milk-allergic individuals vary in their sensitivities to these specific milk proteins. Many allergenic foods contain multiple allergens. For example, peanuts contain three major allergens that are recognized by the majority of peanut-allergic individuals.

Cross-reacting allergens exist in some foods. For example, most individuals with crustacean allergy react to all crustacean species (shrimp, crab, lobster, crayfish). The allergen in crustacean species has been identified as tropomyosin, a major muscle protein. Apparently, the amino acid sequence of tropomyosin is sufficiently similar in all crustacean species to account for the observed cross-reactivity. However, cross-reactions among related species are not universally observed. Peanuts are legumes, but most peanut-allergic individuals are only sensitive to that single legume species and not to other legumes, such as green beans, soybeans, and peas. Allergies to these other legumes also exist but the affected individuals are often sensitive to only one or a few of the many legume species.

B. Pathogenesis

Food allergies are abnormal immunological responses to a food component. Of immune mechanisms that occur in food allergy, the most important is the type I or immediate hypersensitivity reaction (Fig. 1.9). These reactions have very short onset times. The initial event is the production of allergen-specific IgE by B (plasma) cells in response to exposure to an allergenic foodborne protein with the involvement of helper T cells. The IgE attaches to the outer membrane surfaces of tissue mast cells and/or circulating basophils. The mast cells and basophils are thus sensitized; subsequent exposure to the allergen results in the release of allergic mediators from these cells. The allergen cross-links two IgE molecules on the surface of the mast cell membrane, causing the cells to degranulate. The granules of mast cells and basophils contain dozens of mediators of allergic disease, especially histamine. Histamine is released into the bloodstream and reacts with tissue receptors. The symptoms of an allergic reaction (Table 1.6) are dependent upon which tissue receptors are affected. Antihistamines can block allergic reactions by inhibiting the interaction between histamine and its tissue receptors.

A variety of symptoms can be associated with food allergies (Table 1.6), but most reactions involve only a few. Gastrointestinal effects are common. Among the rare life-threatening symptoms are anaphylactic shock, laryngeal edema, bronchoconstriction (asthma), and hypotension.

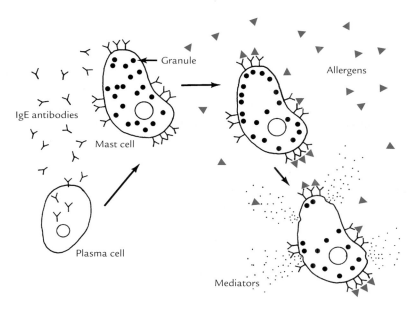

Figure 1.9 Biochemical mechanisms of immediate hypersensitivity. Reprinted with permission from S. L. Taylor (1987). Allergic and sensitivity reactions to food components. *In* 'Nutritional Toxicology' (J. N. Hathcock, ed.), Vol. 2, pp. 173–198. Academic Press, Orlando, FL.

Table 1.6 Symptoms of allergic reactions to foods	
Gastrointestinal symptoms	Respiratory symptoms
Nausea	Rhinitis
Vomiting	Asthma
Diarrhea	Other symptoms
Cutaneous symptoms	Laryngeal edema
Urticaria (hives)	Anaphylactic shock
Eczema or atopic dermatitis	Hypotension
Angioedema	Headache

Because the interaction of allergen with specific IgE bound to mast cells and basophils results in the release of large quantities of histamine and other mediators of allergic disease, the tolerance of food-allergic individuals for the offending food is extremely low. The threshold dose for food allergens is not precisely known and likely varies among individuals. However, milligram amounts of the offending food are sufficient to elicit adverse reactions in the most sensitive individuals. The severity of an allergic reaction is likely to increase in proportion to the ingested dose of the allergenic food.

C. Defense Mechanisms

Individuals with food allergies obviously do not have effective defenses against this type of foodborne disease. However, protective mechanisms do operate in other individuals, mostly in the intestinal lumen. The digestive proteases serve as one protective mechanism, although most food allergens are relatively resistant to proteolysis. The permeability of the gut wall is also a critical factor. Infants are thought to be more susceptible to food allergies because their intestinal walls are more permeable to food proteins. Secretory IgA is another intestinal defense against food allergies. Homologous IgA secreted into the intestinal lumen can react with the allergen before it reaches the sensitized mast cells and basophils. Such blocking antibody responses are among the mechanisms developed by infants as they begin to outgrow their food allergies. The development of oral tolerance to ingested food proteins, mediated by the immune system in the gut, is clearly an important phenomenon because all dietary proteins are foreign to the human immune system and would induce adverse reactions except for the development of oral tolerance.

D. Celiac Disease as a Possible Food Allergy

Celiac disease is characterized by a malabsorption of nutrients resulting from damage to the absorptive epithelial cells of the small intestine. The intestinal damage occurs in susceptible individuals when they consume the protein fractions of wheat, rye, barley, and triticale. The sensitive individuals display an abnormal immunologic response to the grain proteins, but the exact mechanism of the response has not been completely elucidated. An abnormal cell-mediated immune response is definitely involved in the inflammatory and cytotoxic events that lead to damage of the absorptive epithelial layer. The symptoms are diarrhea, bloating, weight loss, anemia, bone pain, chronic fatigue, weakness, muscle cramps, and, in children, failure to grow. Symptoms may begin at any age and do not appear instantaneously in susceptible individuals when they eat the provoking grains, since the intestinal damage requires some time to develop. Remission of symptoms on removal of wheat, rye, and barley from the diet is also slow due to the need to repair the existing damage. Celiac disease is an inherited trait seen in approximately 1 in 3000 individuals in the US. The prevalence of celiac disease is higher in some other parts of the world, including some European countries. Some individuals may suffer from subclinical, undiagnosed celiac disease, which may worsen over time with continuing exposure to the responsible grains.

E. Treatment of Food Allergies

While food allergies can be treated with drugs such as antihistamines, the treatment is only effective after the onset of symptoms. Preventive therapy with pharmaceutical agents does not exist. The major means of treatment of true food allergies is the specific avoidance diet. Simply put, the patients must carefully avoid the food(s) that provoke their reactions. Some products made from the offending food may not contain the allergenic proteins. For example, peanut oil usually contains no protein and will not trigger allergic reactions in peanut-sensitive individuals. Avoidance diets can be complicated by cross-reactions. For example, many individuals with allergies to shrimp will also be allergic to other crustacea, such as crab and lobster, but can often consume other seafoods such as mollusks or finfish.

Table 1.7 Examples of food-associated idiosyncratic reactions

Reaction	Implicated food or ingredient
Migraine headache	Chocolate
Asthma	FD&C Yellow #5 (tartrazine)
	Sulfites
Hyperkinesis	Food-coloring agents
Aggressive behavior	Sugar
Chinese restaurant syndrome	Monosodium glutamate

VII. Idiosyncratic Illness

Like food allergies and metabolic food disorders, idiosyncratic reactions usually affect only certain individuals in the population. Idiosyncratic reactions are simply those individualistic adverse reactions to foods that occur through unknown mechanisms; a few are listed in Table 1.7.

A. Unproven Reactions

Evidence for the existence of some of these reactions is scanty. In some cases, the only evidence is anecdotal reports in the medical literature that are not substantiated by challenge studies or other diagnostic evidence; such reports should be viewed largely as guesswork.

In other cases, the relationship has been largely disproven yet the public may persist in its belief that the reaction is due to some particular food ingredient. A good example is the relationship between hyperkinetic behavior in children and food-coloring agents. Several controlled challenge studies have been conducted on hyperkinetic children. None of the studies has substantiated the original hypothesis. Still, some parents continue to believe that food colors are a common cause of hyperkinetic behavior.

Another example is asthma ascribed to FD&C Yellow #5, tartrazine. Initial studies seemed to demonstrate a relationship between tartrazine and asthma in a small percentage of the asthmatic population. However, other clinical challenge studies have failed to confirm these initial findings, and any role for tartrazine in the causation of asthma seems remote.

Sometimes these relationships are difficult to prove because of the subjective nature of the described condition. One of the most subjective complaints listed in Table 1.7 is Chinese restaurant syndrome (CRS). CRS is characterized by a series of subjective symptoms variously described as burning, tightness, or numbness in the upper chest, neck, and face beginning within minutes after the start of a Chinese meal and lasting only a few hours at most. Occasionally, other symptoms such as dizziness, headache, chest pain, palpitations, weakness, nausea, and vomiting are reported. Of all these symptoms, only vomiting is an objective, readily observable symptom.

While many individuals claim sensitivity to monosodium glutamate (MSG) and provide histories of CRS, attempts to confirm the relationship between MSG and CRS with controlled challenges have failed. The failure may be due in part to the lack of ability to measure the symptoms objectively. However, another possibility is that CRS is due not to MSG, but to some other substance in Chinese meals.

B. Proven Reactions

Some idiosyncratic reactions are well established. However, they fall into the idiosyncratic category because the mechanism of the reaction is not understood. From Table 1.7, the best example is sulfite-induced asthma.

The association between ingestion of sulfites and initiation of an asthmatic response in a small percentage of the asthmatic population has been firmly established by controlled, blinded challenge studies. The asthmatic reactions occur within minutes after the ingestion of a triggering dose of sulfites and can be quite severe on occasion. Several deaths have been attributed to the ingestion of sulfited foods by sulfite-sensitive asthmatics. The mechanism by which sulfite induces an asthmatic reaction in sensitive individuals is not understood. It is known that only a small percentage of the asthmatic population is at risk with estimates varying from 1 to 8% of all asthmatics. Challenges with sulfited foods indicate that some sulfited foods, such as lettuce, are much more likely to trigger asthmatic reactions in sensitive individuals than other sulfited foods, such as shrimp or dehydrated potatoes. The level of residual sulfite and the form of sulfite present in the food (free or bound to other food constituents) are important in determining the likelihood of a reaction.

VIII. Summary

Although food seldom transmits illness, there are several known ways in which foodborne illness can occur. Everyone is susceptible to infections and intoxications, although some portions of the population are at greater risk than others. Infections result from eating food containing a viable organism that multiplies in the body, sometimes causing disease. Intoxications result from eating food that contains a substance that poisons the body. In addition to these more frequent types of foodborne illness, there are others that affect only a select portion of the population. Allergies occur when food contains a substance to which the consumer is hypersensitive (generally an adverse immune mechanism), whereas metabolic disorders (some of which are called intolerances) involve the individual's inability to process food components that are successfully digested by others. Idiosyncratic illnesses are events that may or may not be caused by components of food and that affect rare individuals in ways that are not yet understood. Despite the various hazards of eating that have been described, not eating is still more hazardous, and eating seldom results in illness.

Bibliography

Carpenter, C. C. J. (1982). The pathophysiology of secretory diarrheas. *Med. Clin. North Am.* **66**, 597–609.

Metcalfe, D. D., Sampson, H. A. and Simon, R. A. (eds) (1997). 'Food Allergy – Adverse Reactions to Foods and Food Additives', 2nd ed. Blackwell Scientific, Boston, MA.

Rozman, K. and Hanninen, O. (eds) (1986). 'Gastrointestinal Toxicology'. Elsevier, New York.

Strombeck, D. R. (1979). 'Small Animal Gastroenterology'. Stonegate, Davis, CA.

Taylor, S. L. and Hefle, S. L. (2001). Food allergy. *In* 'Present Knowledge in Nutrition', 8th ed. (B. A. Bowman and R. M. Russell, eds), pp. 463–471. ILSI Press, Washington, DC.

Epidemiology, Cost, and Risk of Foodborne Disease

2

Jill A. Snowdon, Jean C. Buzby, and Tanya Roberts

I. Introduction

A. Overview of Foodborne Disease

1. What is foodborne disease?

Foodborne disease is any illness that results from ingestion of food. Food can contain microbiological or chemical agents that cause infections and intoxications. The sources of these agents range from being an inherent constituent of certain food to inadvertent addition during food production, processing or preparation. Many foodborne diseases can also be transmitted through water or through contact with infected farm animals, pets, and humans. Only a few foodborne diseases are transmitted exclusively via foods. At the other extreme, a few diseases that may not be transmissible via foods (e.g. AIDS) have, nonetheless, raised concern when they occur in food service workers.

2. Risk is inherent to eating

Those who produce and prepare food take steps to minimize the chances that a noxious agent will be transmitted via food. Although the chances of transmitting foodborne disease can be diminished, food will never be free of risk because even sterile food can be recontaminated, some toxins can survive cooking, and some individuals may be sensitive to certain foods (e.g. allergies). However, the risk of eating food from a well-managed food production system is low, especially compared to the risk of not eating. Although a few food items could be dropped from one's diet, eating a variety of foods is necessary to maintain good health.

3. Determining that a disease is foodborne

Determining that an illness is foodborne can be difficult because illness may not develop for days and the most recently eaten meal may incorrectly get the blame. One way to determine that a disease is foodborne is to find the agent that caused the disease in a sample of the food that the ill person has eaten. This is not always possible because leftover food may not be available or no analytical technique may exist to detect the agent. Another way to determine that a disease is foodborne is when a cluster of cases of an illness occur among persons who had nothing else in common than having eaten the same food; such clusters are sometimes defined, in recent years, by 'molecular fingerprinting' of agents isolated from diffuse cases. Also, transmission via food may be inferred because the illness affects the digestive tract; this can be a false clue and the absence of digestive tract symptoms does not prove that a disease was not foodborne. Finally, transmission via food may be suspected when the disease that occurs is one that is known to be conveyed through the food in question.

Some diseases are largely, frequently, occasionally, or never transmitted through food. Some agents *might* cause disease and some agents *do* cause disease. Correctly discriminating between these categories avoids diverting funds to less important areas. For example, many people believe that pesticide residues in food are responsible for cancer, whereas experts maintain residues are an insignificant contributor to cancer. However, in the 1980s, substantial public funds were allocated to the low-risk residue area, while funds to prevent actual foodborne hazards relatively were marginal. Ultimately, scientific evidence, not just theoretical assessment, is necessary to determine that a disease is foodborne.

4. Cultural implications of food

Most food selections are based on considerations other than nutrition or safety. Food choices are based on a myriad of factors including social, cultural, psychological, religious, spiritual, and biological motivations. As almost every culture pays some attention to the aesthetics of food preparation and appreciation, it can be argued that food is a significant nondurable art medium. Food also makes a contribution to life as part of our fun, recreation, reward, and gastronomic delight. The senses of taste and smell that guide our reactions to food are part of our evolutionary heritage and presumably are significant to our survival as individuals and as a species. Personal preference is as much of a factor in determining what people eat as is nutrition or safety.

5. Intrinsic relationship between food and water

There is an intrinsic connection between the availability of safe water and safe food. As water is used in production, processing, and preparation of food, contaminated water can easily affect the food supply. Maintaining safe drinking water (and disposing feces safely) is fundamental to the health of any society. Generally the study of waterborne disease is classified under the study of drinking water. From the viewpoint of the analysis of disease, waterborne and foodborne disease are considered as separate entities and foods are the only vehicle discussed in this chapter.

B. Limitations on Foodborne Disease Data

1. Difficulty in measuring foodborne disease

Several events must occur for a foodborne disease to be reported. First, one or more individuals become ill with a disease that is ascribed to the consumption of food. Second, those individuals need to seek a medical diagnosis and perhaps medical care. The diagnosis must be made correctly. Specimens, typically of feces or blood, need to be collected. The specimens then need to be accurately analyzed, perhaps comprehensively, for foodborne disease agents. If a foodborne disease agent is identified, the results need to be reported to a central agency that maintains records on foodborne disease.

It is uncommon for this entire chain of events to happen. The symptoms of foodborne disease are sometimes so mild and short-lived (a day of loose stools, for example) that the illness is barely acknowledged and medical treatment is not sought. Those individuals who seek medical care may not be correctly diagnosed; foodborne disease is often incorrectly attributed to a '24-hour flu'. Clinical samples are seldom collected. When samples are collected, the analysis is typically limited to a few bacteria or perhaps a parasite with which the physician and the laboratory are familiar. The large variety of possible foodborne diseases – caused by toxicants, viruses, bacteria, or parasites – makes comprehensive analysis and accurate detection time-consuming, expensive, and often infeasible. Even if this information is developed, few foodborne diseases are obliged to be reported or the data report may be incomplete. Finally, limited resources are available for compilation, analysis, and publication of the data that are reported.

Since most foodborne disease is under-reported and direct measurement is not practical (for the reasons listed above), the true incidence of foodborne disease can only be estimated. Correction factors are developed through expert opinion or through research. Scientists can use surveys to determine the difference between those illnesses that are reported and those illnesses that actually occur. Direct measurements of illness, e.g. the number of cases identified by laboratories, are multiplied by the correction factor determined by these surveys. Figure 2.1 provides an example of a model to correct for under-reporting of foodborne illness.

2. Importance of foodborne disease is under-rated

The relative importance of foodborne disease tends to be underappreciated in contemporary society even though there may be millions of illnesses each year even in developed countries. This is in large part because most foodborne diseases in developed nations are seldom fatal, are of relatively short duration, are self-limiting, and typically have commonplace symptoms such as diarrhea and vomiting. The more profound foodborne diseases, such as poliomyelitis and cholera, are considered by some as under control by modern medical and sanitary practices. Additionally, the public health importance of foodborne disease is not well understood, especially for some of

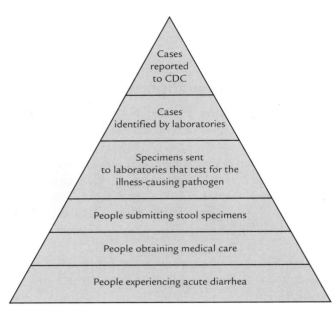

Figure 2.1 A model used to correct under-reporting of foodborne illness. CDC, Centers for Disease Control and Prevention.

the chronic (and expensive) consequences; foodborne illness is a large and complex problem. Having information on the epidemiology, cost, and risk of foodborne disease is key to controlling foodborne disease.

II. Epidemiology of Foodborne Disease

A. Importance of Epidemiology in Foodborne Disease

Epidemiology is the study of the occurrence and distribution of disease in populations and the factors that account for the pattern of disease. In its application to foodborne disease, epidemiology is a tool used to estimate who gets what illness from what food, how often, and under what conditions. Epidemiology measures the incidence and pattern of disease, identifies the factors that result in disease, and leads to recommendations to prevent or minimize disease. Epidemiological data are used to assess the risks of foodborne hazards, establish priorities, allocate resources, increase knowledge and stimulate interest in food safety issues amongst both consumers and policymakers, establish risk reduction strategies, and evaluate the effectiveness of food safety programs.

B. The Study and Measurement of Foodborne Disease

1. Definitions

Pathogens are living or inert agents that cause disease. Epidemiology studies the causes, prevalence, patterns, and determinants of disease, and the prevention of disease

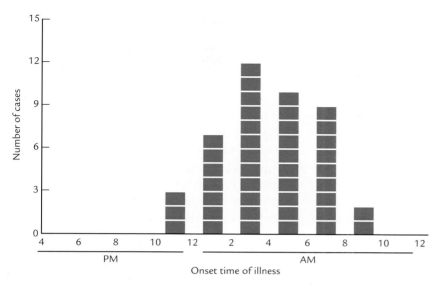

Figure 2.2 Epidemic curve (*span of onsets* = 12 hours).

in populations. It is the study of disease as a mass phenomenon: by examining what happens to the population as a whole, specific recommendations can be made and implemented. Often this study is accomplished through *surveillance* – a system of monitoring and recording disease events.

Descriptive epidemiology examines the features of a disease and asks: who became ill, what were the symptoms, when did the symptoms begin, where did the illness occur, and how many became sick? Plotting this information graphically yields an *epidemic curve* (Fig. 2.2), where the date of onset of symptoms is shown on the *x*-axis and the number of cases on the *y*-axis. The incubation period is the time between exposure and onset of symptoms. If the exposure to a disease-causing agent occurred at a single place and time (a point source), there might be a distinct peak to the epidemic curve. Likewise, the geographic distribution of disease can be portrayed on a map. Finally, the characteristics of the affected individuals, such as age and gender, are included in descriptive epidemiology studies.

Analytical epidemiology compares persons who have been exposed or become ill to those who have not. Other differences between the two groups – particularly the types of foods that have been eaten – are determined and are used in identifying the factors that may have led to disease. Included in this category are case-control studies, cohort studies, and cross-sectional studies.

Molecular epidemiology includes the study of specific attributes of infectious agents at the molecular level. Recently developed techniques include: plasmid profile analysis, restriction endonuclease DNA analysis, and DNA hybridization. These techniques help determine whether a group of individuals became ill with the same or different microbial strain or subtype. They also help provide assurance in correctly identifying the source of the disease under study.

An *outbreak* of foodborne disease occurs when two or more persons experience a similar illness after ingesting a common food. Certain distinct illnesses, such as botulism and chemical poisoning, may be recorded as an outbreak even when they occur as *incidents* (a single occurrence) of foodborne disease. This is in contrast to the *incidence*, or frequency of foodborne disease. A *case* is an individual who becomes ill as evidenced by microbiological, clinical, or epidemiological data. When that illness is an isolated event, as opposed to being part of an outbreak, it is deemed to be a *sporadic* case. The agent (or cause) of a disease is called the *etiology* of the disease.

Risk factors are events, behaviors, and characteristics identified during epidemiological studies that are strongly associated with the development of disease. A risk factor is not necessarily a sufficient cause of the disease, simply something that is associated with the disease. This is not to be confused with risk analysis and hazard analysis, which are described in a later section.

2. Techniques used to study foodborne disease

Ideally, epidemiology can be used to interpret the myriad of forces that contribute to disease. The study of foodborne disease may include: an analysis of what has happened in the past (e.g. case-control studies of outbreaks or sporadic disease); monitoring current events (through a surveillance system); or analysis of an indirectly related event (e.g. causal relationships). Epidemiological data confirmed by laboratory analyses of food or clinical specimens and accompanied by studies of the food preparation practices provides the strongest evidence about the source of disease. In the absence of laboratory-based data, epidemiological data may provide the only available basis for action, such as a governmental response to an outbreak, to protect the public. These decisions need to be made with care, or wholesome food may be unjustifiably condemned.

a. Investigating outbreaks

Investigations of foodborne disease outbreaks in the US are usually begun by a local unit of government and may be joined by state or federal epidemiologists. A final report is to be submitted to the Centers for Disease Control and Prevention (CDC). The likelihood that an outbreak will be investigated depends upon the severity of the illness, the number who are ill, the interest of the public health personnel involved, and the availability of resources.

The organization of outbreak investigations is described in detail in a manual published by the International Association for Food Protection (formerly the International Association of Milk, Food and Environmental Sanitarians). The investigation may include interviews of affected persons (and some who are not), direct observations of food preparation practices, and often laboratory analyses of clinical and food samples.

The shape of the epidemic curve (Fig. 2.2) is used to indicate the possibility of a common exposure and the nature of the disease transmission. A common source and common exposure (a 'point' epidemic) are suggested when the period of time between ingestion and illness is similar to the incubation period typically associated with a particular disease. Person-to-person contact would result in a longer span of onsets (a propagated epidemic). *Predominant symptoms* are compiled on the basis of the percentage of ill persons showing each symptom. These are used to suggest a diagnosis, which should be confirmed by laboratory testing of clinical samples. *Food-specific attack rates* are calculated as the percentage of illness (*attack rate*) among those who ate the food to the attack rate among those who did not eat the food. Most outbreaks involve a complicated array of events that make interpretation difficult; statistical procedures are often used to support the interpretations.

Regrettably, local resources generally are not available for complete and accurate investigations. Consequently, the final data set submitted to CDC is often woefully incomplete. This contributes to the current situation where about 60% of reported foodborne disease outbreaks are 'of unknown etiology', that is to say that no agent is identified as having caused the disease. Incomplete data sets reported to CDC undermine the meaning and value of the outbreak investigations. Foodborne disease outbreak data may not be compiled and reported because the caliber of the data is so erratic. Nonetheless, the data tend to be used to justify opinions or positions.

b. Studying sporadic disease
The number of people involved in sporadic cases of disease differs from those involved in outbreaks. The mode of transmission and the causes of sporadic cases of foodborne disease may differ from those in outbreaks. Studies of sporadic disease are likely to be prospective to compensate for the isolated nature of the event. Studies of these individual occurrences are often based on reports from the laboratory that isolated the disease agent (known as *culture-confirmed*) or on reports from physicians as determined by diagnostic criteria (known as *clinically based*).

c. Case-control studies
A *case-control* study is one in which an individual who has been ill is matched by characteristics such as age, gender, and geographic locale to individuals that were not ill (called *controls*). An interview or questionnaire is used to gather data on the case and the matched control, and the information is compared to determine the factors that resulted in illness.

d. Surveillance of a particular type of disease or a particular population group
Disease surveillance is a system of monitoring distributions and trends of *morbidity* (disease) and *mortality* (death) data. Surveillance systems may monitor selected events such as outbreaks, a particular type of disease or a particular population group. A system of surveillance may be established to monitor a specific situation. This is done sometimes in association with cohort or case-control studies. The purpose of disease surveillance is to understand causes, detect trends in disease, and develop control measures.

e. Studying related events
Events that are related to foodborne disease but are not directly part of the disease may be used to control or study foodborne disease. This includes microbiological monitoring of food animals or foods (e.g. determining the prevalence of *Listeria* in ready-to-eat

Table 2.1 Sources of data on foodborne disease

Foodborne disease outbreak surveillance system in the United States
Records outbreaks investigated through state and local health authorities and reported to CDC on a voluntary basis. Outbreak data may not reflect sporadic disease, and the majority of foodborne disease is thought to be sporadic. However, outbreak data can provide details and insights that serve as the bases for further study.

National notifiable disease surveillance system
Records information from physicians who report cases to state and local health departments, who report to state health departments, who in turn report to CDC on an annual basis. This is mandatory for salmonellosis.

Laboratory-based surveillance system
State laboratories report data to CDC on botulism, campylobacteriosis, cholera, hepatitis A, listeriosis, salmonellosis, shigellosis, trichinosis, typhoid fever, and other communicable diseases.

National *Salmonella* surveillance system
Public health laboratories (federal, state, and some cities) voluntarily report the number of isolates of *Salmonella* detected to CDC. An electronic system (Public Health Laboratory Information System or PHLIS) for data reporting is in use.

Foodborne diseases active surveillance network (FoodNet)
A collaboration between CDC, the US Department of Agriculture, FDA, microbiological laboratories, and selected state health departments designed to monitor certain diarrheal diseases, determine the proportion attributable to specific foods and develop a network to respond to emerging foodborne diseases. The numbers of clinical cultures positive for *Campylobacter*, *E. coli* O157:H7, *Listeria*, *Salmonella*, *Shigella*, *Vibrio*, *Yersinia*, *Cryptosporidium*, and *Cyclospora* within clearly defined and predetermined geographic regions are actively collected in a system described as *active laboratory-based surveillance*. Case-control studies and surveys of clinical laboratories, physicians, and individuals are also carried out, including surveys to measure the proportion of diarrheal diseases that are undiagnosed and unreported so that the true disease incidence can be estimated. This surveillance system provides the most accurate and comprehensive information on foodborne diseases in the US and represents a dramatic advance in foodborne disease surveillance.

PulseNet
A national network for communicating and comparing data on *E. coli* O157:H7, *Salmonella*, and *Listeria*. A single standardized laboratory protocol (using pulsed-field gel electrophoresis) is used to subtype bacterial strains. Each participating laboratory can compare their strains to those isolated in other laboratories for rapid detection of clusters and related cases.

Food-related illness and death in the United States (see Bibliography)
A landmark paper which relied on available data (especially from FoodNet) as well as expert opinion to estimate the incidence of specific and overall foodborne disease. Currently, the only study of its kind to estimate a large variety of foodborne diseases with specificity and precision.

Other sources
Other sources of data include the National Center for Health Statistics, the Health Care Financing Administration, the US Bureau of the Census, state health departments, health maintenance organizations, insurance companies, the United Nations, and the World Health Organization.

foods), developing models that represent estimates of disease or patterns of disease transmission (e.g. risk assessment models), or collecting specialized information electronically (e.g. computer transmission of data on phage types of bacterial isolates).

3. Sources of data on foodborne disease

Most foodborne disease data in the US are based on surveillance systems operated by CDC in conjunction with state health departments (Table 2.1). Results of the studies and annual summaries are often reported in scientific publications such as CDC's journal *Emerging Infectious Diseases* and the *Mortality and Morbidity Weekly Report*. Medical data on individual cases, often published in the literature as a case history, as well as other reports in the scientific literature and risk models based on pathogens' prevalence in foods and on infectious doses, are additional sources of data. Expert opinion also plays an important part in the development of estimates on foodborne disease. Direct measurement of foodborne disease is difficult and databases are incomplete; the opinion of experts is often solicited to help bridge these gaps.

C. Estimates of Foodborne Disease: Incidence and Nature

1. Estimates of foodborne disease in the US

Comprehensive data on foodborne disease in the US were very limited until the 1999 report from CDC was released. This study refined previous estimates and represents the most accurate information available to date. The final estimates of 76 million cases, 325 000 hospitalizations, and 5000 deaths were developed from a multistep process involving the most current data, using conservative assumptions and information specific to individual pathogens whenever available.

The first estimates developed were for diseases resulting in diarrhea and vomiting caused by microorganisms that could be identified (Table 2.2). These numbers were based primarily on reports of bacterial, parasitic, and viral isolates from microbiological laboratories. For each species, these numbers were adjusted for under-reporting by multiplying by a correction factor, which had been developed by scientific study. Next, estimates for diseases whose symptoms are not dominantly diarrhea and vomiting (toxoplasmosis, listeriosis, and hepatitis), but are caused by identifiable microorganisms, were developed. The third set of estimates were for foodborne disease from agents that could not be identified (and which may or may not be microbiological in nature). This was accomplished by estimating the total number of individuals in the US who experience diarrhea and vomiting each year. The number of individuals estimated to be ill with diarrhea and vomiting from known agents was subtracted from the overall number of those ill.

As not all of these agents are transmitted exclusively by food, the estimated proportion of those who became ill was multiplied by the fraction believed to be foodborne. The number of individuals ill with and without diarrhea and vomiting from a known disease agent transmitted by food were added together to estimate cases from known disease agents. Those ill from unknown disease agents were added to those ill from known disease agents to compute the total (Table 2.3). Although no information on trends is available from this study, the data are a reliable benchmark with which to judge ongoing efforts to control foodborne disease.

The surveillance systems and reports mentioned above have limitations and could contribute more to the control of foodborne disease if they were continued, expanded, and refined. The 1999 CDC estimates comprehensively cover bacteria, viruses, and parasites, and estimate the extent and nature of foodborne disease of unknown etiology. However, information on nonmicrobiological foodborne disease (i.e. allergies, naturally occurring toxins, chemical poisonings, and idiosyncratic illness) is limited or nonexistent, and the long-term impact of foodborne disease (e.g. chronic sequelae) is just beginning to be studied. Additionally, direct measurements of more than the eight organisms monitored by FoodNet are needed, as is a better understanding of the unknown agents

Table 2.2 Estimated foodborne illnesses, hospitalizations, and deaths in the US from known pathogens

Pathogen	Illnesses	Hospitalizations	Deaths
Bacterial			
Bacillus cereus	27 360	8	0
Brucella spp.	777	61	6
Campylobacter spp.	1 963 141	10 539	99
Clostridium botulinum	58	46	4
Clostridium perfringens	248 520	41	7
Escherichia coli O157:H7	62 458	1 843	52
E. coli non-O157 STEC	31 229	921	26
E. coli enterotoxigenic	55 594	15	0
E. coli, other diarrheogenic	23 826	6	0
L. monocytogenes	2 493	2 298	499
S. typhi	659	494	3
Salmonella, nontyphoidal	1 341 873	15 608	553
Shigella spp.	89 648	1 246	14
Staphylococcus spp.	185 060	1 753	2
Streptococcus spp.	50 920	358	0
Vibrio cholerae, toxigenic	49	17	0
Vibrio vulnificus	47	43	18
Vibrio, other	5 122	65	13
Yersinia enterocolitica	86 731	1 105	2
Subtotal	*4 175 565*	*36 466*	*1 297*
Parasitic			
Cryptosporidium parvum	30 000	199	7
Cyclospora cayetanensis	14 638	15	0
Giardia lamblia	200 000	500	1
Toxoplasma gondii	112 500	2 500	375
Trichinella spiralis	52	4	0
Subtotal	*357 190*	*3 219*	*383*
Viral			
Norwalk-like viruses	9 200 000	20 000	124
Rotavirus	39 000	500	0
Astrovirus	39 000	125	0
Hepatitis A	4 170	90	4
Subtotal	*9 282 170*	*21 167*	*128*
Grand total	**13 814 924**	**60 854**	**1 809**

Adapted from Mead, P. S., Slutsker, L., Dietz, V., McCaig, L. F., Bresee, J. S., Shapiro, C., Griffin, P. M. and Tauxe, R. V. (1999). Food-related illness and death in the United States. *Emerg. Infect. Dis.* **5**, 607–625.

Table 2.3 Total food-related illnesses, hospitalizations and deaths in the US

Disease agents	Illnesses	Hospitalizations	Deaths
Known	14 000 000	60 000	1800
Unknown	62 000 000	265 000	3200
Total	76 000 000	325 000	5000

Adapted from Mead, P. S., Slutsker, L., Dietz, V., McCaig, L. F., Bresee, J. S., Shapiro, C., Griffin, P. M. and Tauxe, R. V. (1999). Food-related illness and death in the United States. *Emerg. Infect. Dis.* **5**, 607–625.

Table 2.4 Incidence rates of foodborne disease in some European countries (number of cases per 100 000 inhabitants)

Country	Population (million)	1985	1989	1990	1991	1992	1993
Austria	7.8	19.2	61.5	102.7	113.5	137.3	128
Poland	37.8	60.7	93.9	86.7	89.2	75.3	58.9
Spain	38.8	102	116	97	95	–	–
UK (England and Wales)	49.7	37.7	104	104	103	126	137
UK (Scotland)	5.1	38.6	61.6		58.6	64.9	64

Adapted from World Health Organization Surveillance Programme for Control of Foodborne Infections and Intoxications in Europe: sixth report 1990-1992 (1995). Federal Institute for Health Protection of Consumers and Veterinary Medicine (FAO/WHO Collaborating Centre for Research and Training in Food Hygiene and Zoonoses), Berlin.

that cause foodborne disease (comprising 81% of foodborne illnesses and hospitalizations and 64% of deaths). Consistent and expanded funding for the surveillance of foodborne diseases is needed to help ensure a safe food supply because surveillance provides the information that leads to effective control strategies.

2. Estimates of the incidence of foodborne disease on an international basis

a. World Health Organization surveillance data

The most comprehensive international foodborne disease data are described by the World Health Organization (WHO). Nearly 50 European nations report to the Food and Agriculture Organization/WHO Collaborating Centre for Research and Training in Food Hygiene and Zoonoses. Located in Berlin, this center publishes reports on a periodic basis, the most recent covering data from 1990 to 1992. The individual nations have central contact points to collect and evaluate national data on cases and outbreaks, and forward the information to the WHO center. Although most nations have some requirements for notification of certain diseases, the reporting system varies within each nation and an exact comparison of national figures is not possible. Hence, Table 2.4 is presented as an example of European data that are available.

In general, *Salmonella* species are still the most important causative agent reported to WHO with *Salmonella* Enteritidis being the most isolated serotype. Recently, reports of cases of *Campylobacter* are rapidly rising and may surpass *Salmonella* in magnitude. *Staphylococcus aureus* and *Clostridium perfringens* are also frequently reported as foodborne disease agents in Europe. The relative importance of viruses and other unmeasured agents is unknown.

The WHO also summarized the conclusions from 14 391 outbreak investigations from 1990 to 1992. *Salmonella* Enteritidis was responsible for 50.9% of the reported investigated outbreaks where the causative agent was identified. *Salmonella* Typhimurium was responsible for 3.4%, *Staphylococcus aureus* 3.5%, *Clostridium perfringens* 3.0%, trichinellosis 1.5%, and *B. cereus* 1%. The food vehicles in the outbreaks was identified 71.4% of the time (as opposed to 50% in the US): eggs, meats, and sweet foods lead the list with percentages of 25.4, 23.4, and 17, respectively. It is believed that the pathogens generally originate on the farm, with mishandling or cross-contamination occurring in restaurants and homes. Inadequate refrigeration and inadequate heating were involved in 19.3% and 11.3% of the outbreaks, respectively.

b. Other nations

While Canada has an estimated annual foodborne disease incidence of over 2 million cases, data from nations other than those in North America and Europe are not readily available. Many countries do not have reporting systems.

Some data from outbreak investigations are available from Venezuela (1989), Mexico (1981–1990), Cuba (1990), and Thailand (1991–1992). The most frequently identified causes of outbreaks in Cuba were *Staphylococcus aureus*, *Clostridium perfringens*, *Salmonella* spp., *Entamoeba histolytica*, and *Bacillus cereus*. Staphylococcal intoxications were identified as the most frequent foodborne disease in some of the Americas, and biotoxins transmitted by fish or shellfish are also important. Botulism outbreaks, often from homemade preserved foods, still occur in Latin America and the Caribbean.

Parasitic foodborne diseases are very common in South and Central America and likely elsewhere in the world. Responsible agents include *Taenia solium*, *Entamoeba histolytica*, *Giardia lamblia*, and *Cryptosporidium parvum*. These parasites are often found in 2–10% of the population with loose stool. In Thailand, amebiasis, giardiasis, trichinellosis, gnathostomiasis, *Enterobius* infections, angiostrongyliasis, taeniasis, opisthorchiasis, and fasciolopsiasis are parasitic diseases that are believed to occur regularly.

The leading causes of foodborne disease reported in developed countries appear to be *Salmonella* and *Campylobacter*. For example, salmonellosis has been the leading cause of foodborne illness in Japan since 1992 and there are thought to be between 2000 and 3000 cases of campylobacteriosis annually in South Australia alone. *Campylobacter*, *Salmonella*, *Shigella*, and rotaviruses were recovered frequently from children in Thailand; whereas *Vibrio parahaemolyticus* is a common foodborne bacterium in Japan.

Bacteria are considered to be leading causes of foodborne disease in most countries. This may reflect the ability to detect bacteria and the inability or difficulty in detecting viruses and parasites. Bacteria – in contrast to viruses and parasites – may also multiply in foods. Additionally, the nonmicrobial foodborne diseases (i.e. feed contaminated with mycotoxins, chemical food poisoning from heavy metals, and food allergies) are rarely measured.

The pattern of foodborne disease varies greatly in different countries and is dependent upon cultural practices, the size of the population, access to potable water, and sanitary disposition of waste. Individuals in developing nations and in nations in crisis (owing to war or natural disaster) are likely to struggle for enough food to eat. Their struggle to survive is exacerbated by the diarrheal diseases transmitted through their food and water supply. The importance of foodborne disease in these countries cannot be emphasized enough – diarrheal diseases are the leading cause of death in infants and children in undeveloped nations. The entire spectrum of foodborne disease is likely to be of greater consequence in developing nations than in developed nations.

III. Measuring the Cost of Foodborne Disease

A. Importance of Economic Analysis

Economic estimates of diseases illustrate the economic burden of an illness or injury on society, and are important for setting public health priorities. These estimates are developed by analyzing the costs associated with a particular disease or by estimating what consumers would be willing to pay for reduction of risks. Economic analysis permits a comparison of different diseases and different control methods. Likewise, data from different years also can be analyzed.

Economic estimates can be used in three main ways. First, they can be used to evaluate the economic impact of foodborne diseases: which ones create an economic hardship and what is the magnitude of that hardship. Second, they can be used to determine the order in which pathogens should be targeted for control programs. Third, they can be used to compare benefits and costs of pathogen-control programs to determine the level and direction of intervention in both the public and private sectors. In general, economic analyses help identify cost-efficient strategies (i.e. strategies that achieve a given goal for the least cost).

B. Basic Techniques in the Economic Analysis of Foodborne Disease

When estimating the total costs of a foodborne illness, economists create models to represent the different medical outcomes for individuals that develop the disease. Costs for each outcome are then estimated, typically on an annual basis in a technique known as the *cost of illness* method. Assumptions must be made where data are not available; and these assumptions, as well as the rationale for the model, are typically documented along with the results.

These models vary depending upon the symptoms and severity of the disease under study. In general, cases of foodborne illness fall into four categories: those who did not visit a physician, those who visited a physician, those who were hospitalized, and those who died prematurely because of their illness. A fifth category is sometimes added in the study for patients who develop secondary complications from the initial illness.

An example of a model for *E. coli* O157:H7 is presented in Fig. 2.3. The initial acute illnesses (all cases, not just foodborne) are subdivided by the severity of the outcome. The most severe complication in this example is hemolytic uremic syndrome (HUS), which is characterized by red blood-cell destruction, kidney failure, and neurological complications, such as seizures and strokes. While only a few cases develop HUS, the results are severe, long-lasting, and impose a great economic burden.

The cost-of-illness estimates are calculated from: the number of annual foodborne-illness cases and deaths of the particular foodborne illness under study; the number of cases that develop select secondary complications; and the corresponding medical costs, lost productivity costs, value of premature deaths and, where appropriate, other illness-specific costs, such as special education and residential-care costs.

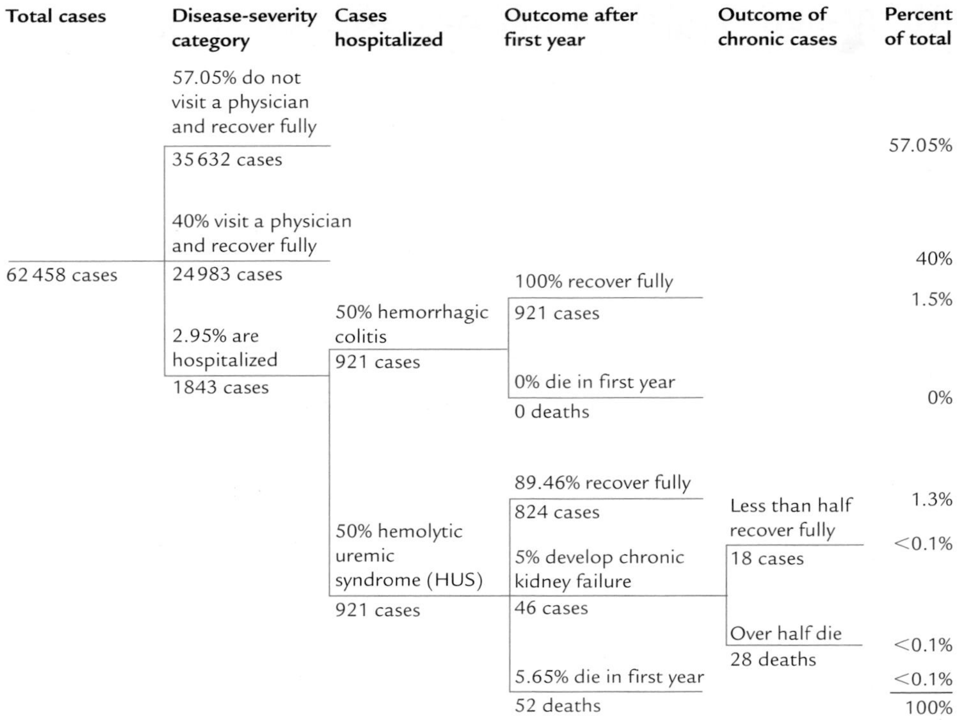

Total cases	Disease-severity category	Cases hospitalized	Outcome after first year	Outcome of chronic cases	Percent of total
	57.05% do not visit a physician and recover fully				57.05%
	35632 cases				
	40% visit a physician and recover fully				40%
62458 cases	24983 cases		100% recover fully		1.5%
	2.95% are hospitalized	50% hemorrhagic colitis	921 cases		
	1843 cases	921 cases	0% die in first year		0%
			0 deaths		
			89.46% recover fully	Less than half recover fully	1.3%
		50% hemolytic uremic syndrome (HUS)	824 cases	18 cases	<0.1%
		921 cases	5% develop chronic kidney failure		
			46 cases	Over half die	<0.1%
			5.65% die in first year	28 deaths	<0.1%
			52 deaths		<0.1%
					100%

Figure 2.3 Distribution of estimated annual US foodborne *Escherichia coli* O157:H7 disease cases and outcomes. Percentages and number of cases may be rounded. Prepared by the Economic Research Service, USDA.

For each category, medical costs were estimated for physician and hospital services, supplies, medications, and special procedures unique to treating the particular foodborne illnesses. Such costs reflect the number of days/treatments of a medical service, the average cost per service/treatment, and the number of patients receiving such service/treatment.

For those people with foodborne illnesses who miss only some days of work, lost productivity is usually approximated by wage rates, published by the Bureau of Labor Statistics. However, some patients die and some develop secondary complications, so that they either never return to work, regain only a portion of their pre-illness productivity, or switch to less demanding and lower paying jobs. Estimating costs for these patients is complicated because there is no consensus on the best way to value these losses (particularly death, in part due to ethical issues). The total cost of lost productivity and premature deaths is the sum for all individuals affected, primarily the patients and, in the case of ill children, their parents or costs for paid caretakers.

The *contingent value* method uses telephone surveys, personal interviews, or mail surveys to elicit what consumers would be willing to pay for a specified risk reduction in a hypothetical scenario. For example, a contingent valuation survey could be used to estimate how much more consumers would be willing to pay for irradiated hamburger over what they would pay for nonirradiated hamburger.

Table 2.5 Estimated annual costs due to selected foodborne pathogens, 2000

Pathogen	Estimated annual foodborne illnesses			Costs (billion dollars)
	Cases	Hospitalizations	Deaths	
Campylobacter spp.	1 963 141	10 539	99	1.2
Salmonella, non-typhoidal	1 341 873	15 608	553	2.4
E. coli O157:H7	62 458	1 843	52	0.7
E. coli non-O157:STEC	31 229	921	26	0.3
Listeria monocytogenes	2 493	2 298	499	2.3
Totals	3 401 194	31 209	1229	6.9

Source: Economic Research Service (2001) (http://www.ers.usda.gov/Emphases/SafeFood/features.htm#start)
Health statistics: CDC (1999) (http://www.cdc.gov/ncidod/eid/vol5no5/mead.htm)

The *experimental auction market* method uses an artificially constructed market situation, such as a laboratory experiment, to determine how much people would be willing to pay for a product that they cannot normally buy through traditional markets. For example, there have been studies that used the experimental auction market method to determine how much people were willing to pay for a chicken sandwich with reduced risk of *Salmonella*. Real people, real money, and real food are used in these otherwise artificial market situations. Both the experimental auction market and the contingent value method measure the value an individual would place on reducing the risk of illness or death.

C. Estimates of the Cost of Foodborne Disease

1. United States

The Economic Research Service (ERS) of the United States Department of Agriculture (USDA) is the leader in estimating costs of foodborne disease in the US. The latest ERS estimates of medical costs, productivity losses, and value of premature deaths for diseases caused by five foodborne pathogens is $6.9 billion per year (Table 2.5). The five bacterial pathogens are: *Campylobacter* (all serotypes); *Salmonella* (non-typhoidal serotypes only); *E. coli* O157:H7; *E. coli*, non-O157 STEC (shiga toxin *E. coli*); and *Listeria monocytogenes*. ERS estimates use the CDC's hospitalization and death estimates for these foodborne pathogens. ERS uses an age-adjusted approach to value premature deaths. The assumed cost of each death ranges from $8.9 million for children who die before their first birthday to $1.7 million for individuals who die at age 85 or older. Because of changes in case estimates, the mix of pathogens analyzed, and the economic valuation of deaths, the ERS estimates are not strictly comparable with earlier ERS estimates of foodborne disease costs.

2. International

Only a few reports are available on the cost of foodborne disease in other countries. The economic impact of 2.2 million annual foodborne illnesses in Canada was estimated as $1.33 billion in 1985 Canadian dollars. Estimated foodborne illness costs in Croatia exceeded US$2 million annually, assuming that the cost per salmonellosis case would be representative of the average foodborne illness and extrapolating cost estimates from studies in other countries to Croatia. It was reported that the estimated 10 million cases of foodborne disease in the UK would be associated with a $1.9 billion dollar expense. The impact of foodborne diseases on travelers alone (if 40% of 500 million tourists are ill at an average cost of $100) could be at least $20 billion.

The main limitation in estimating the economic costs is that estimates of the incidence of disease and the outcomes of disease are lacking. Differences in surveillance systems hinder intercountry comparisons of foodborne illness cases, deaths, and costs. Although precise figures are not known, it is estimated that the economic impact of foodborne illnesses is measured in billions of dollars for each country. The cost of foodborne illness in developing countries may be greater than that in industrialized countries because of high infant mortality, malnutrition, chronic diarrhea, lost work, and child care. However, the economic consequences of foodborne disease are substantial in both industrialized and developing nations.

D. Limitations of the Data

The cost of the impact of all foodborne diseases is not known. Current estimates of foodborne disease in the US undervalue true societal costs for several reasons. First, the estimates provided here focused on only a handful of pathogens whereas over 100 microbiological agents are known to be transmitted through food, and there are numerous foodborne disease agents that are not microbiological. Cost for *all* foodborne pathogens – viral, bacterial, parasitic, fungal, and all the nonmicrobiological agents – need to be included for a complete assessment. Additionally, the cause of most cases of gastrointestinal distress is never identified. Over 60% of the foodborne disease outbreaks that are reported are of undetermined cause and most sporadic cases of foodborne disease go unreported. Hence, the *actual* spectrum of the causes of foodborne disease is unknown and an overall economic analysis is similarly restricted.

Second, economic estimates underestimate the true impact of foodborne disease because the cost estimates tend to focus primarily on medical costs, lost productivity, and the value of premature deaths, while excluding more difficult to measure costs to individual/households, industry, and the regulatory/public health sector. For example, the pain, suffering, and lost leisure time of the victim and the victim's family are not included in these estimates. Neither the loss of business nor liabilities from lawsuits (to the food industry) are included. Likewise the value of preventive action is not included nor are the resources spent by federal, state, and local governments to investigate the source and epidemiology of foodborne disease.

Third, costs for the vast majority of chronic complications associated with foodborne disease are not estimated. The US Food and Drug Administration (FDA) estimates that 2–3% of cases of foodborne illness caused by microbiological pathogens develop secondary illnesses or complications. These complications can be long-term (chronic) and may cause premature death. Examples include arthritis and meningitis, respectively.

Despite these data limitations, economic estimates can be used in the cost–benefit analyses needed in policy making. Economic analyses can assist the allocation of societal resources by identifying data gaps, estimating benefits and costs of alternative control strategies, and helping to set food safety priorities.

IV. Analyzing the Risk of Foodborne Disease

A. Overview of Risk Analysis

Risk analysis is a standardized process to assess, manage, and communicate about risk. Risk analysis originally evolved as a methodical means to make decisions about poorly defined hazards. In particular, the development and use of new chemicals (used as food additives or pesticides) raised the question of their relationship to diseases such as cancer. In many instances there were no data to link the chemical with the disease, nor, in the case of food additives, should disease be expected. Risk analysis was developed to estimate the chances, or risk, of disease *if* low concentrations of the agent under consideration could indeed cause disease.

This emerging science is currently being applied to microbial foodborne disease. Although the cause and effect of microorganisms is better understood than the effect of chemical residues, risk analysis has application to evaluating microbiological hazards as well. The chance of developing a foodborne illness is influenced by a large variety of factors, including food production and preparation techniques, the nature of the pathogen (including its ability to multiply in foods), qualities relating to the individual consuming the food, and the conditions under which the food is consumed (e.g. on a full or an empty stomach). Risk analysis is beneficial in evaluating the impact of these factors and in identifying the means to control them.

B. Basic Concepts

A *foodborne disease hazard* is a biological, chemical, or physical agent in, or a property of, food with the potential to cause illness or injury if not controlled. *Risk* is the probability that a hazard will be manifested; for foodborne disease it is a function of an adverse health effect and the severity of that effect consequential to a hazard(s) in food. A *risk estimate* is a number that describes the probability that the hazard will result. This is different than a risk factor (described earlier in this chapter), which is an element that may contribute to the development of a disease and is typically identified by epidemiological techniques.

Risk analysis consists of the components of risk assessment, risk management, and risk communication. It is a structured process to evaluate the probability of the occurrence and severity of adverse health effects resulting from human exposure to hazards (*risk assessment*), to compare alternatives in light of the results of risk assessment (*risk management*) and to exchange information among interested parties (*risk communication*).

Risk assessment provides a framework for systematically and objectively organizing available information. It evaluates the probability of occurrence and severity of adverse health effects. It is typically divided into four components. *Hazard identification* determines the link between an agent and a human illness. *Hazard characterization* is an evaluation of the nature of the adverse effects associated with a hazard. This may include a dose–response assessment. *Exposure assessment* is the determination of the probability that the hazardous agent (in the instance of foodborne disease) is eaten and the amount of the biological agent likely to be consumed. *Risk characterization* combines information from the hazard identification, hazard characterization, and exposure assessment, and estimates the likelihood and magnitude of adverse health effects (risk estimate).

Risk assessment takes different forms and has different applications. It may be quantitative, with emphasis on numerical expressions of risk, or qualitative (e.g. when expert opinion is the chief tool to evaluate risk). Descriptions of risk assessments also include the assumptions and uncertainties of risk assessment. In addition to providing an estimate of risk for management consideration, risk assessment can also identify gaps in the knowledge base, characterize the most important factors that determine safety in the production-to-consumption pathway of a food, identify efficient strategies to control risks and manage hazards, and provide guidance for research.

C. Applications of Risk Analysis

Risk assessment has great value, particularly in determining the probability that a hazard will occur. Risk assessments have greatest application when they are transparent, flexible, documented, used consistently, and updated as new data become available. Risk assessment of foodborne disease may best be used, similarly to economic analysis, on a relative basis to compare different aspects of disease, and provide guidance for setting priorities and allocating funds.

Some of the first applications of risk analysis to foodborne disease are being carried out by federal governments. For example, the US government studied the relationship between shell eggs and egg products, and the bacterium *Salmonella enterica* serovar Enteritidis. A model was created to represent different aspects of egg production, transportation, storage, preparation, and consumption. Each part of the model was then developed in great detail to represent what was likely to happen to the bacterium, if it was present. The final exposure of humans and the severity of symptoms was then predicted. These types of models are based on statistical principles and use probability distributions to strengthen the accuracy of the predictions. The models can then be used to predict the relative impact of various actions intended to block disease.

A structured risk-analysis process has important implications in international trade. The assessment, management, and communication of risk helps establish equivalent standards and equitable trade between nations while ensuring the safety of the food supply. The processes used must be *transparent* (understood) and have broad acceptance and support.

Risk assessment seeks to identify the likelihood of the hazard actually happening. Hazard Analysis and Critical Control Point (HACCP) is one risk management strategy that seeks to identify and prevent hazards from being transmitted in the food supply. These two techniques may be used concurrently as both are intended to identify and control hazards.

Like epidemiological and economic analysis, risk analysis is a tool that can be applied to identify means to reduce foodborne disease. Understanding of risk can then be applied toward controlling the hazard that leads to disease.

V. The Future of Foodborne Disease

There is likely to be an increased need for attention to the safety of the food supply. As populations grow, both the need for food and the need to dispose safely of fecal and chemical waste (that might contaminate food or water) increase. The steadily increasing need for space for food production and waste disposal, in the face of a finite amount of land, results in a consistent ecological pressure that challenges the ability to provide safe food and water. This pressure may result in the emergence of new pathogens that survive traditional processing and preparation practices, and appear in new foods.

There are other factors currently influencing the incidence of foodborne disease in the US. Incidence may appear to increase as awareness of foodborne disease increases, and as improvements in detection and epidemiological techniques are made. Technological advances are encouraging the mass production and distribution of food; if food is contaminated the chances are greater that it will be in a large batch and distributed widely. New techniques for preserving or making food are being developed. People are more mobile, which increases the probability of exposure to a pathogen as well as the ability to carry (in the intestines) pathogens anywhere that the person goes. The population is changing: more people are living longer and their vulnerability to foodborne disease is increasing as a result of aging, AIDS, cancer, and organ transplants. Influential changes in society result in more food being consumed away from home; food service workers remain important in protecting the food supply. There are new dietary patterns, an interest in new foods, and new merging and mixing of cultures and their cuisines. For example, one trend in the US is greater consumption of 'ready-to-eat' and 'healthy' uncooked foods (fresh fruit, vegetables, and sprouts).

Control of foodborne disease can be enhanced by research. Funds for food safety research, while increasing in the past few years, are still limited. Progress remains to be made in improving our understanding of the ecology of pathogens. Risk assessment models to better estimate the impact of control options from farm to table are still in their infancy. Models of what motivates human behavior change (at home and in industry) to better control foodborne risks are incomplete and studies of how to increase incentives for the private sector to improve its food safety performance are needed.

Other areas affecting the future of foodborne disease include support for programs in public health, sanitary disposition of wastewater, and agriculture. Support for public health waxes and wanes throughout history; currently the public health infrastructure, particularly at local levels, may not adequately cover foodborne disease.

Federal surveillance of foodborne pathogens is better than ever historically. Progress remains to be made, however, in identifying the 'unknown' causes of reported cases of foodborne illness. The infrastructure for the disposal of wastewater may need to be examined for its adequacy under current conditions as should the disposal of animal wastes. Enhanced collaboration between the agricultural, food, regulatory, research, and public health communities may help in identifying effective control measures to ensure food safety.

Epidemiology, economic analysis, and risk assessment are tools that can be used to better understand and reduce the hazards of foodborne disease. More epidemiological data are needed so that economic and risk analyses can proceed with accuracy. The important causes of foodborne disease need to be identified, as do the foods and actions that result in disease. Populations at highest risk for severe foodborne infections need to be determined and the successes of specific intervention efforts need to be assessed. Money is needed for surveillance, research, education, and other intervention programs. Continued commitment from the public and private sectors is needed to ensure protection of the food supply. As food and water are essential to life, efforts to ensure their safety have been and continue to be of fundamental importance to society.

Disclaimer

The views expressed in this chapter do not necessarily represent those of the US Department of Agriculture or the US Government.

Bibliography

Bryan, F. L., Guzewich, J. J. and Todd, E. C. D. (1997). Surveillance of foodborne disease II. Summary and presentation of descriptive data and epidemiologic patterns: their value and limitations. *J. Food Prot.* **60**, 567–578.

Bryan, F. L., Guzewich, J. J. and Todd, E. C. D. (1997). Surveillance of foodborne disease III. Summary and presentation of data on vehicles and contributory factors: their value and limitations. *J. Food Prot.* **60**, 701–714.

Buzby, J. C., Roberts, T., Lin, C.-T. J. and MacDonald, J. M. (1996). Bacterial Foodborne Disease: Medical Costs and Productivity Losses. US Department of Agriculture, Economic Research Service, Commissioned product for FCED, AER No. 741. (http://www.econ.ag.gov/epubs/pdf/aer741/)

Council for Agricultural Science and Technology (1994). 'Foodborne Pathogens: Risks and Consequences'. Task force report no. 122. Ames, Iowa.

Guzewich, J. J., Bryan, F. L. and Todd, E. C. D. (1997). Surveillance of foodborne disease I. Purposes and types of surveillance systems and networks. *J. Food Prot.* **60**, 555–566.

Mead, P. S., Slutsker, L., Dietz, V., McCaig, L. F., Bresee, J. S., Shapiro, C., Griffin, P. M. and Tauxe, R. V. (1999). Food-related illness and death in the United States. *Emerg. Infect. Dis*. **5**, 607–625. (http://www.cdc.gov/ncidod/eid/vol5no5/mead.htm)

Todd, E. C. D., Guzewich, J. J. and Bryan, F. L. (1997). Surveillance of foodborne disease IV. Dissemination and uses of surveillance data. *J. Food Prot.* **60**, 715–723.

WHO (1995). Application of Risk Analysis to Food Standards Issues: Report of the Joint FAO/WHO Expert Consultation. World Health Organization, Geneva.

WHO (1997). Food safety and foodborne diseases. *World Health Stat. Quart.* **50**(1).

www.econ.ag.gov/briefing/foodsafe

www.foodsafety.com

Infections

PART

II

Salmonella

3

Jeffrey T. Gray and Paula J. Fedorka-Cray

I. Overview

Salmonella is a member of the family Enterobacteriaceae that comprises a large and diverse group of Gram-negative rods. Members of the genus *Salmonella* are zoonotic and can be pathogenic in man and animals. Salmonellae are ubiquitous and have been recovered from some insects and nearly all vertebrate species, especially humans, livestock, and companion animals. They can be found free living in nature and as part of the indigenous flora of humans and animals, and are also known to survive well in a desiccated state. They have the ability to grow aerobically and anaerobically, and are metabolically creative, having the ability to utilize a variety of substrates.

One persistent problem encountered with *Salmonella* involves taxonomy and nomenclature. Rather than attempting to explain all of the intricacies of *Salmonella* nomenclature, this chapter will provide some insight on how best to work with the cumbersome manner in which this organism is designated. There have been over 2400 different *Salmonella* serovars described; however, only a few predominate in humans, a given animal population, or a country at any one time. In recent decades we have observed global pandemics of *S. typhimurium* DT104 and *S. enteritidis*; yet we have little knowledge as to the factors that contributed to the pandemics and the differences in characteristics of isolates that have been described between countries. The *S. enteritidis*

Foodborne Diseases 2nd Edn
ISBN: 0-12-176559-8

outbreak induced considerable political concern and has resulted in legislation in many countries to control the prevalence of *Salmonella* in farm animals in order to prevent foodborne infection. Likewise, the emergence of the antimicrobial penta-resistant strain of *S. typhimurium* DT104 (DT104) has induced global concern about the use of antimicrobial drugs in agriculture.

Some *Salmonella* serovars are host-adapted; thus serotypes such as *S. typhi* and *S. gallinarum* are associated with humans and chickens, respectively. Significantly, although *S. choleraesuis* is considered host-adapted in swine, it has been recovered from humans as well as other animal species. The bases for host adaptation are largely unknown; however serovars that are characterized as host-adapted are typically not foodborne pathogens and therefore will not be considered in depth here. In contrast, *S. typhimurium*, one of the most widespread foodborne pathogens, has the ability to infect humans and nearly all animal species. Infection with *Salmonella* is primarily thought to be by ingestion, and large doses are usually required to cause experimental infections. However, epidemiological evidence suggests that aerosols may also play a role in transmission and that the infective dose may be much smaller in many situations.

Although *Salmonella* may colonize the intestine without causing disease, close association with and penetration of the intestinal mucosa are necessary for the induction of diarrhea and systemic disease. Many possible virulence mechanisms have been identified in tissue culture systems *in vitro*, as well as in animals; but there is still much to understand about the *in vivo* role of various *Salmonella* genes that may be involved in invasion. Likewise, evidence for the role of toxins is confusing and often contradictory.

The wider aspects of the ecology and epidemiology of *Salmonella* are, however, of major importance to all those involved in developing methods to control this pathogen. In recent years, persistence of salmonellae has been demonstrated in the environment; in foodstuffs; and in many calf, chicken, and pig production units. The ubiquitous nature of the organism has led some to consider *Salmonella* to be primarily an environmental organism that is pathogenic to animals and humans. A wide range of survival mechanisms have been identified, and further studies regarding the epidemiology and persistence of *Salmonella*, using traditional and molecular approaches, will be important.

II. Characteristics of the Organism

Salmonellae are Gram-negative, rod-shaped bacteria belonging to the family Enterobacteriaceae. They are facultatively anaerobic, nonlactose fermenting, nonspore forming, and most are motile. Salmonellae are not fastidious organisms; they can multiply under a variety of environmental conditions outside of living hosts. They grow readily in many foods, as well as in water contaminated with feed or feces. Complete inhibition of growth occurs at pH <3.8, temperatures $<7°C$, or water activity <0.94. Salmonellae are eliminated by cooking (boiling or equivalent frying) and by pasteurization of milk (71.7°C, 15 seconds) and fruit juices (70–74°C, ≤20 seconds).

Differentiation is based on the presence or absence of capsular antigens, flagellar antigens, or envelope antigens and their reaction to specific antisera, thus classifying them into serotypes. There are currently well over 2400 serotypes. Therefore, *S. typhimurium* can be more accurately described as *Salmonella enterica* serotype (serovar) Typhimurium. However, for utilitarian purposes they are often referred to using a genus-species (or serovar) designation, such as *Salmonella typhimurium* or *Salmonella mbandaka*, when in fact they are each designated *S. enterica*. Serotypes can be further categorized using bacteriophages. Definitive types are based on the pattern of resistance or susceptibility to a set of bacteriophages. For example, *S. typhimurium* definitive type (DT) 104 designates a particular phage type for typhimurium isolates. In contrast, phage types (PT), rather than definitive types, are indicated for *S. enteritidis*, one of the most common being PT4.

Epidemiologic classification of *Salmonella* is based on host preference. One group includes serotypes that infect only humans, for example, *S. typhi* and *S. paratyphi*. Infection in these cases is characterized by chronic fever, bacteremia, and spread to the reticuloendothelial system, involving mesenteric lymph nodes, liver, and spleen. A second group includes serotypes whose host specificity involves primarily animals. For example, *S. pullorum* infects avian species, *S. dublin* infects cattle, and *S. choleraesuis* infects swine. In rare cases, serotypes from this group infect humans, and cause invasive and life-threatening infections with high mortality. The third, and largest, group includes the remaining serotypes, for example, *S. typhimurium* and *S. enteritidis*. These serotypes have a broad host range and infection by these organisms usually results in a mild to severe gastroenteritis that resolves without treatment, although systemic infections, requiring treatment, can occur. A data review by the Foodborne-Disease Outbreak Surveillance System, US Centers for Disease Control and Prevention, indicated that *S. enteritidis* was responsible for the largest number of reported outbreaks, cases, and deaths between the years 1993 and 1997. Typically, *S. enteritidis*, *S. typhimurium*, and *S. heidelberg* are the three most frequent serotypes recovered from humans each year. However, serotype frequency can vary dramatically each year among the different food animals.

III. Characteristics of the Disease

The clinical pattern of salmonellosis can be divided into four disease patterns: gastroenteritis; enteric fever; bacteremia with or without focal extraintestinal infection; and the asymptomatic carrier state. For general food safety applications, all salmonellae should be considered potentially pathogenic. There is difficulty in differentiating one serovar's potential to cause disease over another serovar, as well as difficulty differentiating between strains within serovars. In practice certain serotypes are associated with particular clinical syndromes: *S. typhimurium*, *S. enteritidis*, and *S. newport* with gastroenteritis in humans and animals; *S. typhi* and the paratyphoid species with enteric fever in humans; and *S. choleraesuis* with bacteremia and focal infection with or without antecedent gastrointestinal disturbance in swine. Significantly, *S. choleraesuis* infection in humans results in high mortality, while infection in other animal species may not even induce clinical disease. Differences between infections may be attributed to virulence differences among the isolates, immune status of the host, and ingested dose.

Salmonella gastroenteritis usually follows the ingestion of contaminated food or drinking water and accounts for an estimated 15% of foodborne infection in the US. Typically gastroenteritis in humans begins 24–48 hours after ingestion with fever, nausea, and vomiting, followed by or concomitant with, abdominal cramps and diarrhea. In animals and humans, 24–36 hours postingestion are key times for onset of clinical disease. Subsequently, diarrhea will usually persist as the predominant symptom for 1–4 days and usually resolves spontaneously within 7 days. Fever (39°C) is present in about half of those manifesting clinical illness. The spectrum of disease ranges from a few loose stools to a severe dysentery-like syndrome.

Dehydration is also possible and even life threatening, especially for the very young and elderly. In a case-control study performed in England and Wales, people under the age of 5 years and over the age of 65 years had higher rates of hospitalization. Patients with compromised immune or reticuloendothelial systems were also more severely infected. Studies of the elderly in nursing homes have found that, in foodborne outbreaks, *Salmonella* is responsible for 52% of the outbreaks and 81% of deaths. Additionally, the data indicate that the case-fatality ratio in the elderly is 3.8%, in comparison to a 0.1% for all other *Salmonella* infections.

In humans, antibiotic treatment is usually contraindicated unless the infection has become systemic. Most often, ciprofloxacin is given empirically at the first sign of severe gastroenteritis. Ceftriaxone is often administered intravenously to children with systemic salmonellosis. In production animals (swine), treatment is usually contraindicated but, when necessary, can be given via injection with several treatment alternatives based on numerous considerations, including withdrawal time. Treatment for salmonellosis in poultry is not routinely considered, as *Salmonella* often colonizes birds at hatch but disease is largely absent.

The observed infectious dose is much higher for *Salmonella* than for *Shigella*; ingestion of 10 000 or more *Salmonella* bacilli is required to cause illness in 25% of healthy volunteers. Achlorhydric individuals, or those taking antacids, can be infected by considerably smaller doses. In most cases, disease is mild and resolved by the host; on occasion, however, the microorganisms enter the bloodstream and cause sepsis. Individuals with impaired humoral or cell-mediated immunity, or those with a compromised reticuloendothelial system (as noted above) are at particular risk for severe infection. *Salmonella* bacteremia may be the first manifestation of HIV infection.

The infectious dose in animals tends to be higher than for humans. Experiments have reported that doses ranging from 1 000 000 to 1 000 000 000 bacteria are required before clinical disease is observed. Yet higher doses may be needed for mortality. However, concurrent infection with a virus can dramatically reduce the infectious dose required for both morbidity and mortality. Infected animals show clinical signs similar to those in humans. For instance, the clinical signs in cattle infected with *S. typhimurium* DT104 include diarrhea, fever, and loss of appetite and body weight. Although adult cattle seem to show clinical signs earlier after initial infection, a higher percentage of the calves display clinical symptoms and die. Cattle can also be healthy carriers for up to 18 months. Clinical disease may be more severe, tissue distribution may be more widespread, and mortality may be higher if the animal is concurrently infected with an immunosuppressive pathogen such as a virus. Cattle coinfected with bovine viral diarrhea virus and *Salmonella* can have extensive infections, with *Salmonella* being recovered from the lung, spleen, kidney, and ileum upon necropsy.

Cats infected with *S. typhimurium* DT104 show clinical signs of severe gastroenteritis that include bloody diarrhea, vomiting, fever, anorexia, dehydration, and depression for 4–10 days. Although most cats shed *Salmonella* for about 10 days after recovery, in one clinical case, a cat infected with multiresistant DT104 continued to shed the pathogen in its feces for up to 12 weeks after clinical recovery. Because of the ubiquitous nature of *Salmonella*, one must assume that pets are exposed and, like other animal species, can shed the organism silently. Considering the close contact owners often have with their pets, this makes yet another source for potential human infection.

As indicated above, it is noteworthy that it is very common for animals to be colonized with the organism, especially in organs other than the intestine, and subsequently shed *Salmonella* without showing signs of disease. This is one of the major reasons why it can be difficult to determine when the organism enters the food supply; there may not be overt fecal matter containing *Salmonella* on the carcass, but lymph nodes and other tissues may be infected, which will subsequently transfer to meat, especially ground products. It is also noteworthy that the ubiquitous nature of the genus increases the probability that a certain percentage of humans are exposed to nontyphoid salmonellae and become colonized without manifesting clinical signs. There are few data regarding this occurrence, which could be critically important when considering *Salmonella* epidemiology.

IV. Pathogenesis

The normal architecture of the brush border of intestinal cells is dramatically altered within minutes following *S. typhimurium* infection by the stimulation of membrane 'ruffles' (Fig. 3.1). Ruffles are specialized sites on mammalian plasma membranes where rearrangement of filamentous actin is induced by growth factors, mitogens, or oncogene expression. Evidently, salmonellae have acquired adhesin(s) on their surface that bind to a host cell receptor and stimulate localized membrane ruffling. This ruffling is associated with cytoskeletal rearrangements at the site of bacterial host cell contact and then by a flux of intracellular free Ca^{2+}. The bacteria are engulfed by the cell, presumably by induced pinocytosis associated with the ruffles. Moreover, ruffles directly elicited by invasive *Salmonella* cells facilitate the engulfment of many more nearby *Salmonella*. Thus a single site of *Salmonella*-induced ruffling acts as a portal of entry for many adherent and adjacent bacteria by 'macropinocytosis'. A key feature in this process is the host cellular receptor recognized by *Salmonella*, as well as the bacterial adhesin mediating attachment. This interaction will be an important area for research in coming years. The way in which *Salmonella* subverts the normal cellular

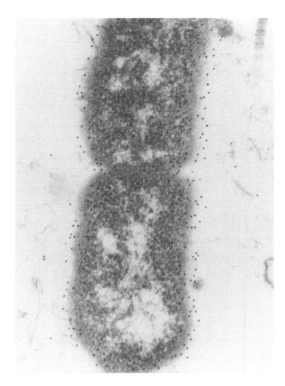

Figure 3.1 Electron micrograph of *Salmonella enterica* serotype typhimurium. Cells are labeled using immunogold particles. Magnification approx. × 21 000.

machinery mediating pinocytosis for its own benefit is of as much interest to mammalian cell biology as it is to understanding bacterial pathogenesis. Because there is no effective model for *Salmonella* gastroenteritis, one can only speculate that the bacteria invade the enterocytes of the large and small bowel, migrate through the cell and pass through the basolateral membrane to enter the lamina propria, and induce a profound inflammatory response.

S. typhimurium, for example, must adapt to various conditions within and outside of the host, such as extreme temperatures and pH, digestive enzymes, detergents, competition with other microorganisms, and host immunity. Through all this, the pathogen must be able to find the right host cells to bind to and begin colonization. *S. typhimurium* transmission to a new susceptible host depends on its ability to interact with host cells.

The Peyer's patches of the distal ileum were identified as the primary invasion sites of *S. enteritidis*, although a few organisms were isolated from the surrounding tissue. About the same time, M cells were discovered. These specialized epithelial cells are located above the follicles on the luminal surface of the Peyer's patches; they maintain tight junctions with adjacent microvilli covering epithelial cells and act as a membrane to separate the lumen from the lymphoid follicles. Directly beneath the follicle epithelium is a dome area that contains B lymphocytes, T lymphocytes, macrophages, and dendritic cells, which activate an immune response to intestinal antigens. M cells function as antigen presenting cells by transporting the bacteria to lymphocytes and macrophages within the dome area.

Shortly after bacterial-epithelial cell contact, *Salmonella* promotes an inflammatory response that is characterized by polymorphonuclear lymphocytes (PMN) surrounding degenerating crypts located between the villi. Bacterial contact with the epithelial cell, rather than bacterial uptake, causes the epithelial cell to produce the cytokine, interleukin-8 (IL-8) and a pathogen-elicited epithelial chemoattractant (PEEC), which guides PMNs to the site of bacterial-epithelial contact. *Salmonella* serotypes that cause enteritis in humans induce PMN migration, whereas *Salmonella* serotypes that do not cause enteritis do not induce PMN migration. IL-8 secretion can also be induced by the translocation of *S. typhimurium* flagellin to the basolateral surface of the epithelial cell, thus initiating an inflammatory response. PMNs play an important role in fluid secretion in the intestine, causing the diarrhea that is associated with the pathogen.

Experiments with tied-off segments of mouse and calf intestines have demonstrated that M cells are invaded within 30 minutes after infection with invasive *S. typhimurium*. Invasion is due to a membrane rearrangement or ruffling mechanism that subsequently leads to columns of actin polymerized into microfilaments that surround and engulf the bacteria. However, only the area of the epithelial cell adjacent to the bacterium undergoes membrane ruffling. At 60–120 minutes, damage to the M cell causes release of the top surface and cell contents into the lumen of the gut, subsequently allowing the bacteria to reach adjacent enterocytes and underlying lymphoid cells. By 180 minutes, the enterocytes have sloughed away, allowing the invasion of significant numbers of bacteria into deeper tissues.

After passing through the intestinal epithelium, invasive salmonellae also enter macrophages via a membrane ruffling mechanism. Unlike epithelial cells, macrophages undergo a generalized membrane ruffling, even in areas where bacteria are not adhered. Survival in macrophages is critical for *Salmonella* because it enables the evasion of the reticuloendothelial system (RES), where the salmonellae reside intracellularly in liver and spleen. In the liver, bacteria reside mainly in Kupffer cells (macrophages) and to a lesser extent in hepatocytes and liver endothelial cells. PMNs kill hepatocytes that contain *S. typhimurium* and also prevent bacterial growth outside of cells. Studies have demonstrated that *Salmonella* has become adapted to living in macrophages in order to evade killing by neutrophils. Salmonellae may also invade through the nasopharyngeal lymph tissue and tonsils, rapidly disseminating via the bloodstream to the intestine and other tissues. Dissemination to other tissues via lymph vessels and blood vessels may be important for long-term colonization without manifestation of clinical illness. Further study is warranted in this area.

V. Virulence Factors of *Salmonella*

Pathogenic *Salmonella* encode virulence factors that are essential for entry, growth, or survival in the host. Once *S. typhimurium*, for example, has been ingested, survival in the stomach depends on the bacterium's ability to tolerate acidic pH. This organism has two acid-tolerance response systems, depending on whether the cells are in log phase or stationary phase. Bacteria that survive the acid in the stomach then enter the small intestine, where they are exposed to high concentrations of bile. When there is a high concentration of bile in the environment, *S. typhimurium* does not invade cells; this aids the organism in its pathogenesis. Once the bacteria reach the distal ileum, the decreased bile concentration acts as a signal to the pathogen to express genes allowing invasion into the epithelial cell. Virulent salmonellae have genes that are essential for invasion of intestinal epithelial cells and for intracellular multiplication. They also trigger engulfment by macrophages, but protect themselves against the normal bactericidal mechanisms inside the macrophages, and may prolong the life of the invaded macrophage.

VI. Transmission

Contaminated food is a common source of infection for humans. A case-control study performed in England and Wales concluded that more humans became infected with DT104 from eating contaminated food than from contact with ill animals. Foodborne illness was most commonly associated with the consumption of chicken, and a higher risk was linked to chicken prepared in a restaurant compared to chicken prepared at home. Additionally, although only 13 of the 83 cases included in the study reported contact with ill animals; this, too, was identified as a significant risk factor. Salmonellae have been recovered from nearly all types of foods including meats, vegetables, and fruits. Prevalence of salmonellae is highest in processed meats, especially ground products. This may be due to the processing procedure, cross-contamination from grinding equipment, or introduction of higher numbers of salmonellae into the product from cheek and other scrap meats, which often contain lymph nodes that are known reservoirs of *Salmonella*.

In addition to transmission by food, salmonellae can also be spread via the environment in a number of ways, indirectly infecting humans and other animals. An important observation is that salmonellae can survive for long periods of time under both wet and dry conditions. The organism is able to survive, and even multiply, in water contaminated with feces. Animals have also become infected after grazing on land that had been washed over by water from an affected farm. Experiments have shown that *S. choleraesuis* can survive in a manure slurry for up to 4 months. Manure lagoons and waste pits serving animal production buildings are often positive for salmonellae, as are litter, dust, and soil. Salmonellae have also been shown to leach through soil and have been recovered some distance from the site of initial wastewater infiltration. Salmonellae can survive for some time during composting and have been recovered from soil for periods >400 days. Survival in excess of 1 year in a dried state has also been observed for both *S. mbandaka* and *S. choleraesuis*. Additionally, following drying, *S. choleraesuis* remained virulent when given to swine.

Transmission to humans can also occur through direct contact with ill or inapparently infected farm animals. Several studies have focused on the relationship between people infected with *S. typhimurium* and their contact with cattle. Human infections with *S. typhimurium* DT104 in the state of Washington were found to be correlated with exposure to cattle and cattle facilities. A greater occurrence of DT104 infections was also observed in Scotland, in people who lived or worked on farms. The incidence in humans was highest during the 2-yearly calving seasons and was correlated with increased morbidity and mortality in calves. An investigation in England and Wales demonstrated that contact with ill cattle was a potential risk factor for acquiring *S. typhimurium*. Strains other than DT104 have similar histories. In 1998, *S. typhimurium* was isolated from a 12-year-old boy who had diarrhea, fever, and abdominal pain. *S. typhimurium* was also recovered from cattle on the father's farm. Investigators believed that the child acquired the pathogen from the cattle because the boy's isolate had an almost identical band pattern on pulsed-field gel electrophoresis. However, the study included no epidemiologic investigation, and the source of both the boy's and the cattle isolates remains uncertain.

Human infections may also result from direct contact with ill pets, such as cats and dogs. Salmonellae are shed in high numbers from the mouths of cats during clinical disease, and grooming habits leave their fur contaminated. Cats also contaminate their paws when scooping dirt over their feces, which can be spread to surfaces in the household. People who handle cats, clean litter boxes, or allow cats free access to food preparation areas, can potentially become infected with the organism.

In 1999, a study conducted in Sweden suggested that wild birds were the source of infection for cats and humans. *S. typhimurium* recovered from sick and dead birds in the area were identical to the *S. typhimurium* isolated from ill cats and people. Recovery of salmonellae from wild geese and turkeys, pigeons, squirrels, and virtually all wild animals have been reported throughout the years on a global scale.

VII. Isolation and Identification

Each bacterial agent of infectious diarrhea has unique pathogenic mechanisms that cause a specific set of symptoms. These symptoms and an adequate patient history are clues that enable diagnosticians to categorize the patient's disease. However, mild forms of gastroenteritis typically resolve without medical intervention. Whether transmitted by food or not, much intestinal distress is caused by viruses.

A. Fecal Samples

Fecal samples are generally the specimen of choice when *Salmonella* is suspected. The fecal sample should not be contaminated with urine; however, if there are portions of the sample that contain blood, pus, or mucus, those should be included for culture along with the fecal matter. Stool samples should contain a minimum of 1–2 g of feces, although more material is preferred. Some collection vials will have a convenient fill line indicating optimal stool quantity. Repeated collection of stools from an individual over 2–3 days can increase the probability of isolation. Preferably, the sample should be submitted to the laboratory within 1–2 hours of collection. If that is not possible, the sample should be put in an appropriate transport media and stored at 4–6°C. The most desirable transport medium for *Salmonella* and *Shigella* is buffered glycerol saline; this is not recommended for *Campylobacter* or *Vibrio* spp., which may be included in the initial differential diagnosis. Therefore, modified Carey–Blair transport medium is recommended as a good universal transport medium and is acceptable for isolation of *Salmonella*. However, it is not as good for isolation of *Shigella* spp. and is inappropriate for *Clostridium difficile*. Unacceptable stool specimens include those that are over 2 hours old that have not been refrigerated, as well as dry rectal swabs. For suspected septicemia, blood culture is indicated.

If the infecting agent is unknown, as it often is, a comprehensive culture process should be followed. The fecal sample should be inoculated on to a routine culture medium, such as blood agar. Additionally, the sample should be inoculated on to an enteric agar medium that has differential qualities, such as MacConkey agar.

If *Salmonella* is suspected, a selective medium should also be used. Selective media for *Salmonella* are designed to inhibit the growth of most members of the family Enterobacteriaciae, while allowing both *Salmonella* and *Shigella*, or just *Salmonella* to grow. Additional differential qualities for the target organism, such as hydrogen sulfide production, will enable a higher frequency of detection. Media such as Hektoen enteric (HE) agar and xylose–lysine–desoxycholate (XLD) are often used for *Salmonella* and *Shigella*. Media such as brilliant green agar with (BGS) or without (BG) sulfadiazine and xylosine–lysine–tergitol (XLT)-4 agar should be used if *Salmonella* is a primary suspect. These agars provide excellent selective and differential properties for *Salmonella*.

For *Salmonella*, the sample should also be inoculated into an enrichment broth to allow low numbers of the bacteria to reach a level detectable following agar culture. One approach is to inoculate the sample into a nonselective enrichment broth such as universal pre-enrichment broth or buffered peptone water for a short period of time, followed by transfer of some of the broth into a more selective medium. This approach works well for recovery of injured cells, which can only begin growth in a permissive environment.

Specific enrichment for *Salmonella* is often warranted and can include inoculating the sample directly into media such as GN-Hajna broth, selenite F broth or tetrathionate broth. After 24 hours of incubation, the broth can be streaked on to agar medium selective for *Salmonella*, incubated for an additional 24 hours, and streaked on to selective agar media. This increases the likelihood that low numbers of *Salmonella* will be detected.

Excellent results have been obtained using the following procedure. The fecal sample is initially inoculated into GN-Hajna (GN) broth and tetrathionate (TET) broth and incubated for 24 hours for GN and 48 hours for TET at 37°C. Following incubation, a 100 µl portion of each of the broths is inoculated separately into 10 ml of Rappaport–Vassiliadis (RV) medium, which is incubated for an additional 24 hours at 37°C. The RV is then streaked to selective *Salmonella* media, most often BGS and XLT4 agars. This process is labor intensive and often too time consuming for most clinical settings; however, for research and epidemiologic investigations, it offers a very sensitive method of determining the *Salmonella* status of a sample.

Agar plates should be examined after 24 hours of 37°C incubation for typical *Salmonella* colony morphology. Colonies having this appearance should be subcultured on to agar and subjected to a screening process for preliminary confirmation. Suspect colonies can be inoculated into triple sugar iron (TSI) agar, lysine iron agar (LIA) and a urea agar slants. Numerous commercial kits are also available for the biochemical screening of suspect colonies, which also provide a reliable preliminary identification of *Salmonella* spp. Any isolate identified as presumptive positive for *Salmonella* should be subjected to complete biochemical identification.

Finally, presumptive positive isolates should be tested for *Salmonella* somatic (O) antigen using slide agglutination methods. It is preferable at this point to submit the isolate to the appropriate state laboratory for complete serological testing (serotyping) and identification. Some laboratories will also phage type the isolate upon request. The serotype is essential in determining isolate origin, and for overall epidemiologic profiling.

B. Food Samples

For food samples, the same media and methods are used as for fecal samples. However, salmonellae in certain foods, such as frozen or dried foods, may be injured; and pre-enrichment in buffered peptone water or other pre-enrichment broth is recommended. Detailed methods for sampling and *Salmonella* testing of foods can be found in the *Compendium of Methods for the Microbiological Examination of Foods* (see Bibliography).

VIII. *Salmonella* Surveillance

PulseNet, the national molecular subtyping network for foodborne disease surveillance, was established by the Centers for Disease Control and Prevention (CDC) and several state health department laboratories to facilitate the molecular subtyping of bacterial food borne pathogens for epidemiologic purposes. The system uses pulsed-field gel electrophoresis (PFGE) DNA fingerprints to compare isolates obtained from various sites. PulseNet, which began in 1996 with 10 laboratories typing a single pathogen (*Escherichia coli* O157:H7), now includes 46 state and two local public health laboratories and the food safety laboratories of the US Food and Drug Administration (FDA) and the US Department of Agriculture (USDA). Four foodborne pathogens (*E. coli* O157:H7; nontyphoidal *Salmonella* serotypes, *Listeria monocytogenes* and *Shigella*) are being subtyped, and other bacterial, viral, and parasitic organisms will be added soon. This system offers a powerful method of comparing isolates obtained from various sources. However, while PFGE is an important epidemiologic tool, molecular typing usually cannot stand alone in determining the source of an isolate, a thorough epidemiologic investigation is needed for isolate source determination. Further information about the PulseNet Molecular Epidemiology System can be found at the CDC website (http://www.cdc.gov).

IX. Treatment and Prevention

The primary therapeutic approach for *Salmonella* gastroenteritis is supportive therapy including fluid and electrolyte replacement, and the control of nausea and vomiting. The illness is typically self-limiting. Antibiotic therapy is contraindicated because it has a tendency to increase the duration and frequency of the carrier state, and may delay resolution of the disease. In patients with underlying risk factors, antimicrobial therapy is applied as a prophylactic measure aimed at preventing systemic spread. For patients with systemic salmonellosis, antimicrobial therapy is essential and antimicrobials are often given continuously for a period of days by intravenous infusion. No human vaccine is available for nontyphoid *Salmonella* infection.

Gastroenteritis usually is self-limited and resolves without treatment. Fluid and electrolyte replacement may be required. Antimicrobial treatment does not shorten the illness or reduce the symptoms; in fact, it may prolong excretion of the organisms. Antimicrobial agents are indicated only for neonates or persons with chronic diseases who are at risk of septicemia and disseminated abscesses. Plasmid-mediated antibiotic resistance is common. Clusters of resistance genes that occur on the bacterial chromosome, such as can be found in *S. typhimurium* DT104, are a more recent phenomenon and present great problems in treatment. If a *Salmonella* infection warrants treatment, antimicrobial susceptibility tests should be performed on the isolate. Empiric treatment may be indicated in some instances. Drugs that retard intestinal motility (i.e. that reduce diarrhea) appear to prolong the symptoms and the fecal excretion of organisms. For humans, the treatment of choice for enteric fevers and septicemia is often ceftriaxone, a third-generation cephalosporin. Ampicillin or ciprofloxacin should be used in patients who are known chronic carriers. For *S. typhi* infections, the gall bladder may need to be removed to eliminate the chronic carrier state. Focal abscesses should be drained surgically whenever feasible.

Salmonella infections in humans can be minimized or prevented through public health efforts and implementation of personal hygiene measures. Proper sewage treatment, a water supply that is chlorinated and monitored for contamination indicated by coliform bacteria, hand washing before food handling, pasteurization of milk and fruit juices, cleaning and disinfection of food preparation surfaces and utensils, and proper cooking of poultry and meat are all important. Hazard analysis and critical control point (HACCP) systems were developed by the food-processing industry several decades ago. More recently, HACCP measures for the postharvest (i.e. slaughter) process have been put in place for *Salmonella* in the US and other countries, and can have an important impact in the reduction of human *Salmonella* cases. Summaries of the incidence of disease can be monitored on the CDC website indicated above. Additionally, USDA meat plant inspection data can be found at www.fsis.usda.gov. Both websites indicate that salmonellosis is declining. Additionally, each commodity group is actively involved in developing preharvest or on-farm HACCP, or other types of monitoring and control programs, which will reduce *Salmonella* at the farm level. The egg industry has already implemented such a plan.

Antimicrobial resistance has become a concern for both human and animal health on a global scale. Scientific groups, such as the US National Institute of Medicine, the American Society for Microbiology and the World Health Organization, have expressed concern about the national and global increase in antibiotic resistance, and the complex issues surrounding the increase in resistance observed in community and institutional settings. In addition to concerns about treatment failures in veterinary and human medicine related to the development of antimicrobial-resistant organisms, there are concerns about the potential transfer of resistant organisms from animals to humans. Although antimicrobials are typically not indicated for the treatment of uncomplicated *Salmonella* infections in humans, their use in treating septic infections is critical. Multiple resistance has also emerged among bacteria, including *Salmonella* species. In addition to resistance developing among salmonellae to the more historical drugs such as the penicillins, resistances to third-generation cephalosporins and the fluoroquinolones have also emerged. These drugs are typically reserved for treatment of complicated *Salmonella* infections and are considered drugs of last resort; if they do not control salmonellae, deaths may result.

Most industrialized countries have implemented both *Salmonella* monitoring systems and antimicrobial resistance monitoring systems. The National Antimicrobial Resistance Monitoring System (NARMS) monitors trends in antibiotic resistance of *Salmonella* from a variety of sources throughout the US. Testing of human *Salmonella* isolates is conducted at the CDC, while testing of animal *Salmonella* isolates is conducted at the USDA laboratory in Athens, Georgia. *Salmonella* isolates from retail meat are tested at the FDA in Laurel, Maryland. The FDA Centers for Veterinary Medicine coordinate the NARMS programs; additional information and data may be found at www.fda.gov website, by linking to the respective NARMS websites.

X. Summary

Salmonella is ubiquitous and can be recovered from humans, animals and throughout the environment. It is incumbent upon every person to understand potential sources of foodborne illness and implement personal hygiene measures, which minimize the risk of infection. Proper cooking and storage of meats, fish, and eggs is essential. Additionally, for those involved in the production of food, implementation of production practices that minimize infection of animals, or presence of salmonellae on fruits and vegetables, will ensure a continued safe and wholesome food supply.

Bibliography

Andrews, W. H., Flowers, R. S., Silliker, J. and Bailey, J. S. (2001). *Salmonella. In* 'Compendium of Methods for the Microbiological Examination of Foods', 4th ed. (F. P. Downes and K. Ito, eds), pp. 357–380. American Public Health Association, Washington, DC.

Anonymous (1994). Enterobacteriaceae. *In* 'Clinical Veterinary Microbiology' (P. J. Quinn, M. E. Carter, B. K. Markey and G. R. Carter, eds), pp. 209–306. Elsevier Science, New York.

Arbeit, R. D. (1995). Laboratory procedures for epidemiologic analysis. *In* 'Manual of Clinical Microbiology' (P. R. Murray, E. J. Baron, M. A. Pfaller, F. D. Tenover and R. H. Yolken, eds), pp. 190–208. American Society for Microbiology Press, Washington, DC.

Curtiss, R. III, Ingraham, J. L., Lin, E. C. C., Brooks Low, K., Magasanik, B., Reznikof, W. S., Riley, M., Schaechter, M., Umbarger, H.E. (eds) and Neidhardt, F. C. (editor-in-chief) (1996). '*Escherichia coli* and *Salmonella*, Cellular and Molecular Biology', 2nd ed. American Society for Microbiology Press, Washington, DC.

Hogue, A., Akkina, J., Angulo, F., Johnson, R., Petersen, K., Saini, P. and Schlosser, W. (1997). *Salmonella typhimurium* DT104 Situation Assessment. US Department of Agriculture, Food Safety and Inspection Service.

Olsen, S. J., MacKinon, L. C., Goulding, J. S., Bean, N. H. and Slutsker, L. (2000). Surveillance for foodborne disease outbreaks – United States, 1993–1997. *Morbid. Mortal. Weekly Rep.* **49**, 1–62.

Pezzlo, M. (1995). Aerobic bacteriology. *In* 'Clinical Microbiology Procedures Handbook' (H. D. Isenberg, ed.), pp. 1.0.1–1.20.47. ASM Press, Washington, DC.

Prescott, J. F., Baggot, J. D. and Walker, R. D. (eds) (2000). 'Antimicrobial Therapy in Veterinary Medicine', 3rd ed. Iowa State University Press, Ames, IA.

Ryan, K. J. and Falkow, S. (1994). Enterobacteriaceae. *In* 'Sherris Medical Microbiology: An Introduction to Infectious Diseases' (K. J. Ryan, ed.), pp. 321–344. Appleton and Lange, Stamford, CT.

Wray, C. and Wray, A. (eds) (2000). '*Salmonella* in Domestic Animals.' CABI Publishing, New York.

Shigella

4

Keith A. Lampel and Anthony T. Maurelli [1]

I. Introduction

The genus *Shigella* is divided into four serogroups based on the somatic O-antigens of the lipopolysaccharide: *Shigella dysenteriae* (serogroup A), *Shigella flexneri* (serogroup B), *Shigella boydii* (serogroup C), and *Shigella sonnei* (serogroup D). There are more than 40 serotypes in serogroups A, B, and C and only one serogroup of *S. sonnei*. *S. dysenteriae*, which causes the most severe symptoms of shigellosis (also known as bacillary dysentery) and is responsible for epidemic outbreaks of dysentery, is found in developing countries in Africa, Latin America, and Asia. *S. sonnei* is the most frequent species isolated in developed countries and causes the least severe symptoms. *S. flexneri* is the most common etiological agent of shigellosis isolated in the developing countries but is being reported more frequently in the developed countries. *S. boydii* is usually found in the Indian subcontinent with more cases emerging in some European countries.

In 1999, data reported to the Centers for Disease Control and Prevention (CDC; www.cdc.gov/mmwr), show that *Shigella* remains as the third leading bacterial agent of foodborne illnesses in the US. The incidence of shigellosis has fallen over the past few years, with the latest figure showing 17 521 cases for 1999. This may reflect better surveillance programs, such as the CDC FoodNet (www.cdc.gov/foodnet), and a

---

[1] The opinions or assertions contained herein are the private ones of A.T.M. and are not to be construed as official or reflecting the views of the Department of Defense or the Uniformed Services University of the Health Sciences.

Foodborne Diseases 2nd Edn
ISBN: 0-12-176559-8

more active role by other federal agencies to monitor the food supply. In the FoodNet program, five sentinel sites are used in an active surveillance program for foodborne pathogens. Data reported from FoodNet for the year 2000 showed that 85% of the isolates were identified as *S. sonnei* and 13% were identified as *S. flexneri*.

The World Health Organization (WHO) tracks foodborne diseases in the world noting the rate of occurrences and distribution of particular pathogens, including several emerging bacterial agents. Data reviewed for 1996 and 1997 estimated that the number of cases of shigellosis worldwide was 164.7 million, with 163.2 million episodes occurring in developing countries. In past reports, such as the WHO Surveillance Programme for Control of Foodborne Infections and Intoxications, many countries noted outbreaks of shigellosis caused by the ingestion of food contaminated with *Shigella*.

Diarrheal diseases are a leading cause of deaths worldwide, and it is estimated that shigellosis is responsible for between 600 000 to 1.1 million deaths each year. *Shigella* spp. cause 5–15% of all cases of diarrhea globally with children 1–4 years old being the most susceptible to these infectious agents. Transmission of the *Shigella* bacilli in individual cases is usually associated with person-to-person contact via the fecal–oral route. In outbreaks, the source of *Shigella* is contaminated food or water. Humans and higher primates are the only known hosts for *Shigella* spp. Incidences of shigellosis are seasonal with higher number of cases occurring during warmer periods of the year.

II. Shigellosis

Clinical symptoms of shigellosis range from mild, watery diarrhea to severe dysentery. The typical incubation period is 1–2 days but may extend up to 7 days. Initial symptoms include abdominal pain, watery diarrhea, and fever. As the illness progresses, severe dysentery occurs as marked by passage of stools with blood and mucus, fever, tenesmus (rectal spasm), and abdominal pain. The presence of leukocytes in stools from patients with dysentery is a characteristic that is used to distinguish this diarrheal disease from diarrhea caused by *Escherichia coli* and rotaviruses. In the elderly, the immunocompromised, and malnourished children, dehydration, acidosis and death can result from loss of fluids and electrolytes.

All of the clinical features of shigellosis can be defined in terms of the life cycle of *Shigella* in the host. Bacterial invasion of colonic epithelial cells eventually leads to death of the infected host cell and the spreading focus of infection leads to the abscess formation seen in patients. The extensive cell death and tissue destruction at this site leads to an intense inflammatory response. Thus, the blood and mucus in dysentery stools as well as the accompanying fever can all be directly related to the invasion, intracellular replication and spread of *Shigella* in the colonic tissue.

In most adults, clinical symptoms persist from a few days up to 2 weeks. During this period of time, the bacteria can still be shed from the recovering patient and potentially spread to others. In many communities, childcare facilities, which are prone to spread of *Shigella* amongst children and attendees, require monitoring of stool samples from infected individuals until results are negative for *Shigella*.

Further complications of shigellosis include intestinal perforation, septicemia, toxic megacolon, seizures, reactive arthritis syndrome, and hemolytic uremic syndrome (HUS). *Shigella*-induced reactive arthritis syndrome (a sterile inflammatory polyarthropathy) and has been correlated with individuals that are HLA-B27 histocompatibility positive. The role of *Shigella* and other foodborne bacteria in triggering reactive arthritis syndrome is described elsewhere (see Smith, J. 1994). HUS is one of the sequelae of *S. dysenteriae* type I infections. Symptoms include hemolytic anemia, thrombocytopenia, and acute renal failure, sometimes leading to death in young children. Shiga toxin produced by *S. dysenteriae* type I is believed to be responsible for causing HUS. Enterohemorrhagic *E. coli* (EHEC), such as serotype O157:H7, produce several shiga-like toxins, one of which is nearly identical to the shiga toxin of *S. dysenteriae* type I. HUS is one of the major clinical manifestations of disease caused by EHEC (see Chapter 5).

S. dysenteriae is associated with the severe form of dysentery that can have a fatality rate of up to 20%. Antibiotic therapy can shorten the duration of the disease and reduce the period of time that the patient is infectious. The other *Shigella* spp. cause illnesses that are rarely fatal and usually self-limiting. In these cases, oral rehydration therapy is usually sufficient to counter dehydration and electrolyte loss.

A great concern to public health officials is the low infectious dose of *Shigella* spp. Volunteer studies showed that the ID_{50} (the infectious dose required to cause disease in 50% of the volunteers) of *Shigella* is as low as 200 shigellae, although it has been reported that the ingestion of as few as 10 organisms is sufficient to cause disease. The low ID_{50} of *Shigella* accounts for its high communicability, particularly in impoverished, crowded populations. The consequence of this feature is that a contaminated food source has the potential to cause explosive outbreaks of dysentery with secondary cases likely to occur among close contacts of infected individuals. Thus, infected food handlers can contaminate food and spread infection among large numbers of individuals.

III. Taxonomy and Characteristics of *Shigella*

A. Taxonomy

Shigella species are members of the family Enterobacteriaceae and are taxonomically identical with the genus *Escherichia* and closely related to *Salmonella*. Shigellae are Gram-negative, facultatively anaerobic, non-motile, rod-shaped bacteria. They are differentiated from other bacteria by their inability to: (1) produce hydrogen sulfide; (2) ferment lactose; (3) utilize citrate as a sole carbon source; and (4) produce gas from fermentation of carbohydrates (with the exceptions of *S. flexneri* 6, *S. boydii* 13 and 14, and *S. dysenteriae* 3). The four species are further distinguished from each other by the fermentation of sugars and sugar alcohols, the presence or absence of arginine dihydrolase or ornithine decarboxylase, and the production of indole. Serological identification of shigellae is accomplished by agglutination tests.

Since *Shigella* and *E. coli* are nearly genetically and biochemically identical, differentiation is sometimes difficult. One class of pathogenic *E. coli* strain, enteroinvasive *E. coli* (EIEC), is also a causative agent of bacillary dysentery and harbors a large plasmid that has the same genetic determinants for virulence as *Shigella*. Shigellae and EIEC are lactose-negative, although some EIEC strains are late lactose fermenters, and some strains of EIEC, e.g. serotype 0124, share common epitopes with the O-antigens of a few *Shigella* spp. *E. coli* that are part of the normal flora of the human gut are differentiated from shigellae by their motility, lysine decarboxylase activity, gas production from carbohydrate utilization, mucate fermentation and alkaline production on acetate or Christensen citrate agar.

B. Growth and Survival of *Shigella* spp.

The survival of *Shigella* in foods is dependent upon temperature, the pH and presence of inorganic acids. Shigellae seem to survive in foods with neutral pH and when stored at room temperature. A review of growth and survival conditions of *Shigella* spp. in foods is presented elsewhere (see International Commission on Microbiological Specification for Foods, 1996).

1. Culture media

Growth of *Shigella* spp. in broth medium is observed over a range of temperatures. *S. sonnei* can grow at temperatures as low as 6°C and *S. flexneri* at 7°C and up to 48°C. Under laboratory conditions, *S. sonnei* and *S. flexneri* grow in culture media with nearly the same pH values, between 4.8 and 9.3. Under certain culture conditions with media supplemented with organic compounds, such as formic or acetic acid, salts (3.8–5.2% NaCl), or nitrite (300–700 ppm), no growth of shigellae is observed.

2. Foods

In foods, survival time is quite different at −20°C, 4°C, room temperature and a brief exposure to 80°C; *S. flexneri* and *S. sonnei* can survive for the longest period of time at room temperature. In acidic foods, such as citrus juices and carbonated soft drinks, the survival time for *S. flexneri* and *S. sonnei* is from 4 hours to 10 days. *S. dysenteriae*, tested in orange juice at 4°C and grape juice at 20°C, survived up to 170 hours and 2–28 hours, respectively. *S. sonnei* and *S. flexneri* can survive at 4°C for 21 days in foods commonly implicated in outbreaks, such as cheese, potato salad, and mayonnaise. The ingestion of contaminated vegetables has been linked to several outbreaks of shigellosis. In one study it was shown that, when *S. flexneri* is seeded into several different vegetables, the number of shigellae that survive remains constant and in sufficient numbers to produce clinical symptoms if ingested. In another study, *S. flexneri* was added to milk, broths (chicken, beef and vegetable), meats, and vegetables, and incubated at temperatures ranging from 12°C to 37°C. Growth was observed in all foods at all temperatures but in milk at 12°C, growth was variable, and in carrots at some temperatures, the pathogen not only did not grow but died.

C. Pathogenesis

After ingestion, *Shigella* passes through the stomach into the small intestine. Many bacterial pathogens fail to survive transit through the stomach and are killed by the low pH environment. However, the remarkable resistance of *Shigella* to extremely low pH contributes to the organism's ability to cause disease at such low doses. The passage of *Shigella* through the small intestine is accompanied by the first signs of diarrhea, though little replication of the bacteria is thought to occur here. Synthesis of enterotoxins by the bacteria is probably responsible for this effect.

The large intestine (colon) is the principal site of *Shigella* colonization and the area which displays the most severe pathology associated with dysentery. The bacteria induce their own uptake into epithelial cells of the colon, perhaps by initially invading the lymphoidal M cells, which are found interspersed throughout the epithelial layer. Genetic studies have demonstrated that the capacity to invade mammalian cells is absolutely required for virulence of *Shigella*. Following invasion of the target cell, the bacteria lyse the membrane of the phagocytic vacuole and begin replicating in the host cell cytoplasm. This step distinguishes the life style of *Shigella* from those of other invasive pathogens such as *Salmonella*, which remain contained within a membrane-bound vacuole.

Once free in the host cell cytoplasm, *Shigella* begin to recruit elements of the host cytoskeletal network and start to display an amazing form of intracellular motility. The bacteria use actin filaments of the host cytoskeleton to propel themselves through the cytoplasm. Even more striking is the observation that *Shigella* use this motility to 'push' out across the host cell membrane into adjacent cells and thus spread the focus of infection without ever leaving the safety of the host target cell.

Much has been learned of the genes and gene products responsible for *Shigella* pathogenesis. Genes implicated in virulence are primarily encoded on a large plasmid, while some accessory genes are encoded on the chromosome. Expression of these genes is coordinately regulated in response to temperature such that virulence genes are optimally expressed at 37°C (the normal temperature of the human host) and expression is repressed at lower temperatures. The ability of *Shigella* to cause disease involves the action of more than two dozen genes. The products of the virulence genes can be divided into three classes: regulatory factors, secreted products, and the Type III secretion apparatus. While the secreted 'invasins' of *Shigella* are unique, the machinery for transport of these products out of the bacteria possesses elements in common with the secretion apparatus of other human pathogens such as *Yersinia* spp., *Salmonella*, and enteropathogenic *E. coli*.

IV. Foodborne Transmission

A. Sources of Contamination

Shigella is primarily passed from person to person by the fecal–oral route. From infected people, *Shigella* can be transmitted by several means including food, fingers, feces, and flies. Examples of foods incriminated in past outbreaks are salads (potato, tossed, egg, tuna), cheese, stewed apples, chicken, shrimp, clams, and milk. Food handlers with poor personal hygiene are the major source for contamination of foods, with improper storage of contaminated foods the second most common factor. Inadequate cooking, contaminated equipment, and food from unsafe sources are other means for spreading shigellae. Houseflies are passive vectors. Shigellae can spread to foods by either passing through the gut of flies or by being transported directly from contaminated feces via the surfaces of the flies.

B. Detection of Shigellae in Food

Shigella spp. are not indigenous to any foods and therefore their presence is indicative of human fecal contamination. The importance of rapid detection and identification of the etiological agent is critical in reducing the spread of shigellae throughout the population. Since shigellae are not associated with any specific food and are introduced into foods by poor hygienic practices, routine surveillance of *Shigella* in foods is not performed.

1. Bacteriological methods

Procedures for isolating *Shigella* spp. from foods take 5–10 days to complete and are often very ineffective. According to the *Bacteriological Analytical Manual* (BAM), food samples are added to *Shigella* broth supplemented with 0.5–3 μg/ml of novobiocin, an inhibitor of many Gram-negative bacteria. This enrichment step is performed anaerobically at 42–44°C. After 20 hours, cells are streaked on to MacConkey agar plates and incubated at 35°C for an additional 20 hours. Lactose-negative colonies are examined by Gram stain and several biochemical properties, e.g. H_2S production, motility, urease activity, lysine decarboxylase activity, citrate utilization. Group-specific *Shigella* antiserum is used to identify the serotype of the isolated bacteria.

Alternative methods of isolating *Shigella* spp. use differential and selective media, such as MacConkey, *Shigella–Salmonella*, eosin–methylene blue, or xylose–lysine–deoxycholate agars. Colonies that are lactose-negative are considered to be presumptive for *Shigella* and are subjected to further analysis on triple sugar iron (TSI) or Kligler iron agar. Colonies that fail to produce H_2S, but produce acid but not gas on Kligler iron agar or are alkaline on TSI agar slants are tested by agglutination using *Shigella*-specific antisera.

The problem of recovering *Shigella* from foods can be illustrated by the report of June *et al.* (1993). In this study, the BAM method was used and was unsatisfactory in recovering *Shigella* seeded in six foods (potato salad, chicken salad, cooked shrimp salad, lettuce, raw ground beef, and raw oysters). The failure to detect *Shigella* in foods using bacteriological methods can be attributed to several factors: (1) shigellae may be present in low numbers or in an injured state making recovery using culture media difficult; and (2) analysis of suspected foods can be compromised by improper storage conditions (abuse by time and/or temperature). Since shigellosis outbreaks are usually initially recognized by clinical laboratory findings several days to weeks after the onset of the disease, loss of the food sample (disposed or eaten) during this time interval further compounds the difficulties in identifying *Shigella* as the cause of a foodborne outbreak.

2. DNA-based assays

The inefficiency and insensitivity of the bacteriological methods for detection of *Shigella* in foods may be overcome by using DNA-based assays. The first gene probes to target *Shigella* spp. were DNA fragments from the large virulence plasmid and were used to detect virulent strains in a colony hybridization format. Later, synthetic oligonucleotides that were derived from DNA sequence of specific virulence genes present on the virulence plasmid or the *Shigella* chromosome were used. Most gene probes are used to detect *Shigella* in clinical samples. In one study using an oligonucleotide (18 bases long) directed to a plasmid-encoded virulence gene, the specificity and sensitivity of a probe assay was tested in foods seeded with *S. flexneri*. The efficacy of this probe was dependent on the food; no shigellae were detected when seeded at 10^2 CFU/g in alfalfa sprouts with an indigenous microbial background of 1.4×10^8 CFU/g. Unless the number of the commensal microbial population is reduced, detection of *Shigella* spp. from foods most likely will not be successful using DNA probes.

Another approach has been the application of the polymerase chain reaction (PCR) to detect *Shigella* spp. in foods. This technique is an *in vitro* amplification system designed to increase the number of target nucleic acid sequences to a detectable level using DNA or RNA as templates. PCR has a significant advantage over other methods since only a relatively small number of shigellae need to be present for PCR-based assays. PCR primer sets, usually short (18–22 bases) synthetic oligonucleotides, which flank the region that is targeted for amplification, have been used to detect shigellae in clinical samples and foods. PCR primers that have been used are targeted to specific, plasmid-encoded virulence genes. Other approaches include combining PCR and dot-blot hybridization with immunomagnetic separation. In this technique, monoclonal antibodies to the O-polysaccharide of *S. dysenteriae* type 1 or *S. flexneri* are attached to magnetic beads to separate these pathogens effectively from stool samples.

There are a few reports of the use of PCR to detect shigellae in foods. Since PCR assays are rapid, sensitive, and do not require selective enrichment, analysis of foods for the presence of *Shigella* spp. can be performed in less than 1 day. However, there are several limitations to using PCR assays for the detection of shigellae in foods: (1) most PCR primers are directed to plasmid-encoded virulence genes – a false-negative result can occur if the strains have lost the plasmid; (2) extraction of template DNA from pathogens lodged in foods can be difficult and can lead to DNA contaminated with inhibitors of PCR or loss of the template; and (3) although PCR assays are sensitive and require very few target molecules, there is a critical minimum number of bacteria needed (in some instances 100 cells) to yield a PCR product. Therefore, if the entire food sample contains a very low number of pathogenic organisms, it is a challenge to isolate sufficient quantity and quality DNA for PCR to be conclusive.

V. Treatment and Prevention

A. Treatment

Antimicrobial resistance of pathogenic bacteria has become a worldwide problem and is a serious concern in the treatment of patients with shigellosis. Developing countries have a significant problem of treating patients, owing to multiresistant bacterial strains, particularly *S. dysenteriae* type I. This problem is compounded by the lack of availability and the cost of newer antibiotics that would be effective. The recommended choice of antibiotics is fluoroquinolones or, when specifically indicated, norfloxacin, with trimethoprim–sulfamethoxazole and ampicillin as alternatives, although strains resistant to these latter antibiotics have been reported.

B. Prevention

A vaccine directed against shigellae is not currently available but the need for one is obvious. The WHO has recognized the importance of vaccine development and have selected *S. flexneri* 2a, *S. dysenteriae* type 1 and *S. sonnei* as targeted pathogens. Until a vaccine is available, the only effective preventive measures are: (1) strict adherence to proper sanitary procedures, such as good personal hygiene (especially hand washing); and (2) identification of carriers who may introduce the pathogen during food preparation.

VI. Summary

Shigella species, the etiological agents of bacillary dysentery, continue to be a major cause of diarrheal diseases worldwide. Transmission of these pathogens is primarily from person-to-person contact via the fecal–oral route, but the organisms are also spread in foods and water. Humans and higher primates are the known reservoirs for *Shigella* and, due to its low infectious dose in humans, rapid and widespread dissemination of this pathogen occurs. Shigellae are the third leading cause of reported foodborne diseases in the US. Since *Shigella* spp. are not indigenous to any food, these cases arise principally due to poor personal hygiene of an infected food handler. Much is known about the host–pathogen interactions, both genetically and physiologically, with the key factors of *Shigella* virulence being the ability to invade colonic epithelial cells, the intercellular spread to neighboring cells and the destruction of the colonic mucosa.

Shigella spp. are a formidable, worldwide, public health threat. Since 30–50% of cases of dysentery worldwide are caused by *Shigella* and nearly 600 000 deaths occur each year from shigellosis, the dissemination of shigellae is a major concern for public health. The WHO has also placed emphasis on food safety since contaminated foods cause 70% of the 1.5 billion cases of diarrhea each year with 3 million deaths of children under the age of 5. Emphasis is placed on personal hygienic practices with hand washing and improvement in safe handling of foods in homes and vendors as a means to reduce the spread of *Shigella*.

Bibliography

Andrews, W. H., June, G. A. and Sherrod, P. S. (1995). *Shigella*. *In* 'Bacteriological Analytical Manual', 8th ed., pp. 6.01–6.06. AOAC International, Gaithersburg, MD.

Bean, N. H. and Griffin, P. M. (1990). Foodborne disease outbreaks in the United States, 1973–1987: pathogens, vehicles, and trends. *J. Food Prot.* **53**, 804–817.

Bennish, M. L. (1991). Potentially lethal complications of shigellosis. *Rev. Infect. Dis.* **13** (suppl. 4), S319–324.

Centers for Disease Control and Prevention (2001). Summary of notifiable diseases, United States, 1999. *Morbid. Mortal. Weekly Rep.* **48**, 1–104.

DuPont, H. L. (1995). *Shigella* species (Bacillary dysentery). *In* 'Principles and Practice of Infectious Diseases' (G. L. Mandell, J. E. Bennett and R. Dolin, eds), pp. 2033–2039. Churchill Livingstone Inc., New York.

Edwards, P. R. and Ewing, W. H. (1972). 'Identification of Enterobacteriaceae'. Burgess Publishing, Minneapolis.

International Commission on Microbiological Specifications for Foods (1996). 'Microorganisms in Foods. 5. *Shigella*.' Blackwell Scientific Publications, Oxford.

June, G. A., Sherrod, P. S., Anaguana, R. M., Andrews, W. H. and Hammack, T. S. (1993). Effectiveness of the Bacteriological Analytical Manual culture method for the recovery of *Shigella sonnei* from selected foods. *J. AOAC Int.* **76**, 1240–1248.

Kotloff, K. L., Winickoff, J. P., Ivanoff, B., Clemens, J. D., Swerdlow, D. L., Sansonetti, P. J., Adak, G. K. and Levine, M. M. (1999). Global burden of *Shigella* infections: implications for vaccine development and implementation of control strategies. *Bull. WHO* **77**, 651–666.

Lampel, K. A. and Maurelli, A. T. (2001). *Shigella* species. *In* 'Food Microbiology: Fundamentals and Frontiers', 2nd ed. (M. P. Doyle, L. R. Beuchat and T. J. Montville, eds), pp. 247–261. American Society for Microbiology Press, Washington, DC.

Sansonetti, P. J., Tran Van Nhieu, G. and Egile, C. (1999). Rupture of the intestinal epithelial barrier and mucosal invasion by *Shigella flexneri*. *Clin. Infect. Dis.* **28**, 466–475.

Smith, J. L. (1987). *Shigella* as a foodborne pathogen. *J. Food Prot.* **50**, 788–801.

Smith, J. (1994). Arthritis and foodborne bacteria. *J. Food Prot.* **57**, 935–941.

WHO Surveillance Programme For Control of Foodborne Infections and Intoxications in Europe. Sixth Report 1990–1992 (1995). Federal Institute for Health Protection of Consumers and Veterinary Medicine, Berlin.

WHO (1997). Vaccine research and development. *Weekly Epidemiol. Rec.* **72**, 73–80.

Escherichia coli

5

Pina M. Fratamico, James L. Smith, and Robert L. Buchanan

I. Introduction

Escherichia coli was isolated and described in 1895 by Escherich, and subsequently, the organism was recognized as part of the flora of the intestinal tracts of warm-blooded animals and humans. The organism is a minority member of the bowel flora and its role is uncertain. The commensal status of *E. coli* obscured the potential pathogenicity of the organism for decades. In the early 1920s, however, *E. coli* was shown to be involved in urinary tract infections, and in the late 1940s, enteropathogenic *E. coli* strains were identified as a cause of infantile gastroenteritis.

In the past 40 years, additional *E. coli* types, which can act as gastrointestinal pathogens, have been isolated. At present, there are six different categories of pathogenic *E. coli* that produce diarrhea (Table 5.1). The mechanisms by which diarrhea is produced vary for each type of pathogenic *E. coli* and include bacterial attachment to host cells, invasion of intestinal cells, and production of toxins. In view of the number of diarrhea-producing *E. coli* types that are now known (Table 5.1), food microbiologists, consumers, and regulatory agencies can no longer regard *E. coli* as merely an indicator of fecal contamination. The purpose of this chapter is to discuss the major diarrheagenic *E. coli* groups, including the pathogenesis of the diarrheal illness, virulence factors, foodborne transmission, diagnosis, and detection.

Foodborne Diseases 2nd Edn
ISBN: 0-12-176559-8

Table 5.1 Types of *Escherichia coli* that can act as diarrheic pathogens

E. coli type	Major O serogroups	Type of diarrhea; virulence factors
Enteropathogenic (EPEC)	18, 26, 44, 55, 86, 111, 114, 119, 125, 126, 127, 128, 142, 157, 158	Acute and/or persistent diarrhea; virulence factors include attaching and effacing lesions, and localized adherence mediated by bundle-forming pili; plasmid and chromosomally mediated
Enterotoxigenic (ETEC)	1, 6, 9, 11, 15, 20, 25, 27, 60, 63, 75, 80, 85, 88, 89, 99, 101, 109, 114, 128, 139, 148, 153, 159	Acute watery diarrhea; virulence factors include adherence and heat-labile toxins; plasmid and chromosomally mediated
	11, 12, 15, 20, 25, 27, 60, 63, 75, 78, 80, 85, 88, 89, 99, 101, 109, 114, 115, 139, 148, 149, 153, 159, 166, 167	Acute watery diarrhea; virulence factors include adherence and heat-stable toxins; plasmid and chromosomally mediated
Enteroinvasive (EIEC)	11, 28, 29, 112, 115, 121, 124, 135, 136, 138, 143, 144, 147, 152, 164, 167, 173	Acute dysenteric diarrhea; virulence factors include cell invasion and intracellular multiplication; plasmid and chromosomally mediated
Enterohemorrhagic (EHEC)	22, 26, 48, 55, 91, 103, 104, 111, 113, 118, 121, 145, 153, 157, 163	Bloody diarrhea with or without hemolytic uremic syndrome; virulence factors include attaching and effacing adherence, Shiga toxins, hemolysin; plasmid and chromosomally mediated
Enteroaggregative (EAEC)	3, 4, 7, 9, 15, 21, 44, 51, 59, 77, 78, 86, 91, 92, 106, 111, 113, 126, 141, 146	Persistent diarrhea; virulence factors include aggregative adherence and heat-stable enterotoxin; plasmid mediated
Diffusely adherent (DAEC)	15, 75, 126	Diarrhea in children; virulence factors include plasmid-mediated and chromosomally mediated fimbrial and nonfimbrial adhesins

II. General Characteristics of *Escherichia coli*

Taxonomically, *E. coli* is located in the family Enterobacteriaceae. Members of the species are Gram-negative straight rods (1.1–1.5 µm × 2.0–6.0 µm) arranged singly or in pairs. They either possess peritrichous flagella and are motile, or lack flagella and are nonmotile. Capsules or microcapsules may be present. The optimum growth temperature is 37°C. *E. coli* is facultatively anaerobic (has both fermentative and respiratory types of metabolism) and chemoorganotrophic (depends on organic chemicals for energy and carbon source).

The biochemical characteristics of the species are listed in Table 5.2. Approximately 95% of *E. coli* strains have an IMViC (indole, methyl red, Voges–Proskauer, citrate) pattern of + + − − (biotype 1), whereas the others have a − + − − pattern (biotype 2). Taxonomists have shown that *E. coli* and *Shigella* form a single species based on DNA relatedness (70–100%), and it is difficult to separate the two biochemically (also discussed in Section III). However, the separate nomenclature is maintained for medical reasons.

Serotyping/serogrouping of *E. coli* is useful for subdividing the species into serovars. Serological classification of *E. coli* involves determining the antigenicity of three surface antigens: the O (somatic lipopolysaccharide); K (capsular); and H (flagellar) antigens. There are currently about 173, 103, and 56 O, K, and H antigens, respectively, in the *E. coli* antigenic scheme. The typing scheme has been expanded to

Table 5.2 Biochemical characteristics of *Escherichia coli*

Biochemical characteristic	Reaction
Indole reaction	+
Methyl red	+
Voges–Proskauer (production of acetoin from glucose)	−
Citrate utilization	−
Gelatin liquefaction at 22°C	−
Acid and gas from glucose	+
Acid and gas from lactose, D-mannitol, D-sorbitol, L-arabinose, maltose, D-xylose, trehalose, D-mannose	+
Acid and gas from L-rhamnose, melibiose	[+]
Acid and gas from sucrose, dulcitol, salicin, raffinose	d
Acid and gas from D-adonitol, myo-inositol, cellobiose, α-methyl-D-glucoside, D-arabitol	−
Nitrate → nitrite	+
ONPG hydrolysis (β-galactosidase)	+
Oxidase	−
Catalase	+

Reaction symbols: +, 90–100% of strains positive; [+], 76–89% of strains positive; d, 26–75% of strains positive; −, 0–10% of strains positive.

include the fimbrial F-antigens, although the fimbrial serogroup is not routinely included in the serotyping formula. Fimbrial serotyping can only be performed in specialized laboratories since one *E. coli* strain may produce multiple fimbrial types and variation in expression of fimbriae may occur. The combination of O and H antigens defines the *E. coli* serotype, and serotyping of isolates is useful for foodborne outbreak and epidemiological studies.

III. Enteroinvasive *Escherichia coli*

Enteroinvasive *E. coli* (EIEC) strains, first shown to cause diarrhea in volunteer studies in 1971, are taxonomically and physiologically more related to *Shigella* than to other *E. coli*. Additionally, the O antigens of many EIEC serogroups are closely related antigenically to those of *Shigella* serovars. *Shigella* spp. and EIEC are non-motile (except for EIEC serogroup O124), late or non-lactose fermenting, and lysine decarboxylase negative. The clinical syndromes caused by EIEC and *Shigella* infection are often similar, and the site of infection is predominantly the colonic mucosa. *Shigella* and EIEC penetrate epithelial cells, lyse the endocytic vacuole, multiply within the cell, then move through the cytoplasm and spread to adjacent epithelial cells. Genes required for invasion are carried on large plasmids (140 MDa) designated pInv and include the *ipa* (invasion plasmid antigen) genes, *ipaA-ipaD*, *ipaH*, and *ipaR*. The invasion plasmid in EIEC possesses virulence genes that are identical to the virulence genes harbored by the *Shigella* invasion plasmid. Spontaneous loss of the plasmid in EIEC correlates with loss of the ability to invade cells. Most patients infected with EIEC develop watery diarrhea indistinguishable from that caused by enterotoxigenic

E. coli. Only a minority of patients go on to develop symptoms typical of *Shigella* dysentery, manifested by: abdominal cramps; blood, mucus, and white blood cells in stool; fever; and tenesmus. The infective dose for EIEC is generally at least 1000-fold higher than that of *Shigella*. Outbreaks caused by EIEC have been linked to imported French cheese, potato salad, tofu, and guacamole. Person-to-person transmission can also occur.

Shigella and EIEC are identified by their ability to invade and form plaques in HeLa cell monolayers and to cause keratoconjunctivitis in guinea pigs (Sereny test). These phenotypic assays correlate with invasion. Other assays for detection include DNA hybridization using probes pMR17 and *ial* (invasion-associated locus) derived from the pInv plasmid, polymerase chain reaction (PCR) assays targeting the *ipaH* and *ial* genes, and an enzyme-linked immunosorbent assay (ELISA) to detect the *ipaC* gene.

IV. Diffusely Adherent *Escherichia coli*

Diffusely adherent *E. coli* (DAEC) strains do not fall into the enteropathogenic *E. coli* serogroups and do not form enteropathogenic *E. coli*-like microcolonies on target cells *in vitro*. Diarrheagenic *E. coli* are assigned to the DAEC group based on their specific adherence pattern to HEp-2 or HeLa cells. Surface fimbriae designated F185 and an approximately 100-kDa outer membrane protein (AIDA-I) associated with the diffuse adherence phenotype have been identified. Additionally, 57-kDa and 16-kDa (CF16K) proteins may also contribute to the diffuse-adherence phenotype. Some DAEC strains induce fingerlike projections extending from infected host cells, which embed or protect the bacteria. A large percentage of DAEC isolates hybridize with a DNA probe derived from the *daaC* gene associated with expression of F185 fimbriae. These organisms are primarily associated with diarrhea in children aged 1 to 5 years. The clinical syndrome associated with DAEC has not been clearly defined. In one study 54% (13 out of 24) of patients had vomiting, 61% (14 out of 24) had diarrhea, and 42% (10 out of 24) had fever; however, compared to controls (children without any potential enteropathogen), there was no association of DAEC infection with diarrhea or fever. Of the 24 DAEC isolates, 13 were *daaC* positive. These data indicate that DAEC are heterogeneous, and that only some strains may be pathogenic for children. In the absence of volunteer studies, the significance of DAEC as enteric pathogens remains uncertain. Studies to identify virulence determinants and to analyze additional strains are needed.

V. Enteropathogenic *Escherichia coli*

A. Pathogenesis

Since first associated with cases of human illness in the 1940s, enteropathogenic *E. coli* (EPEC) remains a major cause of outbreaks and sporadic cases of infantile diarrhea in developing countries. Symptoms of illness caused by EPEC include profuse watery diarrhea, vomiting, and low-grade fever. Central to EPEC pathogenesis is the production of attaching and effacing (AE) lesions in the gut mucosa with localized destruction of microvilli and intimate attachment of bacteria to host cells. The bacteria sit on a pedestal-like structure derived from the host cell. Actin, α-actinin, talin, ezrin, and myosin light chain accumulate beneath the attached bacteria. Genes involved in the formation of AE lesions and signal transduction activity in epithelial cells are encoded on a 35-kb pathogenicity island called the locus of enterocyte effacement (LEE). Genes on the LEE locus include *espA*, *espB*, and *espD* that encode secreted proteins, *sep* and *esc*, which encode a type III secretion system, and *eae* (*E. coli* attachment and effacement), which encodes a 94–97-kDa protein called intimin. Intimin mediates close contact of the bacterium to the eukaryotic cell after interaction of Tir (translocated intimin receptor), a 78-kDa bacterial protein encoded by *tir*, located upstream of *eae*. Components of the type III secretion system and EspA mediate translocation of the Tir protein into the target cell, where it is phosphorylated and inserted into the host membrane. Intimin in the bacterial outer membrane binds to Tir and possibly other receptors, therefore, EPEC inserts its own receptors into the host membrane.

Another feature of EPEC is the ability to adhere in localized microcolonies to target cells *in vitro*. Localized adherence is associated with the presence of 50–70 MDa-sized plasmids designated the EPEC adherence factor (EAF) plasmids. A cluster of 14 genes on the EAF plasmid is required for expression and assembly of bundle-forming pili (BFP), which mediate the localized adherence phenotype. Other types of fimbriae, which may also play a role in adhesion, have been described in EPEC. Some EPEC strains produce the EAST1 toxin that is also produced by enteroaggregative and enterotoxigenic *E. coli* as evidenced by hybridization of EPEC strains with the *astA* gene encoding EAST1 (discussed in Sections VI and VII). Several studies have demonstrated that EPEC are able to invade epithelial cell lines; however, unlike true intracellular pathogens, EPEC do not multiply in these cells and do not escape the phagocytic vacuole.

B. Epidemiology

Infection with EPEC occurs predominantly in children less than 2 years of age. Breast milk and colostrum have factors that inhibit adherence of EPEC to HEp-2 cells and show protection against EPEC infection. Disease due to EPEC occurs predominantly in developing countries, and the route of transmission is fecal–oral, with vehicles including foods or formula, contaminated hands of caregivers, and contaminated fomites such as linen, scales, toys, and carriages. Asymptomatic carriage can occur, thus, healthy children may shed EPEC in their stools. Asymptomatic children and adults are thought to be the reservoir of EPEC infection. The EPEC serotypes found in animals are usually not those associated with human infection.

C. Diagnosis and Detection of EPEC

The characteristics that define EPEC are the AE histopathology, possession of the EAF plasmid, and the absence of Shiga toxin. Shiga toxin-producing enterohemorrhagic

E. coli also produce attaching and effacing lesions on host cells (discussed in Section VIII), and therefore, must be distinguished from EPEC. The fluorescent actin stain (FAS) procedure, in which labeled phalloidin binds to filamentous actin in host cells beneath adherent bacteria, is used to determine the AE phenotype. Localized adherence on HEp-2 and HeLa cells correlates with the presence of the EAF plasmid. An ELISA assay using an antibody to an EPEC strain absorbed with a plasmid-cured derivative was developed to detect strains showing localized adhesion. PCR and DNA hybridization assays targeting sequences on the EAF plasmid, and *bfp* and *eae* genes have also been employed to detect EPEC.

VI. Enteroaggregative *Escherichia coli*

A. Pathogenesis

There is some evidence to indicate that enteroaggregative *Escherichia coli* (EAEC) infection can be foodborne. Strains of EAEC have been isolated from the contents of infant feeding bottles, and EAEC outbreaks have been associated with food. In Japan, more than 2500 children from a number of schools became ill after eating lunches prepared at a central kitchen. A single EAEC serotype (nontypable O:H10) was implicated; however, the serotype was not found in any foods served to the children. Four outbreaks of foodborne EAEC-induced diarrhea have occurred in the UK. Food was implicated in these outbreaks, but the actual food vehicle was not determined.

The EAEC are *E. coli* that do not secrete heat-labile toxin, heat-stable toxin, or Shiga toxin (toxins discussed in Sections VII and VIII) but do adhere to HEp-2 cells in an aggregative or 'stacked brick' (AA phenotype) adhesion pattern. The EAEC strains are isolated most frequently from infants and young children with chronic diarrhea of greater than 2 weeks' duration. In 15 case-controlled studies, the incidence of EAEC colonization in children with diarrhea was 19.4% (range 5.1–51%) compared with 10.4% (range 1.1–22%) in children without diarrhea. Therefore, asymptomatic carriage of EAEC may be common. The EAEC are also associated with diarrhea and weight loss in HIV-infected individuals.

Two prominent histological features present in EAEC diarrhea are a thick mucus gel on the intestinal mucosa and necrotic lesions in the ileal mucosa. Diarrhea caused by EAEC is secretory in nature (usually watery) and mucoid. Other symptoms may include vomiting and dehydration. Oral hydration is an effective therapy. Up to one-third of patients may have grossly bloody stools. There is some indication that EAEC infections may lead to growth retardation in infants and children. A model for EAEC pathogenesis consists of three stages: (1) stage I involves initial adherence to the intestinal mucosa and mucus layer; (2) stage II includes enhanced mucus production leading to a thick EAEC-encrusted biofilm on the mucosal surface; and (3) stage III involves the elaboration of toxin(s) that result in damage to the mucosa and intestinal secretion.

B. Virulence Factors

Information concerning virulence factors in EAEC is limited and confusing. There are several fimbrial types involved with aggregative attachment, and while the pathology suggests that a toxin is involved in EAEC diarrhea, it is not clear that the toxin(s) responsible have been identified. Attachment occurs mainly to the colonic and ileal mucosa, thus, EAEC are large-bowel pathogens that colonize through adhesion mediated by fimbriae. A number of EAEC adherence factors have been demonstrated, including aggregative adherence fimbriae, AAF/1 and AAF/II. However, not all EAEC produce AAF/I and /II fimbriae. Other morphologically distinct types of aggregative fimbriae described as hollow rod, rod, fibrillar, and fibrillar bundles are also associated with EAEC. *In vitro* invasion and internalization of EAEC by HeLa cells has been shown. The invasion process is inhibited by cytochalasin D and other endocytosis inhibitors. *In vivo* invasion by EAEC strains has not been reported in the literature.

The EAST1 toxin, a plasmid-mediated, low-molecular-weight, protease-sensitive, partially heat-stable toxin with an *in vitro* mode of action similar to the enterotoxigenic *E. coli* heat-stable toxin, was first demonstrated in EAEC. It is not clear, however, that EAST1 has a role in EAEC-induced diarrhea *in vivo*. A more detailed discussion of EAST1 can be found in Section VII. A high-molecular-weight (108 kDa), heat-labile protein toxin has recently been isolated from EAEC. A partially purified preparation induced tissue damage, inflammation, and mucus secretion in isolated rat jejunum. The toxin gene (*pet*) is located on the 65-MDa EAEC virulence plasmid, which also contains the genes for the aggregative phenotype, AA. The plasmid-encoded toxin (Pet) belongs to the autotransporter class of secreted proteins and is highly homologous to other autotransporter proteins, the EspP protease of enterohemorrhagic *E. coli*, and the cryptic protein EspC of EPEC. In one study, 14.7% (5/34) of EAEC strains were positive for Pet by colony hybridization.

The EAEC form a very heterogeneous group of more than 50 *E. coli* O serogroups. Fimbriae vary from bundles of fine filaments, to thin fimbriae and bundle-forming fimbriae. However, not all EAEC express fimbriae, and not all strains of EAEC produce the toxins, EAST1 and/or Pet. The EAEC are also heterogenous in their ability to induce diarrhea in adult volunteers. The heterogeneity of EAEC makes identification of strains and diagnosis of EAEC-induced illness difficult.

C. Detection Methods and Diagnosis of EAEC Infection

The most definitive identification of EAEC is the demonstration of aggregative adherence to HEp-2 cells; however, the assay is time consuming and cumbersome, and the type of adherence can be easily misinterpreted. DNA probes and PCR assays utilizing regions of the adherence plasmid of EAEC have been used for the identification; however, there is not always a correlation between probe or PCR positivity and the AA phenotype. Other phenotypic assays, which are useful for identification of EAEC, include pellicle or film formation when cultured in Mueller Hinton broth or in polystyrene culture tubes, respectively.

VII. Enterotoxigenic *Escherichia coli*

A. The Disease

In developed countries, enterotoxigenic *E. coli* (ETEC) is an uncommon cause of diarrhea; however, in developing countries, it is a major cause of diarrhea in infants, young children and the elderly. In addition, the organism is the primary cause of traveler's diarrhea for visitors to developing countries.

Newborn and young calves, lambs, and pigs are susceptible to ETEC-induced diarrhea; however, adult animals are not affected. In human adults and children, ETEC colonize the small intestine by attaching to the enterocytic brush border with the aid of bacterial adherence factors, but do not invade or damage intestinal cells. After colonization, the ETEC secrete toxin(s) that induce a noninflammatory watery diarrhea. Blood, mucus, and leukocytes are not found in stools. The infected individual may show nausea and mild to moderate abdominal cramping, but generally there is no fever. In children, the diarrhea can be prolonged, with severe dehydration, and mortality can be high. In addition, ETEC infections have been associated with serious malnutrition in children. Adults with traveler's diarrhea generally have a mild, self-limited illness, of 1–5 days.

The mode of transmission of ETEC is by ingestion of contaminated food or water. Food handlers infected with ETEC can contaminate food and water, or potable water can be contaminated with ETEC-containing sewage. Studies with human volunteers indicate that the infective dose is about 10^8 organisms/individual. Humans are the major reservoir for human ETEC strains. While young animals are susceptible to ETEC infections, animals are not reservoirs for human ETEC strains. In Table 5.3, a number of foodborne and waterborne outbreaks in which ETEC have been implicated are listed. The spread of ETEC may also occur by direct person-to-person contact; however, in view of the high infective dose, person-to-person transmission is probably uncommon.

Traveler's diarrhea induced by ETEC is generally short term and does not normally require antibiotic treatment. Use of antibiotics may change the intestinal flora as well as promote antibiotic resistance in ETEC. However, in cases where the diarrhea is severe and/or prolonged, trimethoprim/sulfmethoxazole treatment may be appropriate. If dehydration occurs, rehydration therapy must be administered. Antidiarrheal drugs can be effective in treatment of ETEC-induced traveler's diarrhea, since such treatments/agents limit fluid accumulation and intestinal motility.

Breastfeeding of infants is considered an excellent prophylaxis against ETEC infection, as well as against gastrointestinal diseases in general. In ETEC-infected infants and children, the approved therapy is rehydration. Antimotility agents, which interfere with peristaltic removal of pathogens, and antibiotic therapy are not recommended for use in ETEC-infected children. Vaccines for use in humans to control ETEC infections are not available.

B. Basis of Pathogenicity

Virulence mechanisms in ETEC involve various host cell attachment and colonizing factors and enterotoxins. The diarrheic effects seen in ETEC infections are due to the

Table 5.3 Foodborne outbreaks in which ETEC strains were implicated

Date reported	Toxin produced	Number ill	Comments
1976	ST	55	Children's hospital in the US; baby formula implicated
1976	ST/LT	8	Travelers attending a meeting in Mexico City; salads containing raw vegetables implicated
	LT	12	
	ST	5	
1976	ST	67	Cruise ship; food suspected
1977	ST	129	Two separate outbreaks in two different locations in Japan; associated with eating lunch
1979	LT	60	Swedish restaurant; shrimp and mushroom salad implicated
	LT	2	Swedish home; cold, shop-sliced roast beef implicated
1980	LT	349	Outbreaks occurred on two separate trips of same cruise ship; drinking water and crabmeat cocktail implicated
1982	ST/LT	415	Restaurant in Wisconsin; Mexican food implicated (food items included sauces, garnish, flour tortillas, guacamole)
1983	LT	282	Hospital in Texas; associated with eating in hospital cafeteria; no specific food implicated
1985	ST/LT	27	Cold buffet at a school in England; curried turkey mayonnaise implicated
1985	ST	45	Clusters of outbreaks occurred at office parties in Washington, DC; French brie cheese implicated; cheese from same plant (same brand and lot) was implicated in outbreaks in Illinois (75 cases), Wisconsin (35 cases), Georgia (10 cases), and Colorado (4 cases); same brand of cheese caused outbreaks in Denmark, the Netherlands, and Sweden
1994	LT, ST	47	Flight from North Carolina to Rhode Island; garden salad was implicated
	LT, ST	97	Buffet served in a mountain lodge in New Hampshire; tabouleh salad implicated
1998	ST1b	>600	School lunches at four elementary schools in Japan; tuna paste implicated – raw carrots, onions, and cucumbers that were part of the tuna paste was probably contaminated with ETEC
1998	ST	372–645	Banquet in Milwaukee, WI; pan-fried spiced potatoes implicated
2000	ST and LT/ST	97	Cruise ship; ship's tap water and/or beverages served with ice implicated
	ST, LT, and LT/ST	19	Cruise ship; beverages with ice and ice water implicated
	ST, LT, and LT/ST	197	Cruise ship; tap water and beverages served with ice implicated
2000	ST, LT, and LT/ST	229	Military posts and civilian communities in the Golan Heights, Israel; drinking water implicated

organism's ability to synthesize and release a heat-labile toxin (LT), a heat-stable toxin (ST) or both toxins, as well as variants of LT-I. Polyclonal antibodies raised against a particular LT-I toxin will neutralize other LT-I toxins and cholera toxin (CT), but will not neutralize LT-IIs. Thus, CT and LT-I are closely related. Conversely, antibodies to a particular LT-II will neutralize LT-II toxins, but not LT-I or CT. The LTh-I variant is of human origin and the LTp-I is of porcine origin. Unlike the genes for cholera synthesis, which are chromosomally located, the genes for LT-I are plasmid mediated. However, the genes for LT-II are chromosomal.

The LTs, similar to CT, are made up of one A polypeptide subunit noncovalently combined with five B polypeptides. The A subunits of LT-I and LT-II have about 50% amino acid homology, whereas the B subunits of LT-I and LT-II are about 10% homologous.

The B subunits bind the toxin molecule to ganglioside GM1 (for LT-I toxins) on the host cell. The A subunit undergoes proteolytic nicking to produce the A1 and A2 fragments. The A1 fragment is linked to A2 by a disulfide bond and the A2 fragment is bound to the B subunits. The A1 fragment exhibits ADP-ribosyl transferase activity; ADP ribosylation of a GTP-binding protein mediates activation of adenylate cyclase. The resultant increase in cyclic AMP within the intestinal mucosa stimulates chloride secretion and decreases sodium absorption. There is loss of fluid and electrolytes with production of a watery diarrhea. The LT-I-producing ETEC cause a milder disease in humans than CT-producing *Vibrio cholerae*; however, LT-II-producing ETEC strains are not involved in human disease. Proteolytic nicking of the A subunit is necessary for CT activity; however, nicking is not necessary for enzymatic activity of the A subunit of LT-I, though nicking enhances the biological and enzymatic activity of LT. Thus, nicking of the A subunit of LT-I is necessary for optimum activity of the toxin. Other bacteria that produce LT-like toxins are *Klebsiella*, *Enterobacter*, *Aeromonas*, *Plesiomonas*, *Campylobacter*, and *Salmonella*.

The ST enterotoxins are subdivided into STI and STII (also referred to as STa and STb, respectively) families. There are two toxins in the STI family: STp is of porcine origin and is an 18 amino acid peptide, whereas STh is of human origin and contains 19 amino acids. The STI peptides have three intramolecular disulfide bonds. The STI enterotoxins are heat and acid stable, are not denatured by detergents, are water and methanol soluble, and are protease-resistant. Disruption of the disulfide bonds leads to biological inactivity. The STI peptides are poor antigens and must be conjugated to a carrier protein to prepare antisera for diagnostic purposes. Both STI and STII genes are located on transposable elements and are plasmid encoded; the plasmid may also code for LT, colonization factors, colicin, and antibiotic resistance. The STs are homologous to guanylin, a mammalian hormone that aids in regulating fluid and electrolyte absorption in the gut. The hormone and STs bind to the same receptor on intestinal epithelial cells.

Toxins of the STI family cause diarrhea by fluid and electrolyte secretion. The STI toxins produce a short-term effect, which is reversible, quick-acting (within 5 minutes), and mediated by activation of guanylate cyclase, whereas the biological effect of CT and LT is prolonged with a lag phase (*ca* 1 hour), reversible, and mediated by activation of adenylate cyclase. The STIs bind to the extracellular domain of guanylate cyclase C with activation of the enzyme in the enterocytes of the small intestine. There is accumulation of cyclic GMP and secretion of Cl$^-$ and water into the intestinal lumen. Unlike LT-I and CT (which bind to adenylate cyclase from various tissues), STI is quite specific and binds only to intestinal guanylate cyclase. STII appears to be primarily found in ETEC isolated from pigs. Unlike STI, which is methanol soluble, STII is methanol insoluble. The STII is a larger peptide and is immunologically distinct from STI. The STII induces secretion of bicarbonate ions and water into the intestinal lumen, and an increase in intracellular Ca^{2+} is seen in the intoxicated intestinal cell. Generally, STII does not contribute to human disease; however, STII-producing ETEC have been the cause of a few cases of human diarrhea.

Other bacteria produce toxins similar to STI including *Citrobacter freundii*, *Yersinia enterocolitica*, and non-O1 *Vibrio cholerae*. A STI-containing plasmid from

ETEC can be transferred to species of *Shigella*, *Salmonella*, *Klebsiella*, *Enterobacter*, *Edwardsiella*, *Serratia*, and *Proteus* with stable maintenance of the plasmid and expression of toxin.

The *astA* gene, which encodes the EAEC heat-stable toxin 1 (EAST1), has been demonstrated in ETEC strains isolated from both humans and animals. In ETEC isolated from piglets, the gene is found in strains that produce STI, LT, or STI + LT and adhesin factor K88. Interestingly, the K88 and EAST1 genes are on separate plasmids. In human LT-, STI-, or LT + STI-producing ETEC strains, colonization factor antigen (CFA/I, CFA/II, or CFA/IV) is associated with EAST1 but all genes are not located on the same plasmid. Using a DNA probe for *astA*, the gene was found in 100% of 75 *E. coli* O157:H7 strains, in 47% of 227 EAEC, in 41% of 149 ETEC, in 22% of 65 EPEC strains, and in 13% of 70 DAEC. In addition, *astA* was present in nondiarrhea producing *E. coli* strains. Thus, the *astA* gene is common in *E. coli*, but the importance of the EAST1 toxin as a virulence determinant is not clear at the present time. The EAST1 is serologically distinct from STI; however, the toxin exhibits homology with the receptor-binding domains of STI. The enterotoxin is presumed to interact with the same receptor binding site of guanylate cyclase as STI, with elicitation of cyclic GMP secretion. Thus, EAST1 acts as a member of the STI family of heat-stable enterotoxins and produces diarrhea by a similar mode of action.

Attachment of ETEC to host cells is a necessary part of the virulence of the organism. In human strains, CFAs I, II, and IV are major adherence factors and are plasmid mediated. The same plasmid may encode CFA, ST, and LT. The CFAs are found only in ETEC strains that cause diarrhea. The CFA structure may be fimbrial rods, flexible fibrils, helical fibrils, or curly fibrils. Analysis of 241 ETEC isolates obtained from Mexican children showed that 46% of the strains had a CFA. Moreover, 65% of LT/ST strains expressed a CFA, whereas 50% of ST and 25% of LT strains produced a CFA. Children infected with ETEC lacking CFAs had diarrhea similar to those infected with CFA-containing ETEC. Since ETEC lacking CFA can cause diarrhea, unidentified colonization factors may be responsible for diarrhea in those strains. The ETEC preferentially colonize the distal two-thirds of the small intestine, an area highly innervated with adrenergic nerves, which elaborate norepinephrine at the terminals present in the mucosal lining. It is probable that intestinal norepinephrine stimulates induction of CFAs in human ETEC strains.

C. Foodborne Outbreaks

Outbreaks of ETEC have been associated with water and food consumption. The foods that were implicated included salads, dipping sauces, or ready-to-eat items such as hot dogs, cold roast beef, cold turkey, and brie cheese (foods that are served raw or that are cooked but served cold) (Table 5.3). The ETEC may be a more common cause of foodborne illness in the US than suspected. A ST-producing *E. coli* O153:H45 strain was implicated as the agent in a recent foodborne outbreak that occurred in Milwaukee, Wisconsin, in which an estimated 372–645 persons out of 1240 had diarrhea (Table 5.3). Pan-fried spiced potatoes were implicated.

D. Detection of ETEC

There is no satisfactory culturing technique that allows direct discrimination between ETEC and nonpathogenic *E. coli*. Therefore, methods for detection of ETEC have relied on detection of the toxins produced. Detection of colonization factors is difficult, owing to their heterogeneity and large number. Methods for toxin detection from fecal, food, and water samples include the suckling mouse assay (for ST production), the Y-1 adrenal cell assay (for LT production), ELISAs, and radioimmunoassays (LT and ST), latex agglutination (LT), and PCR and colony hybridization procedures for detection of LT- and ST-encoding genes.

VIII. Enterohemorrhagic *Escherichia coli*

Enterohemorrhagic *E. coli* (EHEC) cause the most severe and the majority of reported outbreaks of *E. coli*-associated diarrheal disease in the US and in other developed countries. The term EHEC refers to *E. coli* serotypes that share the same clinical, pathogenic, and epidemiologic features with *E. coli* O157:H7, the serotype that is responsible for the greatest proportion of disease cases (Fig. 5.1). The first reported outbreaks caused by *E. coli* O157:H7 occurred in 1982 and were epidemiologically linked to eating undercooked hamburgers served at a fast-food restaurant chain. Numerous outbreaks have since been documented, and it is estimated that greater than 73 000 cases of illness caused by *E. coli* O157:H7 occur each year in the US.

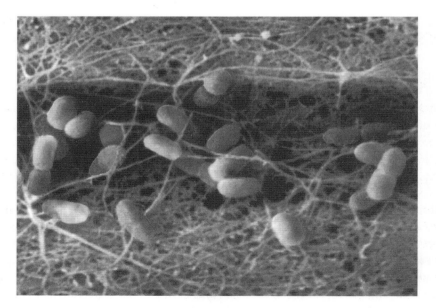

Figure 5.1 *Escherichia coli* O157:H7 attached to collagen fibrils on beef muscle tissue; scanning electron micrograph.

A. Disease Characteristics

E. coli O157:H7 infection can cause a wide spectrum of gastrointestinal illnesses, from asymptomatic infection to mild nonbloody diarrhea and even severe hemorrhagic colitis (HC). Hemorrhagic colitis manifests as a progression from watery to bloody diarrhea, with the absence of fever and pus cells in the stool. The mean incubation period for disease ranges from 3 to 8 days or can be as short as 1–2 days. Illness in patients with nonbloody diarrhea is less severe, and these individuals are less likely to develop systemic sequelae or to die. Symptomatic infections begin with profound abdominal pain and watery diarrhea, with the cecum and the ascending colon as the predominantly affected areas. After 1–3 days, bloody diarrhea develops in 1/4 to 3/4 of patients and lasts 4–10 days. Fever may be mild or absent, and about half of the patients have vomiting. Elevation of blood leukocytes, edema, hemorrhage of the lamina propria, superficial ulceration, pseudomembrane formation, and necrosis of the superficial colonic mucosa can occur with hemorrhagic colitis.

Approximately 7 days after initiation of diarrhea, 5–10% of patients, predominantly those less than 10 years of age, are stricken with hemolytic uremic syndrome (HUS), a severe systemic complication of EHEC infection and the leading cause of acute renal failure in children in the US. The production of Shiga toxins (Stx) by EHEC strains plays a large role in the pathogenesis of hemorrhagic colitis and HUS. With damage and death of endothelial cells of the glomeruli and afferent arterioles caused by Shiga toxin, circulating blood is exposed to the underlying collagen resulting in coagulation, platelet aggregation, and deposition of fibrin. Red blood cells and platelets are mechanically damaged as they pass through the narrowed blood vessels with resultant hemolytic anemia and thrombocytopenia. Narrowing of the capillaries and arterioles of the glomeruli results in reduction of glomerular filtration, and necrosis of kidney tissue occurs with complete occlusion of renal microvessels. The overall effect of HUS is kidney dysfunction.

B. Basis of Pathogenicity

E. coli serotypes that produce HC and HUS, express one or more Shiga toxins, cause AE lesions on intestinal cells, and possess a plasmid of *ca* 60 MDa (pO157) are referred to as EHEC. Although EHEC serotype O157:H7 is the most common cause of HC and HUS in the US, numerous other serotypes produce Shiga toxins and are referred to as Shiga toxin-producing *E. coli* (STEC). Close to 100 STEC serotypes have been associated with disease in humans. However, an STEC serotype is considered in the EHEC group only if it possesses all of the EHEC virulence factors. The most important EHEC virulence factors are the Shiga toxins, also called Verotoxins or Vero cytotoxins because of the cytotoxicity for Vero (monkey kidney epithelial) cells. These toxins were also formerly called Shiga-like toxins, since Shiga toxin 1 differs by only one amino acid from the Shiga toxin produced by *Shigella dysenteriae* type 1 (Stx). The main toxin groups, Shiga toxin 1 (Stx1) and Shiga toxin 2 (Stx2) are encoded on the genomes of temperate lambdoid bacteriophages. Shiga toxin 2 is a heterogenous group of toxins, and several variants have been described (Table 5.4). A single EHEC

Table 5.4 Characteristics of Shiga toxins

Toxin	Cytotoxic to:	Amino acid homology to Stx2		Neutralized by antibody to:	B subunit receptor
		A	**B**		
Stx	Vero and HeLa cells	a		Stx and Stx1	Gb_3
Stx1	Vero and HeLa cells	55	57	Stx and Stx1	Gb_3
Stx2	Vero and HeLa cells	100	100	Stx2	Gb_3
Stx2c	Vero and HeLa cells	100	97	Stx2	Gb_3
Stx2d	Vero and HeLa cells	99	97	Stx2	Gb_3
Stx2e	Vero and HeLa cells	93	84	Stx2	Gb_4
Stx2f[b]	?	63	57	?	?

[a] Stx1 is 98% homologous to Stx, differing by only one amino acid in the A subunit; however, nucleotide and sequence heterogeneities occur in Stx1 operons.
[b] Found in *E. coli* strains isolated from stools of feral pigeons.

isolate may produce Stx1, Stx2, or both toxins. *E. coli* strains that produce Stx2e are responsible for edema disease in swine. The Stx and Stx2e toxins are believed to be chromosomally encoded; however, one study found that Stx2e was encoded on the genome of a Shiga toxin 2e-converting bacteriophage in an ONT(nontypable):H⁻ *E. coli* strain isolated from a patient with diarrhea.

Shiga toxins are composed of one A polypeptide and five B polypeptides that form a pentamer responsible for binding to the eukaryotic cell receptor. The toxin receptor is globotriaosylceramide (Gb_3) or globotetraosylceramide (Gb_4) for Stx2e. The B pentamer binds to Gb_3 (or Gb_4), and the holotoxin is internalized by endocytosis. The toxin moves through the Golgi apparatus and endoplasmic reticulum and into the cytoplasm where the A_1 subunit, which has RNA *N*-glycosidase activity, targets the ribosome. The A subunit is cleaved by a protease in the endoplasmic reticulum and the cytosol to form the 27-kDa A_1 subunit and a 4-kDa A_2 fragment. The mode of action for Shiga toxins is inhibition of protein synthesis in eukaryotic cells by cleavage of an adenine residue from the 28S rRNA within the 60S ribosomal subunit. Additionally, the toxins can stimulate the production of pro- and anti-inflammatory cytokines and induce apoptosis.

Similar to the EPEC, *E. coli* O157:H7 and other EHEC produce the 94–97-kDa intimin (*eae* gene) protein and induce the AE histopathology. However, the *eae* gene of EPEC and *E. coli* O157:H7 (strain 933) shares 94% identity over the first 704 amino acids starting at the N-terminal region and only 49% identity over the remaining 280 amino acids at the C-terminal region, where binding to the enterocytes and Tir occurs. Both the EHEC and EPEC Tir bind intimin; however, the EHEC Tir is not tyrosine phosphorylated. Similar to EPEC, the genes for intimin, the type III secretion system, and the EspA, B, and D secreted proteins are located on the LEE pathogenicity island in EHEC.

A *ca* 60-MDa virulence plasmid is harbored by *E. coli* O157:H7 and other EHEC that have caused human illness. This plasmid varies in size among EHEC serotypes; however, its size is relatively constant within single serotypes. The virulence plasmid of EHEC strain EDL 933, referred to as pO157, has been sequenced, and 100 open reading frames (ORFs) were found. The plasmid encodes the EHEC hemolysin (operon *ehxCABD*), which belongs to the RTX family of exoproteins, KatP, a periplasmic catalase–peroxidase that functions to protect the bacterium against oxidative stress, a serine protease, EspP, and a gene cluster related to the type II secretion pathway of Gram-negative bacteria. There are 13 genes in this operon, *etpC* through *etpO*.

The low infective dose of *E. coli* O157:H7 corresponds to its ability to withstand acid environments. At least three acid-resistance systems function in *E. coli* O157:H7. These are the acid-induced arginine-dependent system and the glutamate-dependent system, which protect the organism against the bactericidal effects of various weak acids such as benzoic, acetic, propionic, and butyric acids, and an acid-induced oxidative system. The oxidative system is dependent on the alternative sigma factor, *rpoS*. Once induced, the acid-resistance systems remain active in *E. coli* O157:H7 during storage at 4°C. The pathogen can survive for several weeks to months in acidic foods such as mayonnaise, sausage, apple cider, yogurt, and cheddar cheese. Additionally, exposure to heat stress may increase the acid resistance of *E. coli* O157:H7.

E. coli O157:H7 releases membrane vesicles containing endotoxic lipopolysaccharide consisting of lipid-A and -O side chains, which may play a role in pathogenesis. Membrane vesicles isolated from *E. coli* O157:H7 may also contain linear DNA and plasmids. There is evidence that vesicles can facilitate the transfer of genetic material including antibiotic resistance genes from *E. coli* O157:H7 to other bacteria, and that the genes are subsequently expressed in the recipient bacteria. Vesicles were also found to contain Stx1 and Stx2, in addition to DNA encoding *eae*, *stx1*, *stx2*, and *uidA*.

There is no evidence that *E. coli* O157:H7 is invasive *in vivo*. An *in vitro* study showed that *E. coli* O157:H7 isolates were engulfed into T24 bladder and HCT-8 ileocecal cells; however, a later report disputed these findings. Quorum sensing, a system of cell–cell communication, which regulates expression of specific genes in a cell density-dependent manner, controls expression of the type III secretion gene transcription and protein secretion in EHEC and EPEC. The *luxS* gene, which is responsible for autoinducer-2 production in *E. coli* and activation of the expression of operons that encode components of the type III secretion system. On the other hand, SdiA, an *E. coli* LuxR homolog, functions as a negative regulator of expression of intimin and EspD in *E. coli* O157:H7. At low cell densities, virulence factors are produced allowing attachment to the colon. At higher cell densities, high concentrations of SdiA results in decreased production of the virulence factors, resulting in elimination from the colon. It is hypothesized that two quorum sensing systems function during different stages of growth of *E. coli* O157:H7.

C. Treatment

There is no established therapy for *E. coli* O157:H7 infection; however, several promising regimens are being tested or proposed. Conjugate vaccines composed of O157 lipopolysaccharide (LPS) conjugated to the Stx1 B subunit or to the *Pseudomonas aeruginosa* exoprotein A, Stx-liposome vaccines, and a vaccine consisting of a

Salmonella landau strain that expresses the O157 antigen have been tested in animals or are undergoing clinical trials. A novel therapy for preventing HUS involves treatment of patients with a Gb_3 receptor analog, called Synsorb Pk, which is designed to bind Shiga toxins in the lumen of the gastrointestinal tract. Synsorb Pk consists of the synthetic trisaccharide (Pk) of the Stx receptor, Gb_3, bound to a calcinated diatomaceous material called Chromosorb. The utility of antibiotics for treatment remains controversial. Studies have shown that administration of antibiotics to children with diarrhea caused by *E. coli* O157:H7 significantly increased the risk of developing HUS compared to children who did not receive antibiotics. Antibiotics may induce the expression of Shiga toxins and/or may cause the release of preformed toxin with bacterial injury caused by the antibiotic. Conversely, treatment with fosfomycin in the early stages of illness (within 3 days of onset) reduced the risk of HUS. Other treatments being evaluated include: administration of humanized Shiga toxin-neutralizing monoclonal antibodies; recombinant bacteria displaying a Shiga-toxin receptor mimic on the surface that neutralizes Shiga toxins; pooled bovine colostrum containing antibodies to Shiga toxins, intimin, and the EHEC–hemolysin; and bovine lactoferrin and its peptides.

D. Infectious Dose

Volunteer studies of EHEC infection in humans have not been performed; thus, the infective dose is not precisely known. However, analysis of *E. coli* O157:H7-contaminated salami eaten by people during an outbreak in 1994 in Washington State indicated an infective dose in the range of 2–45 organisms. The median most probable number of organisms found in raw ground beef patties implicated in an outbreak that occurred in the western US in November 1992 to February 1993 was 1.5 organisms per gram or 67.5 per patty. Furthermore, water and apple cider have been vehicles of infection, and person-to-person transmission of EHEC infection occurs, suggesting a low infective dose.

E. Animal Models

Many animal species have been used as models of EHEC infection, each model having advantages and disadvantages. Postweaning rabbits inoculated with EHEC developed diarrhea and AE lesions in the ileum, cecum, and colon. Rabbits have also been used to study the efficacy of vaccination against *E. coli* O157:H7, and to investigate the effects of Shiga toxin in organs and tissues. Gnotobiotic pigs have been used to study the pathogenesis of EHEC infection and the role of various virulence factors. In the pig model, brain lesions have been induced by oral inoculation of EHEC, and pigs inoculated with strains associated with HUS showed greater evidence of brain lesions than pigs inoculated with strains that were not associated with HUS. In the greyhound, a HUS-like disease manifesting cutaneous and renal glomerular vasculopathy occurs. Typical lesions were observed in dogs from which *E. coli* O157:H7 was isolated and in dogs administered Stx by intravascular inoculation. Lesions can be prevented if anti-Stx antiserum is given. Other models include the chicken, the mouse, which has been used to study the effects of inoculation with Stx and with EHEC, cattle, sheep, deer, goats, and other ruminants. Nonhuman primates may be a suitable model for HUS and other systemic complications of EHEC infection. Animal models will hopefully advance our understanding of the pathophysiology of EHEC-induced HC and HUS, and provide information that will help to prevent, control, and treat EHEC infection.

F. Etiology of Foodborne Cases and Outbreaks

Cattle are the most important reservoir of *E. coli* O157:H7. In the US, surveys of dairy and beef cattle have reported prevalence rates of 1.8–28%. Cattle are generally asymptomatic; however, *E. coli* O157:H7 can cause fatal ileocolitis in newborn calves. Cattle lack Gb$_3$, the Shiga toxin receptor, in the gastrointestinal tract. *E. coli* O157:H7 has also been isolated from deer, horses, goats, sheep, cats, dogs, rabbits, poultry, and from birds such as ravens, doves, and gulls, with prevalence rates of up to 5.2%. Houseflies also can carry O157:H7 in their intestine and on other body parts, and thus, can serve as vectors for dissemination of the organism.

Outbreak investigations have linked a large majority of cases of *E. coli* O157:H7 infection with consumption of undercooked ground beef. Other foods of bovine origin including raw milk and roast beef, and also porcine, avian, and sheep meat have been linked with outbreaks. A large outbreak involving 732 individuals occurred in December 1992 to January 1993 in the western US. Contaminated hamburgers served in a fast-food restaurant chain were implicated as the cause of infection. Numerous other food vehicles such as apple cider, mayonnaise, pea salad, cantaloupe, lettuce, hard salami, and alfalfa and radish sprouts have also been linked to outbreaks. Owing to the ability of *E. coli* O157:H7 to tolerate acidic conditions, the organism can survive in foods of low pH. Radish sprouts were implicated as the vehicle of infection in a large outbreak that occurred in Japan in 1996 that involved thousands of individuals, many of whom were school children. Outbreaks have also been linked to recreational water, well water, groundwater, and municipal water systems. At least seven people died and more than 2300 became ill in Walkerton, Ontario, Canada from drinking contaminated water. Bovine manure from a nearby farm was washed into the town's wells during a flood weeks earlier. In 2000, an outbreak was associated with children visiting a petting zoo in Pennsylvania. Person-to-person transmission of *E. coli* O157:H7 infection in nursing homes and daycare centers also occurs.

G. Diagnosis and Methods for Detection, Isolation, and Identification of EHEC

The Centers for Disease Control and Prevention (CDC) recommends that clinical laboratories culture stools from all patients presenting with bloody diarrhea and HUS on sorbitol MacConkey agar (SMAC); however, since *E. coli* O157:H7 also causes nonbloody diarrhea, it has been recommended that nonbloody stools also be cultured.

Because *E. coli* O157:H7 does not ferment sorbitol, colonies are colorless on SMAC after 24 hours of incubation. Modifications of SMAC medium have resulted in increased selectivity for *E. coli* O157:H7. These include CT-SMAC, which contains potassium tellurite and cefixime or CR-SMAC, in which cefixime and rhamnose are added to SMAC. *E. coli* O157:H7 does not produce the enzyme β-glucuronidase, which hydrolyzes substrates such as 4-methyl-umbelliferyl-D-glucuronide (MUG). *E. coli* O157:H7 colonies do not hydrolyze MUG, therefore, they do not fluoresce. Sorbitol MacConkey agar containing the substrate 5-bromo-4-chloro-3-indoxyl-β-D-glucuronic acid cyclohexylammonium salt (BCIG), referred to as SMA-BCIG, has been used for isolation of *E. coli* O157:H7 from beef. Several commercially available selective and differential media for isolation of the pathogen from foods and other samples include Rainbow Agar O157, CHROMagar O157, BCM O157:H7 Agar, and Fluorocult *E. coli* O157:H7 Agar. Suspect colonies can be tested for the presence of the O157 and H7 antigens using commercially available latex agglutination kits or antisera (Table 5.5). Generally, there are no phenotypic markers shared by all non-O157 EHEC, except for the production of Stxs, that can be used to distinguish them from non-pathogenic *E. coli* rendering detection of these pathogens difficult. Thus, the incidence of disease caused by non-O157 EHEC may be underestimated.

A number of immunoassays and PCR-based assays for detection of *E. coli* O157:H7 are commercially available (Table 5.5). Various multiplex PCR assays employing primers for two or more *E. coli* O157:H7 genes including *stx1*, *stx2*, *eae*, *uidA* (β-glucuronidase), hly_{933}, *fliC* (H7 antigen), and rfb_{O157} (O157 antigen) have also been developed. Shiga toxin can be detected in the supernatants of bacterial cultures, food enrichments, and in fecal samples using immunologic assays including colony immunoblot assays, latex agglutination, and antibody capture or toxin receptor-mediated ELISAs. Several assay kits are commercially available (Table 5.5). Other types of methods are cell toxicity assays using Vero or HeLa cells, and genetic assays such as the PCR targeting *stx* genes or colony hybridization assays using DNA probes.

H. Genetic Fingerprinting and Outbreak Investigation

Differentiation of EHEC isolates below the species level (subtyping or genetic fingerprinting) is an integral part of epidemiologic investigations. Typing methods are used to determine the relatedness of a group of bacterial isolates. Phenotypic methods such as serotyping are gradually being replaced by genetic typing techniques such as ribotyping, random amplified polymorphic DNA, and pulsed-field gel electrophoresis (PFGE). The PFGE technique involves enzymatic digestion of bacterial DNA and analysis of the digestion products (10–20) that are separated by agarose gel electrophoresis with programmed variations in the direction and duration of the electric field (the pulsed field). In an outbreak that occurred in 1993, typing of isolates by PFGE determined that the strain of *E. coli* O157:H7 isolated from patients was indistinguishable from the strain found in hamburger patties served at a fast-food restaurant chain. Because the source of infection was recognized quickly, the patties were recalled, and further illness was prevented. The CDC, in collaboration with the

Table 5.5 Commercially available media and test kits for isolation and detection of *Escherichia coli* O157:H7

Type of assay	Medium/test kit	Supplier
Automated electroimmunoassay system	Detex Pathogen Detection System	VWR Scientific
ELISA	HECO157	3M Canada
	Premier *E. coli* O157	Meridian Diagnostics
	TECRA *E. coli* O157 VIA	Tecra International
Enzyme immunoassay	Assurance EHEC EIA	BioControl Systems, Inc.
Enzyme immunoassay, detection of Stx1 and Stx2	Premier EHEC	Meridian Diagnostics, Inc.
	ProSpectT Shiga Toxin *E. coli* (STEC) Microplate Assay	Genzyme Virotech GmbH
	RIDASCREEN	Lionheart Diagnostics
Enzyme-linked fluorescent assay	VIDAS ECO	bioMérieux, Inc.
Immunoassay	Eclipse *E. coli* O157:H7 Rapid Color Change Test	Eichrom Technologies, Inc.
	ImmunoCard Stat! *E. coli* O157:H7	Meridian Diagnostics
	NOW *E. coli* O157:H7 and O157	Binax, Inc.
	Transia Card *E. coli* O157:H7	GENE-TRAK Systems
Immunochromatography	C QUIC Plus *E. coli* O157 Test	Sun International Trading
	PATH-STICK One Step Rapid *E. coli* O157 Test	Celsis, Inc.
Immunoconcentration	VIDAS ICE	bioMérieux, Inc.
Immunomagnetic separation	Dynabeads anti-*E.coli* O157	Dynal, Inc.
Immunomagnetic separation and electrochemiluminescence	PATH *IGEN E. coli* O157 Test	Igen International, Inc.
Immunomagnetic separation and ELISA	EHEC-Tek for *E. coli* O157:H7	Organon Teknika Corp.
	EIA Foss *E. coli* O157:H7	Foss North America, Inc.
Lateral flow immunoassay	VIP for EHEC	BioControl Systems, Inc.
Latex agglutination	Ecolex O157	Orion Diagnostica
	Microscreen *E. coli* O157	Microgen Bioproducts Ltd
	Prolex *E.coli* O157 Kit	PRO-LAB
	RIM *E. coli* O157:H7 Latex Test	Remel, Inc.
	Wellcolex O157	Murex
Membrane filtration and enumeration on SD-39 agar	ISO-GRID Method for *E. coli* O157:H7	QA Life Sciences, Inc.
Polymerase chain reaction	BAX for Screening/*E.coli* O157:H7	Qualicon, Inc.
	Probelia PCR System	BioControl Systems, Inc.
	TaqMan *E. coli* O157:H7 Detection Kit	PE Biosystems
	TaqMan *E. coli* STX1 and STX2 Detection Kit	PE Biosystems
Reversed passive latex agglutination, detection of Stx1 and Stx2	VTEC-RPLA TD960	Oxoid, Inc.
Sandwich ELISA	Reveal Microbial Screening Test for *E. coli* O157:H7	Neogen Corporation
Selective/differential agar	BCM O157:H7 Agar	Biosynth
	CHROMagar O157	CHROMagar or Hardy Diagnostics
	Fluorocult *E. coli* O157:H7 Agar	Merck
	Rainbow Agar O157	Biolog

Association of State and Territorial Public Health Laboratory Directors created PulseNet, a molecular subtyping network for foodborne disease surveillance in the US. The United States Department of Agriculture (USDA) Food Safety and Inspection Service and the US Food and Drug Administration were integral partners

Table 5.6 Outbreaks and cases of illness associated with Stx1- and Stx2-producing non-O157 *Escherichia coli*

Year reported	Country	Serotype	Disease	Number affected	Stx toxin type
2001	United States	O121:K19	BD	1 (renal and cardiac transplant patient)	2
2000	Ireland	O26:H11	Nonbloody diarrhea	4	1
	United States	O111:H8	HUS (2) BD (20)	58	1 and 2
1999	Japan	O118:H2	BD (9)	126	1
1998	Canada	O121:H11	Thrombocytopenic purpura	1	1
1997	Finland	OX3:H21	BD, HUS	1	2
1996	United States	O103:H2	HUS (urinary tract infection, not gastrointestinal)	1	1
		O111ac:NM	HUS (1)	5	1 and 2
	Australia	OR:H9 (and a strain of *Enterobacter cloacae*)	HUS	1	*E. coli* strain produced 2; *E. cloacae* strain produced 2
1995	Germany	O111:H⁻	BD	1	1
	Australia	O48:H21	BD, HUS	1	1 and 2
		O111:NM	BD (16) HUS (23)	23	1 and 2
	United States	O104:H21	BD (16)	11 (confirmed) 7 (suspected)	2
	Germany	O111:H⁻	HUS (Germany)	1	1 and 2
	Italy	O111:H⁻	HUS (Italy)	1	1 and 2
	Italy	O111:H⁻	HUS (Italy)	1	1
		O120:H19	HUS (Italy)	1	Both Stx1 and Stx2 found in stool, but EHEC isolate produced only Stx1
1994	Italy	O111:NM	HUS	1	1 and 2
		O145	HUS	1	1
		O26:H11	BD, HUS	1	2

Year	Country	Serotype	Syndrome		2e (both serotypes)
1993	United Kingdom	O9ab:H⁻	HUS	1	
		O101:H⁻	HUS	1	2
	Canada	O55:H7	HUS	1	1 and 2
		O111:NM	HUS	?	1 and 2
	Central Europe	O22:H8	HUS	?	2
		O26:H⁻	HUS	?	2
		O26:H11	HUS	?	2
		O55:H⁻	HUS	?	2
		O55:H6	HUS	?	2
		O111:H8	HUS	?	2
	Japan	O18:H⁻	BD	1	1 and 2
	Australia	O111:H⁻	HUS	1	1
	United States	O68:H⁻	BD	1	1 and 2
	France	O103:H2	HUS	6	1
1992	Chile	O26	HUS	2	1 and 2
		O26	HUS	1	1
		O111	HUS	1	1 and 2
		O111	HUS	1	1
	Italy	O111:NM	HUS	1	2
1990	Canada	O91:H21	HUS	1	2e
	United Kingdom	O55:H7	HUS	?	2
		O55:H10	HUS	?	2
		O105ac:H18	HUS	?	1 and 2
		O115:H10	HUS	?	1
		O128ab:H25	HUS	?	2
		O145:H25	HUS	?	2
		O163:H19	HUS	?	2
		O165:H25	HUS	?	2
1989	United Kingdom	O5:H⁻	HUS	?	1 and 2
		O55:H7	HUS	?	2
		O55:H10	HUS	?	2
		O111ac:H⁻	HUS	?	1
		O111ac:H⁻	HUS	?	1 and 2
		O153:H25	HUS	?	2

BD, bloody diarrhea; HUS, hemolytic uremic syndrome.

in the development of PulseNet, and also contribute to the successful operation of the system. Laboratories participating in PulseNet perform PFGE on isolates from humans and the suspected food, and enter the PFGE patterns into an electronic database of DNA 'fingerprints', which are transmitted to the CDC. The PulseNet system is playing an integral role in the surveillance and investigation of outbreaks of foodborne illness caused by *E. coli* O157:H7. It is being expanded to include *Salmonella*, *Listeria monocytogenes*, and *Campylobacter*.

I. Importance of Non-O157 STEC/EHEC

Close to 100 non-O157 STECs have been responsible for cases and outbreaks of HC and HUS (Table 5.6). In Australia and Argentina and in many European countries, non-O157 STEC serotypes are more prevalent than *E. coli* O157:H7. Unfortunately, the non-O157 serotypes do not have an identifiable biochemical marker such as lack of sorbitol fermentation or β-glucuronidase activity to facilitate screening and identification. Detection of these serotypes requires testing for the Shiga toxins or for genes which encode the Shiga toxins.

IX. Emerging Diarrheic *Escherichia coli* Groups

Two new groups of diarrheic *E. coli* have recently been described. In the HEp-2 cell assay, cell-detaching *E. coli* strains cause detachment of the HEp-2 cells from the glass vessel within 3 hours of incubation. The detaching phenotype is closely associated with production of a hemolysin and a cytotoxic necrotizing factor. Enteric colonizing *E. coli* (ECEC) has caused diarrhea in human volunteers. Colonization factors of ECEC are plasmid mediated. There is no reason, given the promiscuity of genetic transfer in Gram-negative organisms, that there will not be more pathogenic types found in the future.

Bibliography

Acheson, D. W. K. and Keusch, G. T. (1995). *Shigella* and enteroinvasive *Escherichia coli*. *In* 'Infections of the Gastrointestinal Tract' (M. J. Blaser, P. D. Smith, J. I. Ravdin, H. B. Greenberg and R. L. Guerrant, eds), pp. 691–707. Raven Press, New York.

Bell, C. and Kyriakides, A. (1998). '*E. coli* – A Practical Approach to the Organism and its Control in Foods'. Blackie Academic Professional, New York.

Cohen, M. B. and Giannella, R. A. (1995). Enterotoxigenic *Escherichia coli*. *In* 'Infections of the Gastrointestinal Tract' (M. J. Blaser, P. D. Smith, J. I. Ravdin, H. B. Greenberg and R. L. Guerrant, eds), pp. 691–707. Raven Press, New York.

Doyle, M. P., Zhao, T., Meng, J. and Zhao, S. (1997). *Escherichia coli* O157:H7. *In* 'Food Microbiology – Fundamentals and Frontiers' (M. P. Doyle, L. R. Beuchat and

T. J. Montville, eds), pp. 171–191. American Society for Microbiology Press, Washington, DC.

Kaper, J. P. and O'Brien, A. D. (eds) (1998). '*Escherichia coli* O157:H7 and Other Shiga Toxin-producing *E. coli* Strains'. American Society for Microbiology Press, Washington, DC.

Law, D. (2000). Virulence factors of *Escherichia coli* O157 and other Shiga toxin-producing *E. coli*. *J. Appl. Microbiol.* **88**, 729–745.

Nataro, J. P. and Kaper, J. B. (1998). Diarrheagenic *Escherichia coli*. *Clin. Microbiol. Rev.* **11**, 142–201.

Nataro, J. P., Steiner, T. and Guerrant, R. L. (1998). Enteroaggregative *Escherichia coli*. *Emerg. Infect. Dis.* **4**, 251–261.

Vallance, B. A. and Finlay, B. B. (2000). Exploitation of host cells by enteropathogenic *Escherichia coli*. *Proc. Natl Acad. Sci. USA* **97**, 8799–8806.

Willshaw, G. A., Cheasty, T. and Smith, H. R. (2000). *Escherichia coli*. *In* 'The Microbiological Safety and Quality of Food' (B. M. Lund, T. C. Baird-Parker and G. W. Gould, eds), pp. 1136–1177. Aspen Publishers, Inc., Gaithersburg, MD.

Campylobacter jejuni and Related Organisms

6

Sean F. Altekruse and David L. Swerdlow

I. *Campylobacter jejuni*

A. History

Campylobacter spp. were first recognized as human pathogens in the 1970s but have undoubtably caused human illness for much longer. An organism now considered to be *C. fetus* was first isolated in 1909 (then called a related *Vibrio*) from fetal tissues of aborted sheep. In the 1970s, microbiologists using modified atmosphere isolation methods tested stools from human patients with diarrheal illness of unknown etiology and discovered that *Campylobacter* spp. were both common and important human pathogens. *C. jejuni,* by far the most prevalent cause of human illness, is now the most commonly diagnosed cause of sporadic foodborne bacterial infection in the US and other industrialized nations. *C. jejuni* is a Gram-negative, curved rod with polar flagella (Fig. 6.1). Other selected *Campylobacter* and related species also cause human disease. Some researchers place the genus in Campylobacteriaceae, which includes *Campylobacter* and *Arcobacter* species, and *Helicobacter pylori* in rRNA superfamily VI. Others place these organisms in the epsilon subdivision of the Protobacteria. Each species has its own reservoir. *C. jejuni,* the most common species associated with human illness, is often found in poultry. *C. coli*, which is a less common cause of human illness, is found in swine, and *C. lari* is found in seagulls and shellfish.

B. Clinical Illness

1. Burden of illness

In the US, approximately two million cases of human campylobacteriosis each year affect nearly 1% of the population. Although *C. jejuni* is the most frequently diagnosed

Foodborne Diseases 2nd Edn
ISBN: 0-12-176559-8

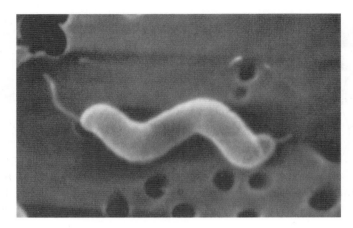

Figure 6.1 Electron microscopic image of *Campylobacter jejuni* cells, demonstrating slender, curved-shaped morphology (0.2–0.5 μm wide and 0.5–5 μm long) with polar flagellum.

bacterial cause of gastroenteritis in the US, *C. jejuni* related deaths are uncommon. Most deaths occur in infants, elderly, or immunosuppressed individuals.

2. Sequelae to infection

Severe sequelae are increasingly recognized as an important part of the epidemiology of human campylobacteriosis. The Guillain–Barré syndrome is a neurologic syndrome occurring about once for every 1000 cases of campylobacteriosis. Early symptoms of Guillain–Barré syndrome include burning sensations and numbness, which can progress to flaccid paralysis. Most patients with this syndrome recover; up to 15% die and 15% experience chronic complications.

Reactive arthritis, or Reiter's syndrome, is a sterile postinfectious process that may be a sequela to acute gastrointestinal campylobacteriosis. Onset of reactive arthritis occurs 7–10 days after onset of diarrheal illness in about 1% of patients with campylobacteriosis. Pain and incapacitation may last for months or become chronic. Septic arthritis and osteomyelitis due to *C. jejuni* infection also occur, generally in patients with underlying disease.

Chronic diarrhea caused by *C. jejuni* is an important clinical complication in patients with underlying diseases such as human immunodeficiency virus (HIV) infection. Patients with HIV infection who develop recurrent campylobacteriosis are at risk for antimicrobial-resistant (e.g. erythromycin) *C. jejuni* infections following long-term therapy. HIV-positive individuals who develop campylobacteriosis have shorter survival times, and higher rates of bacteremia and hospitalization than HIV-positive individuals without campylobacteriosis.

3. Treatment of acute campylobacteriosis

When effective antimicrobial therapy is administered to patients with campylobacteriosis soon after onset of illness, the duration of illness is significantly reduced. In a study of patients with campylobacteriosis in 19 US counties, treatment with antimicrobial drugs within 3 days of onset reduced the mean duration of illness by more than 5 days and treatment within 5 days reduced mean duration of illness by 2 days. Only 38% of patients were treated within 5 days. Later treatment had no impact on duration.

C. Antimicrobial Resistance

When patients with antimicrobial resistant *C. jejuni* infections are treated with an ineffective antibiotic, illness is often prolonged. Among patients with invasive infections, ineffective antibiotic therapy is a potentially life-threatening complication. The rate of antimicrobial-resistant enteric infections is highest in developing nations, where antimicrobial drug use is relatively unrestricted. In the 1990s, however, the emergence of fluoroquinolone-resistant strains of *C. jejuni* was reported in Western Europe, shortly after fluoroquinolones were approved for veterinary use. In parts of Spain, for example, up to 50% of *Campylobacter* isolates from humans were reported to be resistant to fluoroquinolones and an increase in the prevalence of fluoroquinolone-resistant *C. jejuni* strains from humans was reported in Holland, Finland, and other western European nations.

Surveillance is needed to assess patterns of antimicrobial usage in human and animal populations and to monitor changes in the prevalence of antimicrobial resistant *Campylobacter* infections. Reservoirs for resistant organisms that infect humans may include ill persons, animals, foods, and water. Following the US Food and Drug Administration approval of fluoroquinolone use in poultry in 1995, a multiple site surveillance system was developed to monitor the prevalence of fluoroquinolone-resistant *C. jejuni* isolates of human origin. US baseline studies found no resistance to fluoroquinolones in *C. jejuni* isolates obtained from humans in the late 1980s. Nearly 45% of human isolates in the US were resistant ten years later.

D. Pathogenesis

The pathogenesis of enteric *C. jejuni* may involve many factors or virulence determinants. There is, however, little consensus on the relative importance of the factors that have been described. The health of the host and host immunity following repeated exposure affect clinical outcome. Pathogen-specific factors in disease causation may include chemotaxis, motility, mucous colonization, flagellar attachment, adhesion to cells, iron acquisition, host cell invasion, toxin production, inflammation and active secretion, and epithelial disruption with leakage of serosal fluid. Expression of specific virulence determinants may be triggered by host or environmental signals (e.g. temperature, pH, or iron availability).

E. Microbiology

1. Survival in the environment
C. jejuni has a low infectious dose in humans, ranging from 500 to 10 000 organisms. Replication does not occur outside the host intestine, and survival is also poor.

Compared to other enteric bacterial pathogens with low infectious dose, person-to-person transmission of *C. jejuni* is rare. *C. jejuni* is heat tolerant, growing best at 42°C. The organism is microaerophilic. It grows optimally in an atmosphere of approximately 5% O_2, 10% CO_2, and 85% N_2. The organism is also sensitive to stresses including freezing, drying, acidic conditions (pH $\leqslant$ 5.0), and salinity.

2. Sample collection and transportation

Campylobacters may be detected via direct microscopic examination of stools; however, bacterial isolation is more sensitive and specific. Because of the low infectious dose and poor survival of *C. jejuni*, careful specimen handling is important. Specimens should be chilled, not frozen, and submitted to a laboratory without delay. Use of a deep airtight container will minimize exposure to oxygen and desiccation. If a specimen cannot be processed within 24 hours or is likely to contain small numbers of organisms, a suitable transport medium such as Cary Blair is recommended. Diagnostic laboratories can provide guidance on specimen handling procedures.

A variety of procedures are available for isolation of *C. jejuni* from food and clinical specimens. Certain specimens are likely to contain stressed organisms or low numbers of organisms, such as processed foods, stools from convalescent patients, stools from patients that were treated with antibiotics, and swabs that were exposed to oxygen. Pre-enrichment at low temperature may improve recovery. Use of a selective medium containing antimicrobial agents, an oxygen-quenching agent such as charcoal, or a low oxygen atmosphere facilitates *C. jejuni* isolation and decreases the number of colonies that must be screened. However, antimicrobial agents may inhibit the growth of other campylobacters and related pathogens.

3. Typing schemes

There is no standard subtyping scheme for *C. jejuni*. Soon after *C. jejuni* was described, two serotyping systems were developed: a heat stable or somatic O antigen serotyping system; and a heat-labile or flagellar H antigen serotyping system. These typing schemes are labor intensive and require specialized reagents that limit use to reference laboratories. A useful typing scheme should provide consistent results, discriminate isolates into a manageable number of subtypes, and be widely accessible. Genetic typing methods, such as fla A gene polymerase chain reaction (PCR) restriction fragment length polymorphism (RFLP) typing and pulsed-field gel electrophoresis, hold promise.

F. Risk Factors for Human Illness

In the US, outbreaks and sporadic illnesses caused by *C. jejuni* have different epidemiologic characteristics. The two peak seasons for common source outbreaks are May and October, and there are fewer outbreaks during the summer. In contrast, the number of *Campylobacter* isolates reported to the US Centers for Disease Control and Prevention (CDC), which largely represent sporadic illnesses, peak in the summer months. Most outbreaks are associated with drinking raw milk or unchlorinated water, whereas sporadic illnesses are often associated with mishandling or consumption of undercooked poultry or cross-contamination of other foods by raw poultry. Other foods implicated in human illness include sausage and raw shellfish. Travel-associated illness has been documented, as has animal-to-human transmission following contact with immature or diarrheic pets or livestock.

In the US, infants and young adults have the highest rates of *Campylobacter* infection. In infants, this is thought to relate to immature immunity and first exposure to campylobacters in foods. One group with a particularly high incidence of campylobacteriosis is young male adults, probably reflecting poor food handling practices in a population that previously relied on parents to prepare meals.

G. *Campylobacter* Ecology

1. Wildlife reservoirs

Many bird species carry *C. jejuni*. These include migratory birds – cranes, ducks, and geese – and seagulls. The organism is also found in rodents, and other wild and domestic species. Houseflies and beetles carry the organism on their exoskeleton. Further study is warranted to understand environmental sources and reduce the risk that they can 'seed' broiler-type chicken flocks.

2. Poultry

Day-old chicks can be colonized with as few as 40 organisms, and flocks readily pick up *C. jejuni* from the environment. Most flocks are colonized by 4 weeks. Vertical transmission, i.e. from one generation to the next, is unlikely.

Campylobacters in the environment of poultry operations serve as a nidus for reinfection of flocks. Reservoirs within the poultry environment include darkling beetles and houseflies. The use of unchlorinated drinking water facilitates colonization of poultry flocks. Workers who care for other animals before entering poultry house contribute to the colonization of poultry flocks. Feeds are an unlikely source of transmission since their moisture is low and campylobacters are sensitive to drying.

3. Cattle, pigs, and sheep

C. jejuni is present in the intestinal tracts of many cattle, where it often exists as a commensal organism. Young animals are more likely to be colonized than older animals, and feedlot cattle carry campylobacters more often than grazing animals. In one study, colonization of dairy herds was associated with use of unchlorinated drinking water.

Pigs are frequently colonized with campylobacters, principally *C. coli*, generally without evidence of disease. In Norway, *C. coli* was isolated from 58% of pigs that were sampled. Serotypes of *C. coli* found in swine do not always resemble those in humans. *C. jejuni* causes abortion or can occur as an intestinal commensal in sheep.

4. Water

Surface waters often harbor campylobacters, which may originate from fecal contamination by humans, terrestrial animals, or birds. The presence of *C. jejuni* organisms in water is not always detectable by routine culturing methods because under certain conditions, e.g. in the stationary growth phase or upon exposure to atmospheric oxygen, campylobacters become round or coccoid. This morphologic transition may lead to the 'viable but nonculturable state', characterized by uptake of amino acids and maintenance of an intact outer membrane, but inability to grow on selective media. Perhaps organisms in this state are transmitted to poultry via unchlorinated drinking water; however, efforts to test this hypothesis have had variable success.

H. *Campylobacter* in the Food Supply

C. jejuni and other *Campylobacter* species have been isolated from various foods. Surveys of the prevalence of these organisms in foods support epidemiologic evidence implicating poultry, meat, and raw milk as sources of human infection. Most retail chicken contains *C. jejuni* and one study reported an isolation rate of 98% for retail chicken meat. Counts often exceed 10^3 per $100\,g$. Liver and other giblets have high levels of contamination.

Raw milk is a leading cause of outbreaks of *C. jejuni* infections in humans. Bulk tank milk frequently contains the organism. Raw milk is generally contaminated via bovine feces, but *C. jejuni* mastitis with consequent direct contamination of milk may also occur as a result of udder damage. Outbreaks of human infection have occurred after consumption of normal-appearing raw milk obtained from cows with *Campylobacter* mastitis. In England, pasteurized milk in glass bottles with foil lids pecked by birds (jackdaws) has been associated with human campylobacteriosis.

C. jejuni is present in some red meat. The recovery rate is highest for specimens from calves and steers (juvenile cattle). The organism may be present in retail pork and mutton, and is occasionally found in shellfish, fresh vegetables, and mushrooms.

I. Control of *Campylobacter* Infection

1. On-farm controls

Steps are needed to control *Campylobacter* contamination at the farm. Research is needed to establish effective and affordable farm-level controls. Strict management of poultry (and other livestock) production may reduce carriage of campylobacters. Necessary practices might include restricting contact between livestock and other animals, requiring that workers disinfect boots, wash hands, and change into clean clothing before entering animal housing areas. The use of chlorinated drinking water in food animal production holds promise. Treatment of chicks with cecal colonizing bacteria may also reduce *C. jejuni* colonization. Even strict measures may not eliminate carriage by poultry and livestock.

2. Processing controls

Slaughter and processing provide important opportunities for the reduction of *C. jejuni* counts on food animal carcasses. Applied research is needed to enhance the effectiveness of processing controls. Bacterial counts may increase during several points in processing, including transportation from farm to slaughter plant, poultry defeathering, and evisceration. Processing steps that have been reported to reduce bacterial counts include chilling, scalding, and washing. Treatment of chiller water containing sodium chloride with an electrical current may also reduce *C. jejuni* contamination. The use of chlorinated sprays to maintain clean working surfaces and careful processing plant design to reduce surface area in contacts with carcasses may also reduce carcass and retail meat contamination. Lactic acid sprays have been reported to reduce carcass contamination. Carcass irradiation would eliminate most *C. jejuni* on retail meat and poultry. These and other processing level interventions should receive priority for evaluation.

3. Food handling and hygiene

Since campylobacters and other pathogens are commonly found in raw foods of animal origin, careful food handling can reduce the risk of illness. Hand washing, prevention of cross-contamination, and thorough cooking of poultry and meat are necessary controls at the food preparation level. Raw milk and untreated water should also be avoided.

II. Organisms Related to *Campylobacter jejuni*

Other organisms related to *C. jejuni* are human pathogens. Efficient methods are needed for isolation of these organisms. Data on the epidemiology of human illnesses are also needed. Selected pathogens related to *C. jejuni* are described below (Table 6.1).

Table 6.1 Taxonomic position, sources, and disease associations; selected rRNA superfamily VI bacteria

rRNA homology group	Taxon	Known source(s)	Disease associations	
			Human	Veterinary
I	*Campylobacter fetus* subsp. *fetus*	Cattle, sheep	Septicemia, abortion, gastroenteritis, meningitis	Abortion
	C. fetus subsp. *venerealis*	Cattle	Septicemia	Bovine infertility
	C. hyointestinalis	Pigs, hamsters, cattle, deer	Gastroenteritis	Porcine and bovine intestinal hyperplasia
	C. coli	Pigs, poultry, sheep, birds, cattle	Gastroenteritis, septicemia	Gastroenteritis
	C. jejuni subsp. *jejuni*	Poultry, pigs, cattle, other	Gastroenteritis, septicemia, sequelae	Gastroenteritis, avian hepatitis
II	*Arcobacter butzleri*	Pigs, cattle, water, other	Gastroenteritis, septicemia	Gastroenteritis, porcine abortion
	Arcobacter cryaerophilus	Pigs, cattle, sheep, other	Gastroenteritis, septicemia	Porcine, bovine, ovine, equine abortion
III	*Helicobacter pylori*	Humans, primates	Gastritis, ulcers, gastric carcinoma	Gastritis in rhesus monkeys

A. Other *Campylobacter* species

Some *Campylobacter* species other than *C. jejuni*, *C. coli*, and *C. lari* are human pathogens. These include *C. fetus* subspecies *fetus*, *C. hyointestinalis*, *C. upsaliensis*, *C. concisus*, *C. mucosalis*, *C. sputorum*, *C. showae*, *C. cervus*, *C. rectus*, and *C. gracilis*. New pathogens continue to be recognized.

Until recently, *C. fetus* subspecies *fetus* was regarded as an animal pathogen, causing bovine and ovine abortion and sterility. Between 1980 and 1995, *C. fetus* was implicated in at least four outbreaks of human disease in North America, three associated with foods (raw milk, a raw calf liver supplement, and cottage cheese). Infection may result in invasive disease or gastroenteritis. Invasive disease usually affects patients with underlying diseases (e.g. malignancy or HIV).

Between 1979 and 1985, two of four laboratories confirmed *C. hyointestinalis* isolates received by CDC were from stools of homosexual men. Stool isolates were also obtained from an 8-month-old girl who lived on a farm with livestock and a 79-year-old woman who had traveled to Egypt. A small outbreak among family members in Canada may have been associated with drinking raw milk. It is unclear whether this organism causes human disease other than diarrheal disease and whether disease is restricted to persons with compromised immune systems. *C. hyointestinalis*, with or without *C. mucosalis*, is suspected to cause proliferative ileitis in swine and diarrhea in calves.

The first isolation of *C. upsaliensis* was reported in 1983, from stools of healthy and diarrheic dogs. It is a rare human pathogen, but may be under-reported because growth is inhibited by antibiotics present in selective media used for isolation of *C. jejuni* (e.g. cephalothin). Of 11 human cases reported by CDC between 1980 and 1986, three originated from stools and eight from blood. Stool isolates originated from previously healthy patients with acute gastroenteritis and an immunocompromised patient with persistent diarrhea. Blood isolates originated from infants with fever and respiratory symptoms, a woman with an ectopic pregnancy, elderly men with chronic diseases, and immunocompromised adults.

B. *Arcobacter*

Two of four *Arcobacter* species have been associated with human disease, *A. butzleri* and *A. cryaerophilus*. Reports of human illnesses caused by *A. butzleri* include an outbreak of diarrhea in children, severe diarrhea in patients with chronic disease, and bacteremia in a neonate. *A. cryaerophilus* has been recovered from the blood of a uremic patient with pneumonia. The spectrum of reported illnesses may relate to strain differences or the health status of patients. Research regarding mechanism(s) of pathogenesis is needed.

A. butzleri and *A. cryaerophilus* have been found in the environment, but research is needed to identify reservoirs. *A. butzleri*-associated diarrheal illness has been reported in nonhuman primates. *A. cryaerophilus* and *A. butzleri* have been recovered from aborted porcine and equine fetuses. In the Netherlands, *A. butzleri* was present in 53 of 220 (24%) poultry meat specimens as well as in some beef and pork meat specimens. In Germany, the pathogen was detected several times in a drinking water reservoir. *A. cryaerophilus* has been repeatedly found in Bangkok's canals.

C. *Helicobacter pylori*

H. pylori was described in 1982, when it was initially placed in the genus *Campylobacter* because of its structural similarities and its requirement for a microaerobic environment. By 1989, analyses of nucleic acid sequences, ultrastructural features, cellular fatty acids, and growth characteristics placed it in the related genus, *Helicobacter*. *H. pylori* is considered the leading cause of peptic ulcers in humans, a condition once attributed to noninfectious causes such as stress. *H. pylori* may be the most common chronic bacterial infection of humans throughout the world. In developed countries, between 30% and 50% of adults are infected and, in developing countries, 75% of adults may be infected. Incidence of infection increases with age and with lower socioeconomic status.

H. pylori causes gastritis and gastric and duodenal ulcers in chronically infected persons. Infections lasting decades increase the risk of gastric carcinoma and lymphoma. Age-specific stomach cancer rates in industrialized nations have declined with *H. pylori* infection rates. These trends may relate to societal changes in antibiotic usage, hygiene, or sanitation. However, an increase in esophageal cancer has been hypothesized to result from an increase in acidic reflux associated with Helicobacter eradication.

Diagnostic tests for *H. pylori* infection are used for specific purposes. Histologic visualization of the bacterium in biopsy specimens is the gold standard test. Culture of the bacterium from biopsy specimens is possible, but not all clinical centers routinely culture biopsy specimens. Biopsy specimens can be tested for urease and a recently licensed noninvasive test, the urea breath test, is also specific, because *H. pylori* is the only bacterium of the gastric mucosa that elaborates urease, degrading urea to ammonia and bicarbonate. Serologic tests are used to detect IgG and IgA antibodies elicited in response to chronic infection. In most instances, if endoscopy is not indicated, serologic testing is used to screen for infection. If endoscopy is indicated, urease testing is often chosen because it is a sensitive test. The urea breath test is often used to confirm eradication because it is noninvasive, simple, and inexpensive.

The mode of transmission of *H. pylori* is unclear, but people are frequently infected during childhood, perhaps by eating or drinking contaminated food or water, or through person-to-person transmission. Water and raw vegetables have been linked with *H. pylori* transmission in the developing world. Recovery of *H. pylori* from the feces of adults and children suggests the possibility of oral–oral or fecal–oral transmission or fecal contamination of food in transmission. Possible nonhuman reservoirs for *H. pylori* have also been described; these include rhesus macaques, dogs, and cats. The role of these possible reservoirs in human illness is unclear.

Antimicrobial therapy against *H. pylori* is indicated for patients with peptic ulcer disease with evidence of the infection. Regimens vary depending on the age, previous antimicrobial use, and symptoms of patients. The success of current treatment regimens varies and is affected by patient compliance and prevalence of antimicrobial-resistant strains; however, eradication rates of up to 90% can be achieved.

III. Summary

Campylobacter jejuni, first described in 1973, is the most frequently diagnosed bacterial cause of human gastroenteritis in the US. Human campylobacteriosis also causes severe sequelae, including Guillain–Barré syndrome and perhaps reactive arthritis. Improperly prepared or mishandled poultry are common sources of human infection; however, other foods including red meat and raw milk, may also be contaminated. Animal contact and international travel are also risk factors for illness. Each link in the food chain from producer to consumer has a role in preventing illnesses caused by this pathogen, with particular emphasis on poultry. *Helicobacter pylori*, transferred from the genus *Campylobacter* in 1989, is the leading cause of peptic ulcer disease and perhaps the most common chronic infection of humans. Infections lasting several decades increase the risk of gastric carcinoma and lymphoma. Further information on *C. jejuni*, *H. pylori*, and related organisms (e.g. *C. upsaliensis*, *Arcobacter butzleri*) is necessary to define their reservoirs, clinical syndromes, and the number of illnesses they cause each year. These data will assist in defining human illness prevention strategies.

Bibliography

Friedman, C. R., Neimann, J., Wegener, H. C., *et al.* (2000). Epidemiology of *Campylobacter jejuni* infections in the United States and other industrialized nations. *In* '*Campylobacter*', 2nd ed. (I. Nachamkin and M. J. Blaser, eds), pp. 121–138. American Society for Microbiology, Washington, DC.

International Commission on Microbiological Specifications for Foods (1996). *Campylobacter*. *In* 'Micro-organisms in Foods. 5. Characteristics of Microbial Pathogens' (T. A. Roberts, A. C. Baird-Parker and R. B. Tompkin, eds), pp. 45–65. Blackie Academic & Professional, London.

Jerris, R. C. (1995). *Helicobacter*. *In* 'Manual of Clinical Microbiology', 6th ed. (P. R. Murray, E. J. Baron, M. A. Pfaller, F. C. Tenover and R. H. Yolken, eds), pp. 492–498. American Society for Microbiology Press, Washington, DC.

Nachamkin, I. (1995). *Campylobacter* and *Arcobacter*. *In* 'Manual of Clinical Microbiology', 6th ed. (P. R. Murray, E. J. Baron, M. A. Pfaller, F. C. Tenover and R. H. Yolken, eds), pp. 483–491. American Society for Microbiology Press, Washington, DC.

Nachamkin, I. (2001). *Campylobacter jejuni*. *In* 'Food Microbiology: Fundamentals and Frontiers', 2nd ed. (M. P. Doyle, L. H. Beuchat and T. J. Montville, eds), pp. 179–192. American Society for Microbiology Press, Washington, DC.

Newell, D. G., Ketley, J. M. and Feldman, R. A. (eds) (1996). 'Campylobacters, Helicobacters, and Related Organisms.' Plenum Press, New York.

Vandamme, P. and Goossens, H. (1992). Taxonomy of *Campylobacter*, *Arcobacter*, and *Helicobacter*: a review. *Zentralbl. Bakteriol.* **276**, 447–472.

Yersinia enterocolitica

7

Georg Kapperud

I. Introduction

Yersinia enterocolitica has been the focus of growing interest during the past couple of decades. This bacterial species has been isolated with increasing frequency from human patients with acute enteritis, who sometimes exhibit symptoms resembling appendicitis. *Y. enterocolitica* has attracted considerable attention owing to its ability to cause serious postinfectious complications.

The organism has been isolated from humans in many countries of the world, but it seems to be found most frequently in cooler climates. In developed countries, *Y. enterocolitica* can be isolated from 1% to 4% of all human cases of acute enteritis. There appears to have been a real and general increase in incidence. In many countries, *Y. enterocolitica* is not routinely looked for by medical laboratories and is therefore likely to be underdiagnosed.

II. Characteristics of the Disease

Illness caused by *Y. enterocolitica* is referred to as yersiniosis. *Y. enterocolitica* is associated with a spectrum of clinical syndromes in man, which are described below.

Foodborne Diseases 2nd Edn
ISBN: 0-12-176559-8

A. Acute Intestinal Infections

Acute noncomplicated enteritis is by far the most frequently encountered manifestation. In 3–15% of cases, the infection causes mesenteric lymphadenitis, terminal ileitis or both, which give rise to symptoms resembling appendicitis. The incubation time of *Y. enterocolitica* enteritis ranges from 1 to 11 days and clinical disease typically persists for 1–2 weeks, but it occasionally may last for several months. The minimal infective dose has not been determined. The organism may be excreted in the stools for a long time period after symptoms have resolved. It is generally unnecessary to treat acute noncomplicated enteritis with antibiotics. However, patients with systemic or extraintestinal infections should be treated. For such cases, therapy with doxycycline or trimethoprim–sulfamethoxazole has been recommended.

B. Extraintestinal and Systemic Infections

Septicemia and localized extraintestinal infections are rare manifestations that almost exclusively are seen in patients with underlying illness.

C. Postinfectious Sequelae

Although a range of postinfectious sequelae has been reported, reactive arthritis and cutaneous manifestations like erythema nodosum are the most common. The two latter complications occur mainly in adults and are caused by serogroups O:3 and O:9. Reactive arthritis following *Y. enterocolitica* infection typically persists for 1–4 months, but follow-up studies indicate that prolonged symptoms may occur in a significant proportion of cases.

The ability of the bacterium to proliferate at low temperatures poses a problem in blood transfusion. *Y. enterocolitica* present during transient bacteremia in blood donors may multiply in blood products stored at 4°C and produce septic shock upon transfusion.

III. Characteristics of the Organism

Yersinia forms a genus within the family Enterobacteriaceae. The cells are small rods, sometimes coccoid in shape, and Gram-negative. *Y. enterocolitica* has been divided into more than 70 serogroups, of which only a few have been conclusively associated with human or animal disease. Somewhat simplified, *Y. enterocolitica* may be divided into three groups according to clinical significance; each group comprises different serogroups as listed below.

- *The human pathogens.* Serogroups O:3, O:5,27, O:8, and O:9 are the most important causative agents in man. Although other serogroups may occasionally cause infection, these variants are completely dominant.

- *The animal pathogenic strains* also belong to particular serogroups. O:2 has been associated with disease in goats, sheep, and hares, while O:1 caused widespread epizootics among chinchillas in the early 1960s. With few exceptions, O:1 and O:2 have not been incriminated in human disease.
- *The environmental strains* usually lack clinical significance and comprise a wide range of variants, which are ubiquitous in terrestrial and freshwater ecosystems. A number of closely related *Yersinia* species are also frequently encountered in nature (*Y. frederiksenii, Y. kristensenii, Y. intermedia, Y. aldovae, Y. rohdei, Y. mollaretii*, and *Y. bercovieri*), all of which are apathogenic.

There are appreciable geographic differences in the distribution of the pathogenic serogroups. O:3 is the most widespread in most parts of the world, including Europe, Japan, and Canada. Until recently, the most frequently reported variants in the US were O:8 followed by O:5,27. In recent years, serogroup O:3 has been on the increase in the US and now accounts for the majority of isolates in certain states.

Y. enterocolitica is able to multiply at temperatures approaching 0°C, a circumstance which means that it can grow in properly refrigerated foods. However, some results indicate that *Y. enterocolitica* competes poorly with other psychrotolerant organisms. *Y. enterocolitica* can survive in frozen foods for long periods. The heat resistance, salt tolerance, and pH tolerance are comparable to that of other Enterobacteriaceae.

Y. enterocolitica causes enteritis by adherence to and penetration of the epithelial cells in the terminal ileum, followed by invasion of the intestinal mucosa, and multiplication in the lymphoid tissue of the intestine. Virulent strains harbor a particular plasmid of 40–50 MDa in size. The plasmid encodes a series of proteins, several of which are important virulence determinants. At least two chromosomal gene loci are also necessary for expression of virulence.

IV. Transmission via Food

Y. enterocolitica is frequently encountered in healthy carriers among warm- and cold-blooded animals, in foods, and in the environment. However, the vast majority of the strains isolated from these sources are apathogenic variants. Although pets may occasionally be fecal carriers, the pig is the only animal consumed by man that regularly harbors the pathogenic serogroups O:3 and O:9. In addition to being fecal commensals, these serogroups inhabit the oral cavity of swine, especially the tongue and tonsils. As a result of present slaughter techniques, they are also frequently encountered as surface contaminants on freshly slaughtered pig carcasses. Pathogenic *Y. enterocolitica* have only infrequently been recovered from pork products at the stage of retail sale. This phenomenon might be explained by the lack of proper selective methodology for isolation of pathogenic strains. Studies using genetic probes or the polymerase chain reaction (PCR) have indicated that such strains are more common in pork products than previously documented.

Epidemiologic investigations have supported the role of pork as a vehicle for *Y. enterocolitica*. Case-control studies of sporadic cases, conducted in Belgium and Norway, have identified consumption of pork as an important risk factor for infection. Following an outbreak due to serogroup O:3 among children in Atlanta, Georgia, a case-control study showed that household preparation of chitterlings (raw pork intestines), was significantly associated with illness.

In contrast to O:3 and O:9, serogroup O:8 appears to be rare in swine. O:8 may have an entirely different reservoir and ecology. In Japan, small rodents have been identified as a reservoir for O:8. Outbreaks and sporadic cases due to this serogroup have been traced to ingestion of contaminated water, water used in manufacturing or preparation of food (e.g. bean sprouts, tofu), and milk products, which probably became contaminated subsequent to pasteurization. Consumption of untreated water was also identified as a risk factor for infection with serogroup O:3 in a case-control study conducted in Norway.

V. Isolation and Identification

In food, one can expect to find a broad spectrum of yersiniae, the vast majority of which have no medical importance. However, the development of isolation media and procedures that clearly differentiate pathogenic from non-pathogenic variants has been difficult. A number of isolation procedures are currently in use. Most methods require time-consuming resuscitation and enrichment, and no single method provides optimal isolation of all pathogenic serogroups. Cefsulodin–irgasan–novobiocin (CIN) agar, a differential selective medium, is more effective than routine enteric media for the recovery of *Y. enterocolitica* from food.

Identification of *Y. enterocolitica* is based on cultural–biochemical characterization, including biotyping. Serotyping is conducted by slide agglutination against specific O-antigen sera. Since a majority of the strains capable of causing disease belongs to only a few serogroup–biotype combinations, serotyping and biotyping are sufficient to differentiate pathogenic strains from nonpathogens for practical purposes. In addition, a series of *in vitro* virulence assays has been described. Methods based on genetic probes or the PCR enable rapid, sensitive, and specific detection of all pathogenic variants.

VI. Control and Prevention

Preventive measures, which reduce contamination and improve hygiene, during all stages of pig production and pork processing, are essential to reduce infection with serogroups O:3 and O:9. *Y. enterocolitica* may be difficult to control efficiently at the farm level, and its detection during the routine meat inspection is practically impossible,

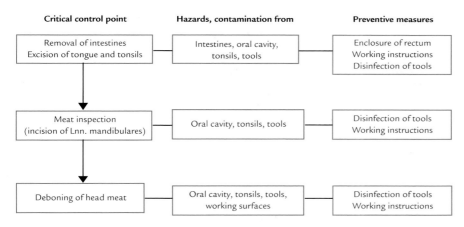

Figure 7.1 Preventive measures at critical control points during pig slaughter, aimed at reducing contamination with *Y. enterocolitica*.

since pigs are healthy carriers. During the slaughtering process, bacteria from the oral cavity or intestinal contents may easily contaminate the carcasses and the environment in the slaughterhouse. Improved hygiene at critical control points should be attempted (Fig. 7.1). Special attention should be paid during:

* circumanal incision and removal of the intestines;
* excision of the tongue, pharynx, and particularly the tonsils, which ideally should be left on the head of the pig;
* postmortem meat inspection procedures, which involve incision of the mandibular lymph nodes;
* deboning of head meat.

Changes in slaughtering procedures, including technological improvements, may be required during these activities. Enclosing the anus and rectum in a plastic bag after rectum loosening, markedly reduces contamination of the carcass.

Preventive and control measures should also focus on informing of all categories of people involved in production, processing, and final preparation of food, about the importance of good hygienic practices. Strict hygiene is particularly necessary because *Y. enterocolitica* is able to propagate at refrigeration temperatures. Therefore, chilling of food products should not be considered as an effective control measure for this microbe. Consumption of undercooked pork should be discouraged. The need to adhere to preventive measures such as pasteurization of milk and avoidance of recontamination and cross-contamination after treatment, should be emphasized. Avoidance of contact with feces from pigs or domestic pets may also reduce transmission. Although *Y. enterocolitica* isolated from water are not usually among the types pathogenic for humans, a risk of transmission through water is present. Since *Y. enterocolitica* is sensitive to chlorination, proper treatment of drinking water and water used for food processing should eliminate the risk of infection from this source.

VII. Summary

Yersinia enterocolitica has emerged as a worldwide pathogen associated with a range of clinical entities. Although acute enteritis is by far the most frequent manifestation, the bacterium may also cause serious postinfectious complications. *Y. enterocolitica* encompasses a spectrum of variants of which only a few have been conclusively associated with human disease. The great majority of the strains involved in human infection belong to a few distinct serogroups, which show appreciable geographic differences in distribution. Serogroup O:3 is the dominant pathogen in most parts of the world, followed by O:9, O:8 and O:5,27. The development of isolation procedures that clearly differentiate pathogenic from nonpathogenic variants has proven problematic. Of special significance in food hygiene is the ability of *Y. enterocolitica* to grow in properly refrigerated foods. There is strong indirect evidence that pigs and food products of porcine origin are the major sources for human infection with serogroups O:3 and O:9. The pig is the only animal consumed by man that regularly harbors pathogenic *Y. enterocolitica*. Preventive and control measures should focus on the improvement of hygiene during slaughtering of swine and on education of people involved in food processing and preparation.

Bibliography

Bottone, E. J. (1997). *Yersinia enterocolitica*: The charisma continues. *Clin. Microbiol. Rev.* **10**, 257–276.

Cornelis, G. R. (1994). *Yersinia* pathogenicity factors. *Curr. Top. Microbiol. Immunol.* **192**, 243–263.

Cover, T. L. and Aber, R. C. (1989). Yersinia enterocolitica. *N. Engl. J. Med.* **321**, 16–24.

Kapperud, G. (1991). *Yersinia enterocolitica* in food hygiene. *Int. J. Food Microbiol.* **12**, 53–66.

Kapperud, G. and Slome, S. B. (1998). *Yersinia enterocolitica* infections. *In* 'Bacterial Infections of Humans', 3rd ed. (A. S. Evans and P. S. Brachman, eds), pp. 859–873. Plenum Medical Book Company, New York.

Robins-Browne, R. M. (2001). *Yersinia enterocolitica*. *In* 'Food Microbiology. Fundamentals and Frontiers', 2nd ed. (M. P. Doyle, L. R. Beuchat and T. J. Montville, eds), pp. 215–245. American Society for Microbiology Press, Washington, DC.

Schiemann, D. A. (1989). *Yersinia enterocolitica* and *Yersinia pseudotuberculosis*. *In* 'Foodborne Bacterial Pathogens' (M. P. Doyle, ed.), pp. 601–672. Marcel Dekker, Inc., New York.

Clostridium perfringens[1]

8

Ronald G. Labbe and Vijay K. Juneja

I. Introduction

Historically *Clostridium perfringens* is best known for its association with gas gangrene (primarily due to alpha-toxin) as became apparent during the First World War. It remains the primary cause of this medical condition. The first report of its association with food and diarrhea dates from the turn of the century, but there was little evidence to confirm this since the organism was well known to be present in human stools, certain foods, and the environment. More conclusive evidence for its role in foodborne illness appeared in the 1940s in England and the US when large numbers of *C. perfringens* were isolated from outbreaks in which gravy and chicken were vehicles. The classic studies by Hobbs and coworkers in the 1950s firmly established the role of the organism in foodborne illness. They noted that the organism could be divided into two broad categories: heat-resistant (HR) (and weakly or nonhemolytic) or heat sensitive (HS) (and hemolytic). This was based on the survival of the spores for 1 hour at 100°C. It was initially believed that only the HR strains were involved in foodborne illness. However, in the 1960s both types were recognized as causing foodborne illness. The enterotoxin responsible for food poisoning was demonstrated in 1969 by its ability to induce fluid accumulation in ligated ileal loops of rabbits. The food poisoning role of the organism was soon proven conclusively by human feeding experiments.

---

[1] Mention of brand or firm name does not constitute an endorsement by the US Department of Agriculture over others of a similar nature not mentioned.

Foodborne Diseases 2nd Edn
ISBN: 0-12-176559-8

II. Characteristics of the Disease

C. perfringens food poisoning is neither a true intoxication nor an infection. The illness results typically from ingestion of large numbers ($>10^8$) of viable vegetative cells in temperature-abused foods. The cells survive stomach passage and subsequently sporulate in the small intestine. The enterotoxin known as *C. perfringens* enterotoxin (CPE) is produced during sporulation and is released together with the mature spore during sporangial autolysis.

The illness consists of diarrhea and abdominal pain with an incubation period of 8–24 hours. Fever is rare, which is consistent with the absence of a true infection. Symptoms subside within 24 hours without treatment, although residual cramping may persist an additional 1 or 2 days. The brief nature of the illness is presumably due to normal turnover of intestinal epithelial cells to which the enterotoxin has bound and removal of unbound toxin as a result of diarrhea. Fatalities do occur occasionally, mainly limited to elderly or hospitalized individuals.

In animal models, the enterotoxin induces a reversal of net transport in the small intestine, resulting in the secretion of water, sodium, and chloride. The brush border of villus tip epithelial cells is the enterotoxin's primary site of action. In animal models, desquamation of epithelial cells occurs within 15 minutes of exposure to the enterotoxin with the loss of large quantities of membrane and cytoplasm to the lumen. Scanning electron micrographs reveal localized damage to villus tips with a covering of rounded blebs of the tips (Fig. 8.1). In this regard CPE differs from cholera toxin (CT), which causes little tissue damage to the gastrointestinal tract. It also differs from CT, as well as *Escherichia coli* labile toxin in that the CPE does not bind to GM_1 ganglioside of membranes but rather to 50 000 and/or 22 000 molecular weight membrane proteins whose normal physiological roles are unknown.

The use of tissue culture cells has also provided information on the mode of action of the toxin at the cellular levels. Vero (African green monkey kidney) cells have been particularly useful in this regard. The enterotoxin initially affects the permeability (influx and efflux) of small molecules such as ions and amino acids. This leads to the loss of macromolecular precursors, in turn leading to a loss of viability. Loss of fluids and electrolytes in such *in vitro* systems presumably corresponds to the diarrheal symptoms associated with the illness. It remains unclear whether the enterotoxin affects the human large intestine. It does bind to rabbit colonic cells but causes little effect on the rabbit colon.

Sporadic cases of a particularly severe foodborne illness, often fatal, associated with *C. perfringens* type C have occurred throughout the world; however, the disease is common in Papua New Guinea. Medically it is a necrotizing, hemorrhagic jejunitis, commonly called 'Pig-Bel' (necrotizing enteritis) and associated with pig feasting in which pigs are insufficiently cooked. The disease may progress to gangrene of parts of the small intestine leading to severe toxemia and shock. Consumption of local foods, such as sweet potatoes, containing trypsin inhibitors as well as malnutrition are also apparent factors leading to low levels of intestinal proteases, which may otherwise degrade the responsible beta-toxin produced by type C. As a prophylactic measure, vaccination is carried out using a toxoid preparation of type C cultures

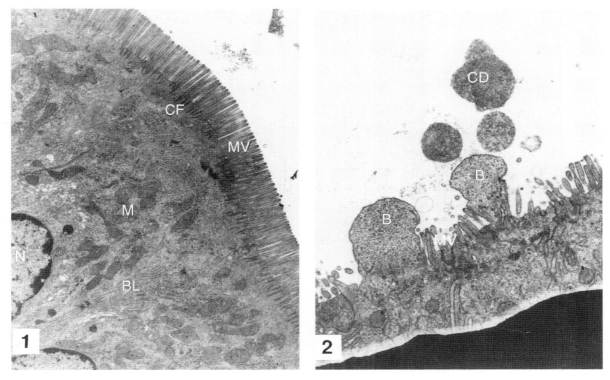

Figure 8.1   Effect of C. perfringens enterotoxin on rat ileum. (1) Electron micrograph of control showing normal villus cell morphology. (2) Specimen treated with enterotoxin for 90 minutes, showing bleb formation. B, bleb; BL, lateral membrane; CD, cytoplasmic bleb; CF, central filaments; M, mitochondria; MV, microvilli; N, nucleus. Reprinted with author's and publisher's permission from McDonel (1979).

of *C. perfringens*. Recently, necrotizing enteritis caused by *C. perfringens* type A has also been reported.

III. Characteristics of the Organism

A. Vegetative Cells and Spores

C. perfringens is a relatively large, nonmotile, Gram-positive, anaerobic rod of the family Bacillaceae. Most strains sporulate poorly in laboratory media, an important aspect, since the presence of the enterotoxin can only be demonstrated in sporulating cultures. The organism is somewhat aerotolerant, which facilitates manipulations associated with analysis of food. A typical sporulating culture is shown in Fig. 8.2.

The organism produces some dozen exotoxins as well as an enterotoxin. The presence or absence of the major lethal toxins is used as a basis to assign isolates into five types, A–E (Table 8.1). Alpha-toxin (lecithinase) is common to all toxin types but is produced in greatest amounts by type A. It is largely responsible for the breakdown of

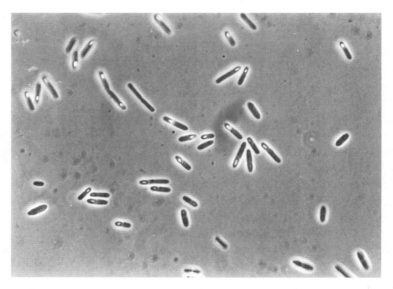

Figure 8.2 Phase-contrast micrograph of sporulating cells of *C. perfringens* showing mature refractile spores.

Table 8.1 Distribution of the major lethal toxins among types of *Clostridium perfringens*

Type	Alpha	Beta	Epsilon	Iota
A	+	−	−	−
B	+	+	+	−
C	+	+	−	−
D	+	−	+	−
E	+	−	−	+

tissue in gangrene, lecithin being an important membrane component. Alpha-toxin is a phospholipase C that splits the lecithin molecule to produce phosphorylcholine and a diglyceride as shown below.

$$CH_2O\ CO\ R$$
$$|$$
$$CHO\ CO\ R$$
$$|$$
$$CH_2O - P - O - CH_2N\ (CH_3)_3$$
$$|$$
$$OH$$

Lecithinase C

It was initially thought that this enzyme could hydrolyze lecithin in food to form phosphorylcholine as the active agent in foodborne illness. However, human feeding studies and the subsequent isolation of the enterotoxin showed that this hypothesis was incorrect.

The morphology of vegetative cells of *C. perfringens* varies with the growth rate. Rapidly growing cells, e.g. in media containing readily fermentable monosaccharides or disaccharides, are noticeably smaller than cells grown in sporulation media, which typically contain starch as the carbohydrate source. A capsule is produced by this organism and is the basis of a serotyping scheme. It can be observed by the addition of India ink to a wet mount of the organism.

Vehicles of foodborne disease caused by *C. perfringens* are usually meat and poultry products. This reflects the organism's demanding nutritional needs; it requires more than a dozen amino acids and several vitamins. In the laboratory the organism grows well in complex media containing meat or casein digests as a source of amino acids, a fermentable carbohydrate, and a complex source of vitamins and nucleotides such as yeast extract. Lactose fermentation to produce H_2 and CO_2 is so vigorous that a milk medium is used as a diagnostic test. The casein clot produced from acid formation is disrupted by the gas produced.

In addition to its ability to sporulate, the other notable characteristic of *C. perfringens* is its potential for rapid growth. At its optimum growth temperature, 43–45°C, and in rich media, generation times of less than 10 minutes have been demonstrated, the shortest for any known bacterium. The growth limits are 15–50°C, but between 15 and 20°C growth is slow and strain dependent. In addition to enterotoxin formation, spore formation and rapid growth at relatively high temperatures are invariably involved in foodborne disease incidents caused by this organism. The organism is not noted for its ability to grow at extremes of pH, low water activity, or in the presence of high salt concentration.

As mentioned above, earlier workers divided *C. perfringens* into two biotypes based on spore heat resistance. The HR types survived 100°C for 1 hour, while the HS strains survived less than 15 minutes. While it is now known that both types are involved in foodborne illness, spores of this organism are still categorized based on their D (decimal reduction) value, i.e. the time required to inactivate 90% of the spore population at a given temperature. The HS group have D_{90C} values of 3–5 minutes as compared to the HR group values of 15–145 minutes. Spores of the HR group require a heat shock (e.g. 75°C for 10 minutes) for optimal germination. An unusual phenomenon, which has been noted for spores of this species, is that lysozyme, when included in the plating medium, resuscitates and promotes the recovery of heat-injured spores, thereby increasing the measured or apparent heat resistance. The affected population of heated spores that are permeable to lysozyme is small, less than 1%. Lysozyme in the plating medium replaces the thermally inactivated spore-germination enzymes, which, during germination, cause the disruption of the spore cortex, allowing core hydration and, consequently, spore germination.

B. The Enterotoxin

As mentioned above, the enterotoxin forms during sporulation of the organism in the small intestine following ingestion of large numbers of vegetative cells in temperature-abused foods, which may not appear spoiled. In some strains a substantial amount

(10–20%) of the soluble protein (cell extract) consists of enterotoxin. The reason for its formation remain unclear but it is known to be formed during stage II–III of sporulation, the levels being distinctly strain dependent.

The enterotoxin has been isolated and characterized. It consists of a single polypeptide of approximately 35 000 Da, with a unique amino acid sequence, except for some limited homology with a nonneurotoxic protein made by *Clostridium botulinum*, and with an isoelectric point of pH 4.3. As with most proteins of this size, it is readily destroyed by heating, e.g. 60°C for 10 minutes. Human feeding studies indicate that relatively large oral doses (8–10 mg) are required to induce symptoms. By comparison, only about 1 μg of staphylococcal enterotoxin induces typical symptoms. As in the case of *Clostridium botulinum* type E neurotoxin, limited treatment of the enterotoxin with trypsin or chymotrypsin increases its biological activity (2–3-fold), suggesting that intestinal proteases may activate the enterotoxin during food poisoning.

In cases of foodborne outbreaks *C. perfringens*, enterotoxin can be detected in the stools of ill individuals but not of controls. Methods for quantifying the enterotoxin are described in Section V.

IV. Transmission via Food

The outbreaks that have been investigated have invariably implicated cooked meat and poultry as the transmission vehicles. A long cooling period at nonrefrigerated storage temperature is a contributing factor. Typically these foods are prepared well in advance and served to large numbers of people. The spores survive the cooking process, which also serves to inactivate competitors and drive off oxygen. Numerous surveys indicated the presence of the organism on about half of beef, pork, and lamb carcasses. Adequate rest and starvation (feed withdrawal) of animals can reduce contamination of carcasses. These early surveys did not determine whether isolates possessed the gene encoding CPE. It is becoming increasingly apparent that, in the case of nonsymptomatic animals, only a minority of isolates possess the enterotoxin gene.

Virtually all human adults harbor the organism in their large intestine, with median counts in the range of 10^3–10^4/g feces. These are typically of the enterotoxin-negative biotype.

V. Isolation and Identification

The presence of characteristic symptoms, incubation period, type of food, and preparation method will indicate the possible role of *C. perfringens* in an outbreak. Confirmation depends either on the isolation of $>10^5$ organisms per gram of implicated foods, the isolation of $>10^6$ spores per gram of feces of ill individuals, or the direct detection of the enterotoxin in stool samples. Serotyping of isolates, which is based on the presence of specific capsular antigens, can provide useful information for epidemiological studies. Unfortunately, many suspect isolates are not typeable and the antisera are not commercially available. Serotyping is less common in the US than in the UK, where public health officials have a large collection of capsular antisera.

The resistance of *C. perfringens* to certain antibiotics, its relative aerotolerance, and its rapid growth rate makes it one of the easiest clostridia to isolate. The currently recommended media is TSC (tryptose–sulfite–cycloserine) agar. It contains iron and sulfite. When grown anaerobically on TSC agar pour plates, *C. perfringens* reduces the sulfite to sulfide, which produces a black precipitate with iron. In cases where low numbers are expected, such as in market or environmental surveys, a most probable number (MPN) procedure using an iron–milk medium has been shown to be useful and does not require anaerobic incubation. Fecal spore counts are easily performed by heating diluted samples at 75°C for 20 minutes and plating on TSC agar.

Because certain other, though rare, clostridia can produce black colonies on TSC media, it is necessary to confirm isolates. Typically 10–12 isolated colonies are picked and inoculated into motility-nitrate and lactose-gelatin tubes. *C. perfringens* is non-motile, hydrolyzes gelatin, reduces nitrate, and ferments lactose. Details of the procedures are found in manuals published by the US Food and Drug Administration and the American Public Health Association (APHA).

Over the years, numerous procedures have been described for detecting the presence of enterotoxin in sporulating broths and stool samples. All require the availability of specific antiserum and some require specialized equipment. Two serological methods are commercially available. One is the enzyme-linked immunosorbent assay, the other a reversed passive latex agglutination. Recently DNA probes have been used to detect *C. perfringens* strains that have the potential to produce the enterotoxin. This avoids the necessity of inducing sporulation of isolates to verify their enterotoxigenicity. On the other hand, positive results obtained by the use of such probes assume that all isolates containing the enterotoxin gene are able to express the toxin, a point which is yet to be demonstrated. More recently, a polymerase chain reaction technique was developed for direct detection and enumeration of enterotoxigenic *C. perfringens* strains in beef. The technique is sensitive enough to detect less than 10 colony forming units (CFU)/g of *C. perfringens* strains harboring the enterotoxin gene in the presence of heterogeneous background bacterial flora present in raw ground beef.

VI. Treatment and Prevention

As with most cases of foodborne illness, those caused by *C. perfringens* are generally self-limiting in otherwise healthy individuals, and antibiotic therapy is not recommended. Only supportive therapy is usually needed; however, severe cases and occasional deaths have occurred in the elderly, in whom dehydration is an important concern.

Because vegetative cells and spores of *C. perfringens* can be present in many raw foods, its presence alone is not of major importance. However, temperature abuse of prepared foods is a major concern and is invariably involved in outbreaks. Rapid chilling and proper reheating are important aspects of control. Hot foods should be held at the APHA recommended holding temperatures for hot foods [$\geq$60°C (140°F)]. Foods to be reheated should reach 71°C (160°F) before consumption to kill vegetative cells.

Bibliography

Katahira, J., Inoue, N., Horiguchi, Y., Matsuda, M. and Sugimoto, N. (1997). Molecular cloning and functional characterization of the receptor for *Clostridium perfringens* enterotoxin. *J. Cell Biol.* **136**, 1239–1247.

Kokai-Kun, J. F. and McClane, B. A. (1997). The *Clostridium perfringens* enterotoxin. *In* 'The Clostridia: Molecular Biology and Pathogenesis' (J. Rood, B. McClane, J. G. Songer and R. Titball, eds), pp. 325–358. Academic Press, London.

Labbe, R. (1989). *Clostridium perfringens*. *In* 'Foodborne Bacterial Pathogens' (M. P. Doyle, ed.), pp. 191–223. Marcel Dekker, New York.

Labbe, R. (2001). *Clostridium perfringens*. *In* 'Compendium of Methods for the Microbiological Examination of Foods', 4th ed. (F. P. Downes and K. Ito, eds), pp. 325–330. American Public Health Association, Washington, DC.

McClane, B. (1997). *Clostridium perfringens*. *In* 'Food Microbiology: Fundamentals and Frontiers' (M. P. Doyle, L. R. Beuchat and T. J. Montville, eds), pp. 305–326. American Society for Microbiology Press, Washington, DC.

McDonel, J. (1979). The molecular mode of action of *Clostridium perfringens* enterotoxin. *Am. J. Clin. Nutr.* **32**, 210–218.

Rhodehamel, E. J. and Harmon, S. M. (1995). *Clostridium perfringens*. *In* 'The Food and Drug Administration Bacteriological Analytical Manual', 8th ed., pp. 16.01–16.05. AOAC International, Gaithersburg, MD.

Smith, L. D. S. and Williams, B. L. (1984). *In* 'The Pathogenic Anaerobic Bacteria', 3rd ed., pp. 101–136. Thomas, Springfield, IL.

Vibrio

Riichi Sakazaki[1]

9

I. Introduction

Until 1960, cholera was considered as a disease caused only by *Vibrio* species and it was not thought that marine organisms produce human diseases. However, since *V. parahaemolyticus* was defined as an etiologic agent of gastroenteritis, medical microbiologists have taken great interest in the relationships between marine organisms and human disease. Subsequently, it has been demonstrated that at least 12 *Vibrio* spp. cause intestinal or extraintestinal infections. Of the 12 *Vibrio* spp., *V. cholerae*, *V. parahaemolyticus*, and *V. vulnificus* are described here according to their importance in foodborne or waterborne diseases.

II. *Vibrio cholerae*

A. Disease

There are two serological groups of *V. cholerae* strains, O1 and non-O1. *V. cholerae* O1 is the causative agent of cholera. Cholera was originally endemic in eastern India. Up to 1960, it had extended from India in six pandemics over the world. There was another focus of cholera-like disease, which was called El Tor cholera in Indonesia, and this illness began to spread throughout the world after 1961. The causative agent

[1] Deceased.

Foodborne Diseases 2nd Edn
ISBN: 0-12-176559-8

of this seventh pandemic was a different biovar in the same serogroup (O1) as the original cholera vibrio. On the other hand, in 1992, there appeared an epidemic of cholera caused by a 'new cholera vibrio', O139 of *V. cholerae* non-O1, in the Indian subcontinent, which has since spread to neighboring countries.

V. cholerae O1 produces either a severe or mild syndrome, but mild diseases are much more common in the seventh pandemic. The ratio of severe to mild or asymptomatic cases is generally between 1 : 5 and 1 : 10 for classical cholera, but only 1 : 25 to 1 : 100 for El Tor cholera. Typical cholera has a sudden onset of vomiting and painless watery diarrhea, which quickly assumes the characteristic 'rice-water' appearance. If the illness is not treated, death can occur very quickly after the onset of symptoms because of severe dehydration. Gastric disturbances, particularly subacidity and gastrectomy, are risk factors for inducing severe cholera symptoms. The incubation period of cholera ranges from 1 to 5 days.

V. cholerae non-O1 can cause a cholera-like diarrhea or gastroenteritis, and occasionally produces extraintestinal infections.

B. Organism

1. Phenotypic characteristics

V. cholerae is a motile rod with a single polar, sheathed flagellum. It can grow in peptone water with 0–6% NaCl. It has high alkaline tolerance. Phenotypical characteristics of *V. cholerae* are summarized in Table 9.1, together with other human pathogenic *Vibrio* spp. Strains of *V. cholerae* O1 are divided into two biovars, classical and El Tor. The differential characteristics of the two biovars are shown in Table 9.2. Although the differentiation between the El Tor and classical biovars was originally based on hemolytic activity, many El Tor isolates from recent epidemics may be nonhemolytic.

V. cholerae is divided serologically into many O groups, in which O group 1 is assigned to cholera vibrio. The numbers of O serogroups have now extended to 140. In O1 strains, there are three O antigenic forms, named Ogawa, Inaba, and Hikojima: the Ogawa strains are the original form; the Inaba strains are mutants that have lost Ogawa-specific fraction; and the Hikojima strains are regarded as a stable intermediate form between the Ogawa and Inaba forms. The conversion from the Ogawa to Inaba is irreversible.

2. Virulence mechanisms

Cholera toxin (CT) is the most important virulence factor responsible for the cholera symptoms. CT comprises one A subunit and five B subunits. Excreted CT binds to G_{M1} ganglioside on host mucosal cells with the B subunit, then the A subunit is translocated into the host cells, where it activates adenylate cyclase and subsequently increases intracellular levels of cAMP. The increased intracellular cAMP concentrations cause hypersecretion characteristic of cholera diarrhea. CT is produced by human isolates of O1 and some non-O1 *V. cholerae*, including O139. CT-negative strains do not cause cholera.

V. cholerae also produces two other toxins, Zot (zona occludens toxin) and Ace (accessory cholera toxin), which possibly collaborate with CT. Another enterotoxin

Table 9.1 Phenotypic characteristics of *Vibrio* species associated with human disease

Test (substrate)	*V. cholerae*	*V. mimicus*	*V. parahaemolyticus*	*V. vulnificus*	*V. fluvialis*	*V. firnissii*	*V. alginolyticus*	*V. metschnikovii*	*V. hollisae*	*V. cincinnatiensis*	*V. carchariae*	*V. damsela*[a]
Oxidase	+	+	+	+	+	+	+	−	+	+	+	+
Nitrate reduction	+	+	+	+	+	+	+	−	+	+	+	+
Growth in peptone water with:												
0% NaCl	+	+	−	−	−	−	−	−	−	−	−	−
6% NaCl	d	d	+	+	+	+	+	d	d	+	+	+
8% NaCl	−	−	+	−	d	d	+	−	−	−	−	−
10% NaCl	−	−	−	−	−	−	+	−	−	−	−	−
Swarming	−	−	d	−	−	−	+	−	−	+	−	−
Susceptibility to O/129												
10 μg	S	S	R	S	R	R	R	S	R	R	S	S
150 μg	S	S	S	S	S	S	S	S	S	S	S	S
Flagellation on solid medium	M	M	P	M	P	P	P	M	M	M	M	M
Indole	+	+	+	+	d	d	+	d	d	−	+	−
Voges–Proskauer	+	−	−	−	−	−	+	+	−	+	d	+
Urease	−	−	d	−	−	−	−	−	−	−	−	−
Lysine decarboxylase	+	+	+	+	−	−	+	d	−	d	+	d
Ornithine decarboxylase	+	+	+	d	−	−	d	−	−	−	−	−
Arginine dihydrolase	−	−	−	−	+	+	−	d	−	−	−	+
ONPG	+	+	−	+	d	d	−	d	−	+	−	−
Gas from glucose	−	−	−	−	+	−	−	−	−	−	−	−
Acid from:												
L-Arabinose	−	−	d	−	+	+	−	+	+	−	−	−
Cellobiose	−	−	−	+	d	d	−	−	−	+	d	−
Lactose	−	d	−	+	−	−	−	d	−	−	−	−
Sucrose	+	−	−	d	+	+	+	+	−	+	d	−
Salicin	−	−	−	+	−	−	−	−	−	+	−	−
Growth on TCBS agar	Y	G	G	G, Y	Y	Y	Y	Y	−	Y	G, Y	G

+, 90–100% strains positive; −, 0–10% strains positive; d, 11–89% strains positive; S, susceptible; R, resistant; M, monotrichous; P, peritrichous; Y, yellow; G, green.
[a] *V. damsela* is presently classified in the genus *Photobacterium*.

Table 9.2 Biovars of *Vibrio cholerae* O1

Test	Biovar	
	Classical	El Tor
Hemolysis of sheep erythrocytes	−	+[a]
Voges–Proskauer	−	+
Chicken erythrocyte agglutination	−	+
Polymyxin B, 50 IU	−	+
Mukerjee's phage IV	Susceptible	Resistant

[a] Recent isolates are sometimes negative.

named 'new cholera toxin' was reported in CT-negative O1 strains, which was demonstrated to cause diarrhea in human volunteers.

Of the adhesins reported, TCP (toxin coregulated pili), which is long filamentous, is the essential adhesin for the pathogenesis of cholera. ACF (accessory colonization factor) and a hemagglutinin, which are outer membrane proteins, have also been suggested to be important in colonization.

Non-O1 strains are much more commonly isolated from environments than are O1 strains, even in epidemic areas. In addition, environmental isolates of *V. cholerae* O1 are almost always CT-negative outside of epidemic areas. Those CT-negative *V. cholerae* may cause not only diarrhea but also extraintestinal infections. In extraintestinal infections, capsule may also be a virulence factor, which mediates adhesion to epithelial cells.

C. Transmission

The most important mode of spread of cholera is through the environment, particularly by contaminated water. Recent investigations have strongly suggested that *V. cholerae* has a free living cycle in which the vibrio appears to have a 'viable but nonculturable' state with a natural reservoir in the environment. Furthermore, *V. cholerae* preferably colonizes the surface of zooplankton and shellfish by its production of chitinase. Also, *V. cholerae* colonizes the digestive tract of shellfish as a result of ingestion of zooplankton and persists for many weeks in such shellfish. Those shellfish may serve as an important vehicle in the transmission of cholera. It is clear, therefore, that the presence of the vibrio is not always associated with fecal contamination from cholera patients. Use of water contaminated with *V. cholerae* during household storage for cooking, bathing, or washing is a significantly increased risk of infection. Beverages, in particular those containing ice made from contaminated municipal water, are a potential vehicle of cholera transmission in epidemic areas.

D. Laboratory Methods

1. Isolation and identification

Although a variety of media have been devised, thiosulfate citrate bile salts sucrose (TCBS) agar for direct plating and alkaline peptone water for enrichment culture are most widely used. TCBS agar is highly selective for pathogenic vibrios including *V. cholerae*. Alkaline peptone water supports good growth of *V. cholerae*, but the incubation period is best limited to 8 hours to prevent overgrowth by other organisms.

Stool specimens are inoculated directly on to TCBS agar plates. For formed feces, approximately 1.0 g of the specimen is inoculated into 10 ml of alkaline peptone water. If there is any possibility of delay before culture, specimens should be collected in alkaline peptone water.

For water samples, several liters are passed through membrane filters and then the filter disks are placed in 50–100 ml of alkaline peptone water. Moore's swabs are also used successfully to isolate *V. cholerae* from water samples. Food samples are homogenized and triturated in alkaline peptone water and then treated in the same way as stool samples. However, shellfish should be cut, but not be homogenized, into small pieces. After adding the pieces to the enrichment broth and shaking vigorously, they are removed from the broth, as some parts of the shellfish may inhibit and adsorb the vibrios.

2. Direct antigen detection

There are enough *Vibrio* cells in watery diarrheal stools of cholera patients to be agglutinated with O antibodies. A coagglutination test using monoclonal antibody to an O1-specific antigen factor has recently become available.

3. Toxin assays

Various modifications of an enzyme-linked immunosorbent assay (ELISA), using purified G_{M1} ganglioside receptor as the capture molecule, are commonly used to assay CT. A latex agglutination assay to detect CT was also reported to be less complicated and less time-consuming than the ELISA. In addition, a variety of molecular approaches to the detection of CT has been developed. The polymerase chain reaction (PCR) has been used for the detection of the *ctx* gene, which is more sensitive than ELISA.

E. Prevention

The important vehicles for the spread of cholera are water and seafoods. The most effective prevention, therefore, is elimination of all sources of contamination that may endanger the safety of public and private water supplies. Special attention should be directed to the disposal of excreta. Processing of food, especially seafoods, must also be subjected to strict supervision.

Cholera usually affects the lower socioeconomic groups, and is associated with poor hygiene and inadequate sanitary facilities. The provision of satisfactory water supplies, together with better sanitation in epidemic areas will ultimately eradicate cholera. In industrial countries, there is no longer any need for cholera to be regarded as a dreaded disease, although it causes fear even among the medical profession and has international economic repercussions.

III. *Vibrio parahaemolyticus*

A. Disease

Since a marine organism was found in an outbreak of food poisoning in 1951 by Fujino and coworkers, *V. parahaemolyticus* has been recognized as a potential enteropathogen all over the world. *V. parahaemolyticus* causes gastroenteritis with severe abdominal pain and diarrhea. The diarrhea is usually watery but can sometimes be bloody. Symptoms usually occur about 12 hours after consumption of infected food. Recovery is usually complete within a few days. The mortality rate is low. Occasionally, it causes extraintestinal infections.

B. Organism

1. Phenotypic characteristics

The organism was given several names, but the name *Vibrio parahaemolyticus* proposed by Sakazaki and coworkers in 1963 has been widely accepted. *V. parahaemolyticus* is a halophilic rod, which can grow on or in ordinary media containing 1–8% NaCl. It has a single polar, sheathed flagellum in broth culture, but young cultures on the surface of solid medium may have unsheathed peritrichous flagella. The phenotypic characteristics of *V. parahaemolyticus* are summarized in Table 9.1. They are divided into many serogroups with combination of 11 somatic (O) and 70 capsular (K) antigens.

2. Virulence mechanisms

Almost all strains isolated from patients with gastroenteritis are hemolytic, whereas the majority of those from sea fish and the marine environment are nonhemolytic on a special agar containing human blood. This hemolytic phenomenon, in which human but not horse erythrocytes are lysed, is called the Kanagawa reaction. The hemolysin responsible for the Kanagawa reaction is known as the thermostable direct hemolysin (TDH). In human volunteers, it was observed that Kanagawa-positive strains produce gastroenteritis, whereas the Kanagawa-negative strains failed to induce any clinical sign. It is possible that the TDH plays a role in the pathogenesis of gastroenteritis due to *V. parahaemolyticus*. Some outbreaks of gastroenteritis are associated with Kanagawa-negative strains. Those Kanagawa-negative strains produce a TDH-related hemolysin (TRH), which is thermolabile. TRH-positive strains are also associated with gastroenteritis.

Several possible adhesins have been reported, but no substantial studies have yet implicated any of the candidate adhesins.

C. Transmission

Gastroenteritis due to *V. parahaemolyticus* is always associated either directly or indirectly with seafoods, though the lack of correlation of the Kanagawa reaction between isolates from patients and implicated seafoods is still one of the more puzzling aspects. Raw fish or shellfish are the most important sources of gastroenteritis caused by this organism in Japan, where the high incidence is undoubtedly due to the national custom of eating raw fish. In contrast, vibrio infection may be less important in European countries, where seafoods are usually cooked shortly before consumption. Nevertheless, cases of gastroenteritis caused by this vibrio have been reported in many countries. Crabmeat and shrimp, which are the seafoods most involved in the incidence of the vibrio infection in those countries, become contaminated from other sources after cooking. Also, vibrio gastroenteritis is sometimes associated with consumption of cured vegetables, which have been contaminated through the kitchen utensils.

Vibrio infection is confined mostly to the warmer months of the year. This seasonal variation of infection is closely associated with the numbers of this vibrio in the estuarine environment. The vibrio is not found in sea fish and sea water in the winter. Oysters are often contaminated with *V. parahaemolyticus* in the summer, but they seldom cause vibrio gastroenteritis because they are usually eaten in the cold season. It is probable that *V. parahaemolyticus* is a frequent cause of infection in developing countries, where waterborne infections with the vibrio could be considered.

D. Laboratory Procedures

1. Isolation

TCBS agar is recommended for the isolation of *V. parahaemolyticus*. However, the routine use of TCBS agar for plating of stool specimens may not be cost effective. Alternatively, MacConkey agar containing an additional 0.5% NaCl is convenient for routine culture of diarrheal stools.

Polymyxin salt broth, containing 2% of NaCl, and 50 μg/ml of polymyxin B is used for enrichment culture of *V. parahaemolyticus* from foods and marine sources. For enrichment of vibrios from shellfish, the same method as that described in the section on *V. cholerae* should be observed. In outbreaks of gastroenteritis due to *V. parahaemolyticus*, however, enrichment cultures of and serotyping of isolates from possibly incriminated seafoods may be largely redundant, as it is seldom possible to detect the identical organism to that from the patients in these foods.

For determination of the Kanagawa reaction, colonies have been tested with a special agar containing human blood (Wagatsuma agar). However, it may not be possible to carry out the test because of the difficulty in using human blood. An ELISA using monoclonal antibodies has been developed recently to demonstrate TDH.

2. Molecular approach

Several molecular approaches to detecting Kanagawa-positive *V. parahaemolyticus* have been developed. DNA probes and oligonucleotide probes specific for the genes encoding TDH or TRH, and a PCR technique using oligonucleotide primers from nucleotide sequences of the *tdh* genes to detect Kanagawa-positive strains in test samples have recently been reported.

E. Prevention

Because *V. parahaemolyticus* is a natural inhabitant in estuarine environments, it is impossible to protect seafish against contamination with this organism. The most important means of controlling infection lay in simple hygienic measures to prevent multiplication of the organism in seafoods and cross-contamination of cooked foods from raw seafood. These measures should be applied equally in markets, shops, and the home environment. Refrigeration or freezing is the most effective method for preventing multiplication. Adequate heating effectively eliminates *V. parahaemolyticus* from seafoods.

IV. *Vibrio vulnificus*

A. Disease

In humans, *V. vulnificus* is associated with two disease syndromes, primary septicemia and wound infection. Septicemia caused by this vibrio is serious, with a fatality rate of 50%. Progression of the illness can be very rapid from asymptomatic to death within 24 hours. About 75% of patients with this *Vibrio* septicemia have pre-existing hepatic diseases. Wound infections with *V. vulnificus* usually develop after trauma and exposure to the marine environment and progress rapidly. The wound infections most commonly present as cellulitis, of which the fatality rate is about 7%. About one-third of patients with wound infection may have some underlying disease. The etiological role of *V. vulnificus* in diarrhea disease has not been proved. An indole-negative biogroup of this vibrio, which is pathogenic to eels, is not thought to be associated with human infections.

B. Organism

1. Phenotypic characteristics

Most strains of *V. vulnificus* are encapsulated. They can grow on or in ordinary media containing 1–6% NaCl. Phenotypic characteristics of *V. vulnificus* are shown in Table 9.1. This species is divided into two biogroups by means of indole production. Serologically, 18 O groups were defined.

2. Virulence mechanisms

Although a variety of potential substances have been suggested as virulence factors of *V. vulnificus*, the role of the capsule in its pathogenesis appears to be the most clearly established. Encapsulated cells are virulent to mice, are resistant to the bactericidal activity of human serum and to phagocytosis, and are able to grow in iron-deficient medium, whereas unencapsulated cells lack these capabilities. In individuals, especially those with hepatic cirrhosis, hepatoma, and hemochromatosis, this vibrio induces septicemia. It is suggested that lack of complement, functional defects of the reticuloendothelial system, the presence of free Fe^2 in serum, or all three are important factors in developing septicemia. Cytolysin, collagenase, and protease are also important virulence factors, particularly in wound infections, as tissue-damaging substances.

C. Transmission

Principal reservoirs for *V. vulnificus* are estuarine seawater and brackish water. Although the vibrios cannot be cultured in the cold season, it is likely that they are not absent but are in a 'viable but nonculturable' state. Foodborne *V. vulnificus* infections have been associated almost exclusively with raw seafood, especially oysters. *V. vulnificus* in oysters increases in number at 18°C or above.

D. Laboratory Procedure

V. vulnificus grows well on TCBS agar, but in a clinical microbiology laboratory this vibrio is isolated on blood agar. To distinguish the vibrios from shellfish and environmental samples, enrichment culture with alkaline peptone water is required before plating.

V. vulnificus has been detected by PCR. A labeled DNA probe and a sandwich ELISA have also been reported to detect the cytotoxin of this vibrio in environmental samples.

E. Prevention

To prevent the risk of septicemia following *V. vulnificus* infection, individuals with hepatic disorders and other underlying diseases should not eat raw oysters and should take care to avoid being injured in marine environments.

V. Summary

V. cholerae is divided on the basis of serology into two groups: O1, which cause cholera; and non-O1, which are not cholera vibrios, except O group 139. The main virulence factors of the O1 and O139 strains are cholera toxin, which is responsible for producing cholera symptoms, and TCP pili, which mediate colonization. The most important mode of transmission is through water contaminated with the vibrio. Shellfish are also important reservoirs, because this vibrio can colonize shellfish for many weeks. CT-negative O1 and the majority of non-O1 strains are avirulent, but sometimes produce diarrhea and occasionally extraintestinal infections.

V. parahaemolyticus causes gastroenteritis in the summer. Most clinical isolates produce a toxin named the thermostable direct hemolysin. In contrast, most food isolates are TDH-negative. TDH-negative vibrios sometimes cause gastroenteritis. These vibrios produce another toxin named the TDH-related hemolysin. This vibrio is an estuarine inhabitant and infection is always associated directly or indirectly with seafood.

V. vulnificus often causes septicemia, which is associated exclusively with the consumption of oysters. Individuals with pre-existing hepatic disease are uniquely susceptible. This vibrio also causes wound infection that develops after exposure to seawater. Principal reservoirs are estuarine seawater and shellfish.

Bibliography

Barua, D. and Greenough, W. B., III (1993). 'Cholera'. Plenum Medical Book Co., New York.
Colwell, R. R. (1984). 'Vibrios in the Environment'. John Wiley & Sons, New York.

Farmer, J. J., III and Hickman-Brenner, F. (1991). The genera *Vibrio* and *Photobacterium*. *In* 'The Prokaryotes', Vol. III, 2nd ed. (A. Balows, H. G. Triiper, M. Dworkin, W. Harder and K.-H. Schleifer, eds), pp. 2952–3011. Springer-Verlag, New York.

Kaper, J. B., Morris, J. G., Jr and Levine, M. M. (1995). Cholera. *Clin. Microbiol. Rev.* **8**, 48–86.

Nishibuchi, M. and Kaper, J. B. (1995). Minireview – Thermostable direct hemolysin gene of *Vibrio parahaemolyticus*: a virulence gene acquired by a marine bacterium. *Infect. Immun.* **63**, 2093–2099.

Sakazaki, R. and Balows, A. (1981). The genus *Vibrio. In* 'The Prokaryotes' (M. P. Starr, H. Stolp, H. G. Truper, A. Balows and H. G. Scheleger, eds), pp. 1273–1301. Springer-Verlag, New York.

Sears, C. L. and Kaper, J. B. (1996). Enteric bacterial toxins: mechanisms of action and linkage to intestinal secretion. *Microbiol. Rev.* **60**, 167–215.

Wachsmuth, I. K., Blake, P. A. and Olisvic, O. (1994). 'Vibrio cholerae and Cholera'. American Society for Microbiology Press, Washington, DC.

Listeria monocytogenes

10

Linda J. Harris

I. Introduction

L. monocytogenes was first described in 1923. Before 1982, *L. monocytogenes* was recognized as a cause of abortions and encephalitis in many animals (particularly cattle and sheep), and was thought to be associated with contaminated animal feed or silage. While it was recognized as a cause of human illness, it was not until 1981 that a foodborne association was widely accepted.

The organism, its epidemiology, mechanisms of virulence, occurrence, and methods of detection in foods have been extensively reviewed; the reader may wish to refer to the documents listed in the Bibliography for a more detailed description of the organism. This chapter will provide a general overview of *L. monocytogenes* and its associated illnesses.

II. *Listeria monocytogenes*

A. Characteristics

L. monocytogenes is a Gram-positive, ovoid to rod-shaped bacterium that is widespread in the environment. It is a facultative anaerobe, and acid but not gas is produced

Foodborne Diseases 2nd Edn
ISBN: 0-12-176559-8

Table 10.1 Growth characteristics of *L. monocytogenes* under otherwise optimum conditions. Adapted from ICMSF (1996)

Limit	Temperature (°C)	pH	NaCl (%)	Water activity
Minimum	−0.4	4.4		0.92
Optimum	37	7.0		
Maximum	45	9.4	10	

from glucose. The organism is capable of multiplying at temperatures between approximately 0° and 45–50°C (it is a psychrotroph with optimum growth at 37°C); it is relatively resistant to NaCl (growth at 10%; survival at 20–30%) and low pH (growth at pH 4.4), is not inhibited significantly by carbon dioxide, and can survive many processing techniques such as freezing and drying. Table 10.1 indicates limits for growth of *L. monocytogenes* under ideal laboratory conditions. Limits for growth may be more restrictive in food systems, where a number of factors may be less than optimum for growth. The Canadian regulatory policy on *L. monocytogenes* includes the following parameters that will not support growth in ready-to-eat products: (1) pH 5.0–5.5 and water activity <0.95; (2) pH <5.0 regardless of water activity; (3) water activity of ≤0.92 regardless of pH; and (4) frozen foods.

B. Behavior in Foods

Fully cooked, modified atmosphere or vacuum packaged foods, including savory fully prepared meals, fully cooked side dishes, dairy and dessert items, deli meats, and sandwiches, are becoming increasingly common. Use of low or oxygen-free and elevated carbon dioxide atmospheres in some of these packages (modified atmosphere packaging; MAP) has been shown to extend the shelf-life considerably by inhibiting the Gram-negative aerobic spoilage bacteria. However, the growth of *L. monocytogenes* may occur with no apparent signs of spoilage. Preservation of MAP foods relies mainly on proper refrigeration (≤4°C). Unfortunately, this barrier is difficult to control after the food leaves the processing plant and will not completely prevent the growth of *L. monocytogenes*.

The survival of *L. monocytogenes* in both fluid milk products and fermented dairy products has been thoroughly studied. *L. monocytogenes* does not survive pasteurization of milk and if present in pasteurized milk it is as a postprocess contaminant. *L. monocytogenes* grows well in liquid dairy products from 4° to 35°C. When *L. monocytogenes* is present in milk used to make cheese, listerial growth is suppressed but not completely prevented by the presence of lactic acid starter cultures. *L. monocytogenes* is concentrated in the cheese curd, with only a small portion of the cells appearing in the whey. Once in the curd, the behavior of the pathogen ranges from growth (feta cheese) to death of almost all of the cells (cottage cheese). During cheese ripening, numbers of *L. monocytogenes* decrease gradually (Cheddar or Colby cheese), decrease rapidly during the initial ripening periods and then stabilize (blue cheese) or increase markedly (Camembert cheese). The manufacture of mold surface-ripened soft cheeses, such as Camembert and Brie cheeses, requires a considerable amount of handling after the initial fermentation, making these cheese types particularly prone to recontamination. In addition, the pH at the surface of the cheese increases as the lactate is degraded by fungi such as *Penicillium camemberti*.

Smoked mussels were associated with an outbreak of listeriosis in New Zealand in 1992 and cold-smoked rainbow trout were implicated in an outbreak in Finland in 1998. The organism can be found in freshwater and seawater from coastal areas. *L. monocytogenes* has been isolated from a wide variety of fresh, frozen, and processed seafood products. Although retail surveys do indicate a high incidence of the organism in seafoods, the levels are often less than 100 per gram. With hot-smoked products, populations of *L. monocytogenes* are reduced during smoking and postprocess contamination is the primary concern, especially in those products with water activities high enough to support the growth of the organism during refrigerated storage. Cold-smoked fish undergo a smoking procedure at 20–30°C for up to 18–20 hours, and sometimes longer. Even though smoke itself may have some antilisterial effect, these cold-smoking conditions may also allow multiplication of *L. monocytogenes* under some circumstances.

Deli meats, paté, and hot dogs have all been associated with outbreaks of listeriosis. Nitrite levels in many of these products may reduce but not prevent the growth of *L. monocytogenes*. The high water activity and pH of the products, vacuum or modified atmosphere storage, and relatively long shelf-life contribute to the risk associated with these products. Although *L. monocytogenes* does not survive the cooking process applied to these products, significant handling during postprocess preparation and packaging can result in contamination. In addition to rigorous postprocessing sanitation and hygiene, postpackaging pasteurization has been suggested as a means of controlling *L. monocytogenes* in these products.

L. monocytogenes is widely distributed on raw fruits and vegetables, and other plant material. However, several studies with relatively large sample sizes have failed to detect the organism, and factors affecting its presence or persistence have yet to be determined. Plants and plant parts used as salad vegetables play a key role in disseminating the pathogen from natural habitats to the human food supply. This role may also be indirect, for example, by contaminating milk via forage or silage, or direct in the form of raw produce.

Documented outbreaks associated with this organism and linked to fresh produce have been limited, and the significance of finding this organism in raw produce is currently unclear. Growth at refrigerator temperatures on a wide variety of fresh-cut fruits and vegetables has been reported. Modified atmospheres do not appear to influence growth rates in these products.

C. Taxonomy

As with other bacteria, the definition of the genus *Listeria* has undergone significant changes in the past two decades, as techniques to distinguish microorganisms become increasingly sophisticated. The genus *Listeria* comprises six species: *L. monocytogenes*, *L. innocua*, *L. welshimeri*, *L. seeligeri*, *L. grayi*, and *L. ivanovii*. On rare occasions, *L. seeligeri* and *L. ivanovii* have been implicated in human infections. However, *L. monocytogenes* is the only species considered to be of public health significance.

D. Serology

All 13 serotypes of *L. monocytogenes* may cause human listeriosis; however, 95% of human isolates are 1/2a, 1/2b, or 4b. Serotype 4b strains are responsible for 33–50% of human listeriosis worldwide, which has led to a hypothesis that clones of this serotype are more virulent than other strains. Serotype 4b has also been responsible for most recorded foodborne outbreaks; the reasons for this are not understood. Pulsed-field gel electrophoresis, ribotyping, multilocus enzyme electrophoresis, restriction fragment length polymorphism analysis, and random amplification of polymorphic DNA are among the methods that have been used to subtype *L. monocytogenes*. Three evolutionary lines of *L. monocytogenes* are generally identified by these techniques. In most studies, serotypes 1/2a, 1/2c, 3a, and 3c are placed in division I, 1/2b, 3b, 4b, 4d, and 4e in division II and 4a forms a distinct relatively homogenous group. Typing methods have been successfully used to identify multistate outbreaks of listeriosis that would not have been recognized until the mid-1990s. None of the current typing methods can be used to distinguish nonpathogenic or less virulent strains.

E. Reservoirs

L. monocytogenes is widely distributed in the environment. It can be found in decaying vegetation, in soils, animal and human feces, sewage, silage, and water. Approximately 2–6% of healthy people are thought to be asymptomatic fecal carriers of the organism, and wild and domestic animals may also be carriers. *L. monocytogenes* has been isolated from a wide range of retail foods. Although numbers are often very low in these products, multiplication of *L. monocytogenes* can potentially occur during refrigerated retail and home storage. The psychrotrophic nature of *L. monocytogenes* makes it of particular concern in refrigerated foods with an extended shelf-life.

III. Listeriosis

Listeriosis can be transmitted through contact with animals, by cross-infection of newborn babies in hospitals, and via food vehicles.

A. Infective Dose

The infectious dose is unknown but thought to be highly strain- and host-dependent. It is likely that <1000 colony-forming units (CFU) is of no concern to healthy adults.

However, it is assumed that this level will cause illness in susceptible persons. Populations of *L. monocytogenes* in foods responsible for outbreak or sporadic cases are often more than 100 CFU/g, although in one major outbreak associated with frankfurters, the levels were less than 0.3 CFU/g.

B. Incubation Period

The incubation period for listeriosis depends upon the susceptibility of the individual and the dose ingested, but is documented to range from 24 hours to 91 days. It is not known why this range is so large other than being related to the number of organisms ingested and the host immune system. Extremely long incubation periods contribute to the difficulty in determining the source of infection.

C. Symptoms

Symptoms range from flu-like vomiting and diarrhea to septicemia and meningitis (Table 10.2). Listeriosis refers to the more serious life-threatening illnesses, while gastroenteritis is the mild illness experienced by healthy adults. Persons with suppressed T-cell-mediated immunity are more susceptible to listeriosis. Pregnant women, newborns, and elderly and immunocompromised individuals are most commonly afflicted and experience a more severe illness. Case fatality rates for these groups range from 13% to

Table 10.2 Illnesses caused by *L. monocytogenes*. Adapted from Bell and Kyriakides (1998)

Type of infection	Source of organism	Symptoms	Time to onset
Zoonotic	Local infection of skin lesions	Mild and self-resolving	1–2 days
Neonatal (listeriosis)	Newborn babies are infected from the mother during birth or as a result of cross-infection from one neonate in the hospital to other	Can be extremely severe, resulting in meningitis and death	1–2 days (early onset) usually from infection during birth; 5–12 days (late onset) from neonate cross-infection
During pregnancy (listeriosis)	Follows consumption of contaminated food	Mother experiences mild flu-like illness or is asymptomatic. Spontaneous abortion, fetal death, stillbirth, and meningitis may occur with unborn infant. Third-trimester infection more common	1 day to several months
Nonpregnant adults and children >1 month (listeriosis)	Follows consumption of contaminated food	Asymptomatic or mild illness. In immunocompromised or elderly individuals, illness may progress to central nervous system infections such as meningitis.	1 day to several months
Gastroenteritis	Consumption of food with exceptionally high levels of *L. monocytogenes*, >10^7 per ml or g	Vomiting and diarrhea, usually self-resolving	<24 hours after ingestion

34%. Patients with acquired immunodeficiency syndrome (AIDS) have been reported to be as much as 280 times more susceptible than the general public to listeriosis.

In pregnant women, the illness can result in abortion, stillbirth, or birth of a severely ill infant. *L. monocytogenes* is one of the few bacteria capable of crossing the placenta and able to gain direct access to the fetus. Newborn babies may also acquire infection after birth from the mother or from other infected infants. In immunocompromised and elderly adults, the illness typically involves infection of the tissues surrounding the brain (meningitis) and infection of the bloodstream (septicemia).

Healthy adults are thought rarely to suffer from listeriosis. However, at least one outbreak has demonstrated that high levels of this organism can cause symptoms in healthy individuals that are similar to flu-like symptoms of vomiting, nausea, and diarrhea.

D. Virulence Factors

L. monocytogenes is an intracellular parasite. Abnormalities in T-cell immunity increase the risk of listeriosis. The T-cell response in the first few days following infection is important to the subsequent outcome of the disease. Considerable progress has been made in the past decade in understanding the pathogenesis of *L. monocytogenes*. Eight genes clustered on the chromosome are associated with virulence. *L. monocytogenes* cells cross the intestinal barrier via intestinal epithelial cells or the M cells of Peyer's patches. The organism is internalized by phagosomes. Surface proteins internalin and p60 are thought to aid internalization of *L. monocytogenes*. Once internalized within a phagosome, the vacuole membrane is lysed and *L. monocytogenes* is released to the cytoplasm where it can multiply. Listeriolysin O and a phosphatidylinositol–phospholipase C are involved in the lytic process. An essential component of virulence is the ability of *L. monocytogenes* to spread directly from cell to cell. To do this the organism uses the host cell actin machinery continuously assembling an actin tail at a pole of the bacterial cell surface. This serves to propel the bacterium across the cytoplasm, pushing the organism against the host cell membrane, thus forming a protrusion that can be ingested by an adjacent cell. Three genes, *mpl*, *actA*, and *plcB*, have been linked to this process. The resulting vacuole is lysed, releasing *L. monocytogenes* into the cytoplasm of the newly infected cell.

The phagosomes are transported via the blood to the lymph nodes, liver, and spleen. Further dissemination of the organism via the bloodstream to the brain, or placenta in the pregnant woman, occurs, giving rise to the various forms of the illness.

IV. Foodborne Transmission of *Listeria monocytogenes*

L. monocytogenes has been responsible for a relatively small number of foodborne outbreaks but is of concern because of its high case-fatality rate. Outbreaks of listeriosis have been associated with vegetable, dairy, seafood, and meat products. It is estimated that 80–90% of listeriosis cases are linked to ingestion of contaminated food; however,

Table 10.3 Examples of foodborne outbreaks of listeriosis

Year	Country	Cases (deaths)	Food	Outbreak serotype	Food isolate?
1980–1981	Canada	41 (18)	Coleslaw	4b	Yes
1983	USA	49 (14)	Pasteurized milk	4b	No
1983–1987	Switzerland	122 (31)	Vacherin cheese	4b	Yes
1985	USA	142 (48)	Mexican-style soft cheese	4b	Yes
1987–1989	UK	>350 (>90)	Belgian paté	4b	Yes
1992	New Zealand	4 (2)	Smoked mussels	1/2a	Yes
1992	France	279 (85)	Pork tongue in aspic	4b	Yes
1994	USA	45 (0)	Chocolate milk	1/2b	Yes
1995	France	20 (4)	Raw-milk soft cheese	4b	
1998–1999	Finland	11(4)	Butter	3a	Yes
1998–1999	US	101 (21)	Frankfurters	4b	Yes
2000	US	29 (7)	Turkey deli meat	1/2a	

demonstration of foodborne listeriosis is relatively rare and most diagnosed cases are sporadic. Recent FoodNet data indicate that total diagnosed infections range from 0.3 to 0.6 cases per 100 000 population in the US, or approximately 900–1800 cases per year. With improvements in medical technology, the proportion of the population susceptible to foodborne listeriosis continues to rise. There are an estimated 1500–1700 foodborne cases in the US annually, with an estimated 400–500 deaths. Annual estimated medical costs and productivity losses are 0.2–0.3 billion dollars. Foods most commonly associated with listeriosis are: those where a listericidal process has not been applied; processed product, which is susceptible to postprocess pre-package contamination; product formulation, which allows growth; those that are stored refrigerated and have a shelf-life of greater than 10 days; and ready-to-eat products.

Sporadic foodborne cases of listeriosis have been associated in case-control studies with hot dogs consumed without reheating, consumption of chicken meat that was still pink, soft cheeses, and foods from delicatessen counters.

A. Foodborne Outbreaks of Listeriosis

Foods most commonly associated with outbreaks of listeriosis are ready-to-eat foods that are stored under refrigerated conditions and have a shelf-life of more than 10 days. Similar foods have been associated with sporadic listeriosis in case-control studies. Frankfurters, deli meats, paté, soft cheeses, and smoked fish are among the ready-to-eat foods that have been associated with listeriosis (Table 10.3).

B. Case Studies of Representative Outbreaks of Listeriosis

1. Coleslaw (1981, Maritime Provinces of Canada)

- *Product type*: Chopped cabbage and carrot, sold premixed at retail outlets
- *Levels*: Not known

- *Serotype*: 4b
- *Extent*: 41 cases – 18 deaths, 2 adults and 16 fetal or newborn
- *Comments*: Cabbage was suspected to be contaminated in the field with *L. mono-cytogenes* from uncomposted sheep manure. Two sheep were suspected to have died from listeriosis, one in 1979 and the other in early 1981. Cabbage was stored in a cold-storage shed for extended periods of time before shredding. No anti-listerial processes were applied.
- *Control options*: Apply good agricultural practices. Control use of manure. Storage at temperatures to prevent the growth of *L. monocytogenes* ($<1°C$) or storage for short periods of time (<10 days).

2. Mexican-style white cheese (1985, Los Angeles County, CA)

- *Product type*: Soft cheese (Jalisco), pH 6.6
- *Levels*: Not known
- *Serotype*: 4b
- *Extent*: 142 cases – 48 deaths
- *Comments*: Environment and equipment were grossly contaminated with *L. monocytogenes* even after clean up. Raw milk deliveries allegedly exceeded pasteurization capacity of plant and cheese samples tested positive for phosphatase (an indication of inadequate pasteurization).
- *Control options*: Adequate pasteurization and sanitation.

3. Chocolate milk (1995, Illinois)

- *Product type*: Pasteurized chocolate milk
- *Levels in implicated food*: 10^7–10^9 CFU/ml
- *Extent*: 45 cases – gastroenteritis, no deaths
- *Comments*: Outbreak strain was isolated from stool samples of infected individuals, from a tank drain at the manufacturing plant and from unopened packs of the implicated milk. Milk was pasteurized and passed to a holding tank before filling. The jacket of holding tank was in poor state of repair and refrigerant could not be used. Lining of the holding tank jacket was not intact. Milk leaked into the jacket where it stayed until the tank emptied. Product could re-enter the vessel and contaminate the remaining product being filled. Sanitizer 'spray-balls' were blocked and it was likely that insufficient cleaning took place. Cartons were taken to the picnic – 2 hours unrefrigerated, then refrigerated, then unrefrigerated for several hours.
- *Control options*: Maintain equipment in good repair. Have an adequate sanitation program in place. Maintain temperature control on perishable products.

4. Hot dogs and deli meats [August 1998–January 1999, Multi-state (CDC, 1998, 1999)]

- *Product type*: Hot dogs and deli meats
- *Levels*: Unknown
- *Extent*: Over 100 cases – 15 deaths, 6 miscarriages

- *Comments*: Construction dust at the plant is believed to have contaminated the product in the packaging room.
- *Control options*: Adequate sanitation during packaging of product. Postpackaging pasteurization (not yet commonly practiced or commercially feasible for all products).

V. Control in Foods

A. Temperature

L. monocytogenes can survive for long periods of time in frozen foods. Although *L. monocytogenes* can multiply under refrigerated conditions, lag and generation times are significantly affected as the temperatures fall below or rise above 10°C (Table 10.4).

B. pH and Water Activity

Inhibition of growth or inactivation of *L. monocytogenes* by acids is dependent upon pH, and acidulant type and concentration. Growth of the organism has been observed at pH 4.3. The minimum water activity allowing growth of *L. monocytogenes* is also dependent to a certain extent upon the solute. However, lower limits are typically reported at or near water activities of 0.90–0.93.

C. Other Factors

Increased carbon dioxide or other changes to package atmosphere do not inhibit *L. monocytogenes*. Modified-atmosphere packaged foods often support the growth of lactic acid bacteria, which in many cases do not interfere with the growth of *L. monocytogenes*. A great deal of research has focused on the inhibitory effects of lactic acid bacteriocins against *L. monocytogenes*. Bacteriocins are small peptides produced by some bacteria that are bactericidal to others. The antilisterial bacteriocin nisin (produced by *Lactococcus lactis* subsp. *lactis*) is permitted as a food additive in some countries and has been shown to help reduce populations of *L. monocytogenes* in some

Table 10.4 Approximate lag and generation times for *L. monocytogenes* generated using the USDA Pathogen Modeling Program Version 6.0 for a pH of 6.0 and water activity of 0.98

Temperature (°C)	Lag time (days)	Generation time (hours)
4	124.9	19.6
6	89.1	13.2
8	64.6	9.1
10	47.6	6.4
12	35.6	4.6
14	27.1	3.4
16	20.9	2.5

products and to decrease its heat resistance in others. Some starter cultures produce antilisteria bacteriocins, which may help to control growth of *L. monocytogenes* in fermented products. Some products marketed as flavor enhancers that are derived from lactic acid bacteria fermentations contain antilisterial bacteriocins that have been shown to influence lag phase and growth rates of *L. monocytogenes* in some products.

Combinations of suboptimal conditions will significantly affect the ability of *L. monocytogenes* to survive or grow in a particular product. When assessing the ability of any given food product to support the growth of *L. monocytogenes*, pathogen growth-modeling programs can be used to assist in product modifications, but when practical experience is lacking, challenge studies may be necessary.

L. monocytogenes resistance to electron beam and gamma radiation is similar to that of other Gram-positive bacteria.

D. Control in the Food-processing Environment

Because of the widespread prevalence of *L. monocytogenes*, its control in the food-processing environment has been particularly difficult. For sensitive foods, those that have an extended chilled shelf-life, the food processor must carefully assess each operation for points of potential contamination and proliferation of the organism.

Elimination of the organism from a food-processing environment has been shown to be very difficult, as *Listeria* is constantly being reintroduced into facilities by employees and incoming raw materials. Proliferation of *L. monocytogenes* in food-processing facilities is promoted by relatively high humidity and high levels of nutrient-laden waste. Controlling *L. monocytogenes* requires both reducing the number of bacteria that might contaminate the final product and equipment surfaces via physical means, and preventing the growth and proliferation of the bacteria by managing the environment. Where products are cooked, there should be strict separation of raw product and processing areas, including restriction of employee movement from one area to the other or even interaction of these two groups of employees during breaks.

Cleaning and sanitizing treatments given to the equipment, walls, and drains should be adequate to destroy or remove *L. monocytogenes*, especially in areas designated for handling final products. In order to verify control of *L. monocytogenes*, plants should consider implementing a monitoring program that tests for an indicator bacterium, such as *Listeria* spp. This type of monitoring system detects the presence of all *Listeria* spp. with the assumption that, if any are detected, *L. monocytogenes* may also be present. The frequency of sampling, location of samples, and the corrective action to be taken are tailored to the plant's operation. In addition to plant sanitation, training and hygiene should be important components of a *L. monocytogenes* control program.

VI. Regulatory Control

In 1985, the US Food and Drug Administration (FDA) began monitoring dairy products for *L. monocytogenes* after a particularly large outbreak of listeriosis in

Los Angeles associated with cheese. In 1989, the US Centers for Disease Control and Prevention (CDC) published a report of a case of listeriosis from the consumption of turkey frankfurters. In response, the US Department of Agriculture initiated a microbiological surveillance program for *L. monocytogenes* in ready-to-eat meats and initiated a zero-tolerance policy, prohibiting the sale of ready-to-eat meat products contaminated with *L. monocytogenes*. Shortly thereafter, this policy was expanded to include all ready-to-eat foods; in practice, this is interpreted as 'processed', not raw, ready-to-eat foods.

This policy designates *L. monocytogenes* an 'adulterant'. Any ready-to-eat food that contains this organism (in a 50-g sample) can be considered adulterated and subject to a Class I recall and/or seizure. A Class I recall is considered a situation in which there is a reasonable probability that the use of, or exposure to, a product will cause serious adverse health consequences. In the US from October 1, 1991 to September 30, 1992, 16% of all recalled products and 57% of class I recalls could be attributed to *L. monocytogenes*. It has been suggested that these procedures, as well as consumer educational efforts, led to a reduction in illness and death associated with this organism. When 1989 and 1993 data were compared, a reduction in illness and death of 44% and 48%, respectively, was observed. However, single outbreaks such as the one associated with hot dogs and deli meats that occurred in 1999 can significantly impact statistics and draw attention to the problems in the food industry that still exist with this organism.

Other countries, including Canada, have adopted more moderate approaches to controlling *L. monocytogenes* that are based on risk (Table 10.5). The International

Table 10.5 Canadian Compliance Criteria for *Listeria monocytogenes* in ready-to-eat (RTE) foods

Category	Action level for LM	GMP status	Immediate action	Follow-up action
1. RTE foods causally linked to listeriosis. This list presently includes: soft cheese, liver pate, coleslaw mix with shelf life >10 days, jellied pork tongue	>0 CFU/50 g	N/A	Class I recall to retail level Consideration of Public Alert Appropriate follow-up at plant level	
2. All other RTE foods supporting growth of LM with refrigerated shelf-life >10 days	>0 CFU/25 g	N/A	Class II recall to retail level Health alert consideration Appropriate follow-up at plant level	
3. RTE foods supporting growth of LM with refrigerated shelf-life ≤10 days and all RTE foods not supporting growth	≤100 CFU/g	Adequate GMP	Allow sale	Appropriate follow-up at plant level
	≤100 CFU/g	Inadequate or no GMP	Consideration of Class II recall or stop sale	Appropriate follow-up at plant level
	≥100 CFU/g	N/A	Class II recall or stop sale	Appropriate follow-up at plant level

Commission on Microbiological Specifications for Foods has developed risk-based guidelines for international trade of foods. These guidelines differentiate foods intended for immunocompetent and immunocompromised individuals.

In addition to industry guidelines, consumer education programs are an important part of controlling listeriosis. Education efforts targeted to susceptible individuals, through the health care systems and by other mechanisms, have been demonstrated to be effective in several countries. The education programs include information on handling foods with an emphasis on adequate cooking and sanitation as well as instructions to avoid or cook many ready-to-eat foods.

VII. Isolation and Identification of *Listeria monocytogenes*

Early methods of detection and identification of *L. monocytogenes* were developed in clinical laboratories. With the first recognized outbreak of foodborne listeriosis in 1981 came a need for adaptation of these methods to isolation from foods.

In a food industry setting it is more common to test for *Listeria* spp. than for *L. monocytogenes* and more common to test environmental samples than food samples. When food or environmental swab samples are tested, methods to determine presence/absence or detection in 25–50 g samples (food) or per swab are most commonly employed. These methods use a selective enrichment in broth followed by isolation of colonies on selective agar. When the organism might be injured as in frozen or dried foods, a nonselective enrichment might be employed prior to selective enrichment. Colonies on selective media exhibiting morphology typical of *Listeria* spp. can be further identified to the species level using biochemical characteristics. Because several types of *Listeria* spp. may be isolated from a food sample, it is common to select more than one colony per plate (often three or more) for confirmation.

Methods available for detection and identification of *L. monocytogenes* have increased significantly in the past decade. Enzyme-linked immunosorbent assays and assays based on DNA probes or polymerase chain reaction (PCR) tests are readily available. These methods can, in some cases, eliminate or reduce the time required for enrichment or, after enrichment, quickly provide a positive or negative sample evaluation. Rapid identification methods include miniaturized multitest bioassays that allow confirmation of presumptive *Listeria* isolates to the species level. Additionally, DNA probes are commercially available, and PCR-based tests or published PCR primers have been successfully used in the identification of *L. monocytogenes*.

A great deal of controversy exists regarding the value of end-product testing for *Listeria* spp., especially when processing methods are adequate to eliminate *Listeria* spp. from the product. In these cases, contamination occurs postprocess/pre-package and tends, when it happens, to be unevenly distributed and sporadic in nature. For this reason, it is more common to test for *Listeria* spp. in the processing environment from floors to food contact surfaces. When the organism is detected, sanitation programs can be adjusted and product can be held if necessary when the organism is detected.

VIII. Summary

L. monocytogenes is a significant cause of foodborne disease in the US and other developed countries. It is particularly a threat to pregnant women and to other people whose immunity is impaired. The associated medical costs related to the severity of the disease, as well as costs of food recalls, are enormous. The ubiquitous nature of the organism makes control very difficult, but industry programs that accomplish this are in use. High levels of skill and dedication are required, and occasional contamination may still occur. Strict maintenance of storage temperatures along the distribution chain to and including the consumer will reduce the opportunity for *L. monocytogenes* to reach high levels in foods prior to consumption. End-product testing for *L. monocytogenes*, as a sole means of control, is not a reliable way of ensuring a safe product. However, environmental testing – as a means of verifying that appropriate good manufacturing practices and HACCP plans are in place – does have value in controlling this organism.

Bibliography

Bell, C. and Kyriakides, A. (1998). '*Listeria:* A Practical Approach to the Organism and Its Control in Foods'. Blackie Academic & Professional, London.

Buzby, J. C., Roberts, T., Lin, C.-T. J. and MacDonald, J. M. (1996). Bacterial Foodborne Disease: Medical Costs and Productivity Losses. USDA, Agricultural Economic Report No. 741.

Centers for Disease Control and Prevention (1998). Multistate outbreak of listeriosis – United States, 1998. *Morbid. Mortal. Weekly Rep.* **47**, 1085–1086.

Centers for Disease Control and Prevention (1999). Update: Multistate outbreak of listeriosis – United States, 1998–1999. *Morbid. Mortal. Weekly Rep.* **47**, 1117–1118.

Farber, J. M. (1993). Current research on *Listeria monocytogenes* in foods: an overview. *J. Food Prot.* **56**, 640–643.

Farber, J. M. and Harwig, J. (1996). The Canadian position of *Listeria monocytogenes* in ready-to-eat foods. *Food Control* **7**, 253–258.

Farber, J. M. and Peterkin, P. I. (1991). *Listeria monocytogenes*, a food-borne pathogen. *Microbiol. Rev.* **55**, 476–511.

International Commission on Microbiological Specifications for Foods (ICMSF) (1996). 'Microorganisms in Foods. 5. Characteristics of Microbial Pathogens'. Blackie Academic & Professional, London.

Johnson, J. L., Doyle, M. P. and Cassens, R. G. (1990). *Listeria monocytogenes* and other *Listeria* spp. in meat and meat products, a review. *J. Food Prot.* **53**, 81–91.

Lammerding, A. M. and Doyle, M. P. (1990). Stability of *Listeria monocytogenes* to non-thermal processing conditions. *In* 'Foodborne Listeriosis' (A. J. Miller, J. L. Smith and G. A. Somkuti, eds), pp. 195–202. Society of Industrial Microbiologists.

Linnan, M. J., Mascola, L., Lou, X. D., Goulet, V., May, S., Salminen, C., *et al.* (1988). Epidemic listeriosis associated with Mexican-style cheese. *N. Engl. J. Med.* **319**, 824–828.

Pearson, L. J. and Marth, E. H. (1990). *Listeria monocytogenes* – threat to a safe food supply: a review. *J. Dairy Sci.* **73**, 912–928.

Pinner, R. W., Schuchat, A., Swaminathan, B., Hayes, P. S., Deaver, K. A., Weaver, R. E., *et al.* (1992). The role of foods in sporadic listeriosis. II. Microbiolgic and epidemiologic investigation. *J. Am. Med. Assoc.* **267**, 2046–2050.

Rocourt, J., Jacquet, C. and Reilly, A. (2000). Epidemiology of human listeriosis and seafoods. *Int. J. Food Microbiol.* **62**, 197–209.

Ryser, E. T. and Marth, E. H. (ed.) (1999). *Listeria*, Listeriosis and Food Safety, 2nd ed. Marcel Dekker, New York.

Schleck, W. F., III, Lavigne, P. M., Bortolussi, R. A., Allen, A. C., Haldane, E. V., Wort, A. J., *et al.* (1983). Epidemic listeriosis – evidence for transmission by food. *N. Engl. J. Med.* **308**, 203–206.

Shank, F. R., Elliot, E. L., Wachsmuth, I. K. and Losikoff, M. E. (1996). US position on *Listeria monocytogenes* in foods. *Food Control* **7**, 229–234.

Smith, J. L. (1999). Foodborne infections during pregnancy. *J. Food Prot.* **62**, 818–829.

Tappero, J. W., Schuchat, A., Deaver, K. A., Mascola, L. and Wenger, J. D. (1995). Reduction in the incidence of human listeriosis in the United States. Effectiveness of prevention efforts? *J. Am. Med. Assoc.* **273**, 1118–1122.

Tompkin, R. B., Scott, V. N., Bernard, D. T., Sveum, W. H. and Sullivan Gombas, K. (1999). Guidelines to prevent post-processing contamination from *Listeria monocytogenes*. *Dairy Food Environ. San.* **19**, 551–562.

Venugopal, R., Tollefson, L., Hyman, F. N., Timbo, B., Joyce, R. E. and Klontz, K. C. (1996). Recalls of foods and cosmetics by the US Food and Drug Administration. *J. Food Prot.* **59**, 876–880.

Suggested Web Sites

N.B. Always look for dates on materials
http://www.fsis.usda.gov/OA/topics/lm.htm – A government website devoted to *L. monocytogenes*. See recent risk assessment – http://www.foodsafety.gov/~dms/lmrisk.html
http://www.foodsafety.gov

Economics of Listeriosis

http://www.ers.usda.gov/briefing/FoodborneDisease/listeria/index.htm
For consumers: *Listeria* and Food Safety Tips (May 1999). PDF Brochure.
For the meat industry: *Listeria* Guidelines for Industry (May 1999).

FDA Bacteriological Analytical Manual

http://www.cfsan.fda.gov/~ebam/bam-toc.html
http://www.cdc.gov/mmwr/preview/mmwrhtml/mm5013a1.htm#tab1

Infrequent Microbial Infections

11

Dean O. Cliver

I. Scope

This chapter will cover only infectious bacteria, with a few exceptions in Section V. Other infrequent infectious agents (e.g. some of the viruses) are discussed in their respective chapters. Bacteria that cause infrequent intoxications are described in Chapter 14.

This is intended as a brief survey, rather than an all-inclusive collection. Categories of bacteria included are as follows.

- Historic – agents that were once a major threat via foods in the US and other developed countries. Although these agents are less significant now in affluent parts of the world, they may still be important in less developed countries.
- Sometimes foodborne – agents that are occasionally transmitted via foods, but represent a greater threat as transmitted by other routes.
- Questioned foodborne pathogens – sometimes present in foods, not surely pathogenic, or pathogenic only in especially susceptible populations. It should be noted that, although some of these agents are discussed in the 1994 CAST report (Foodborne Pathogens: Risks and Consequences) and other references, only *Brucella*, *Plesiomonas*, and *Streptococcus* Group A are reported to have caused food- or water-associated outbreaks during the most recent reporting periods.

II. Bacteria Less Commonly Foodborne than Formerly

A. *Bacillus anthracis*

B. anthracis is a Gram-positive, encapsulated, spore-forming, aerobic, nonmotile rod. It achieved notoriety in the US in November 2001 by transmission through the mail, causing inhalation anthrax (often fatal), as well as serious skin infections. The bacterium can also infect the digestive tract after being ingested. Foodborne anthrax is now virtually unknown in the US but is fairly common in developing countries, where dead or dying animals are often butchered and eaten for want of other sources of food. Because the spores of *B. anthracis* persist well in soil, enzootic anthrax occurs worldwide, but infrequently.

B. *Brucella* spp.

Brucella spp. are Gram-negative coccobacilli that are nonmotile and do not form spores. They measure 0.5–0.7 μm by 0.5–1.5 μm. They are apparently labile outside the host's body. They do not colonize the environment and are readily killed in food by cooking or pasteurization. In settings where milk is regularly pasteurized, persons most at risk are those in animal-handling occupations – farmers, veterinarians, and abattoir workers.

    Brucella spp. were once widely transmitted via unpasteurized milk and milk products, although other animal products might also be vehicles on occasion. The milk-related species are *B. abortus* from cows and *B. melitensis* from sheep or goats. Carcasses of infected animals [including swine (*B. suis*) and buffalo] are also infectious, but do not typically lead to consumer infections. *Brucella* spp. are now rare in Northern Europe and North America, but still a great threat elsewhere.

    Incubation in human illness ranges widely, from 5 days to 2 months (rarely, more). The illness is a recurrent, prolonged, febrile, systemic infection. *B. melitensis* infections, in particular, may be fatal. Treatment with combination of rifampin or streptomycin, and doxycycline is useful, but recurrences due to sequestered organisms are common and may be accompanied by arthritis. Control in cattle once involved vaccination, but test-and-slaughter programs are now the norm and have been quite successful. Continued vigilance is necessary, as *Brucella* spp. sometimes must be eradicated more than once.

C. *Corynebacterium diphtheriae*

This bacterium is a Gram-positive rod that may be slightly curved and enlarged in the middle or at one end. It produces a toxigenic infection, usually of upper respiratory tract in humans. The disease can be life threatening, and is treated with antitoxin (barring hypersensitivity to horse serum) and erythromycin or penicillin. It has been transmitted by raw milk, but the principal reservoir is humans; in the community, it is controlled by vaccination of humans. Even before vaccination had virtually eradicated the agent in the human population, transmission via milk had essentially ceased, owing to the advent of machine milking and milk pasteurization. This suggests that, like the poliomyelitis virus, its source in milk was infected milkers.

D. *Mycobacterium bovis*

The agent of bovine tuberculosis, like its human counterpart (*M. tuberculosis*), is an acid-fast (i.e. once stained, the bacteria are not readily decolorized with acid–alcohol), rod-shaped bacterium. The principal reservoir is cattle, although other large ruminants are also susceptible. Food vehicles are principally unpasteurized cows' milk, and products made from it. Bovine tuberculosis is now rare in Northern Europe and North America, but still a great threat elsewhere. The original specifications for time and temperature of milk pasteurization were selected to kill *M. bovis* in milk.

Tuberculosis in humans that is caused by *M. bovis* is said to be indistinguishable from that caused by *M. tuberculosis*; but since *M. bovis* is most likely to infect *consumers* via the digestive tract, extrapulmonary tuberculosis is more likely. Isoniazid is still the drug of choice in most instances, but resistance is so common worldwide that the World Health Organization (WHO) recommends treatment with combinations. The disease is controlled in cattle by tuberculin skin testing and slaughter of positive reactors. Although this is effective, reintroductions occur from time to time in areas from which bovine tuberculosis has been eradicated, so continued surveillance – and pasteurization of milk – are indicated.

III. Bacteria Rarely Transmitted via Foods

A. *Clostridium difficile*

Like the rest of its genus, this agent is a Gram-positive, anaerobic, spore-forming rod. It is a free-living organism in soils and sediments. It can contaminate foods, but has not specifically been shown to be foodborne. However, food transmission seems to have been suspected principally because this agent causes diarrhea, especially in persons whose intestinal flora has been disturbed by antibiotics, and in infants.

B. *Coxiella burnetii*

This agent is a rickettsia – an obligate intracellular parasite. It is rod-shaped, 0.2–0.4 μm by 0.4–1.0 μm. It occurs worldwide, with a reservoir in sheep, cattle, and goats, but infects some companion and wild animals, as well as birds and ticks. It is most commonly airborne, but is shed in the milk of infected cattle; control of this mode of transmission is by pasteurization of milk. The principal sign of infection is fever. It is treated with tetracyclines or chloramphenicol.

C. *Streptococcus pyogenes* (=Group A)

These bacteria are Gram-positive, spherical or ovoid, generally <2 μm in diameter, and grow in chains. They are nonmotile and do not form spores. They are distributed worldwide. Historically, the most common vehicle has been unpasteurized cows' milk (also true of β-hemolytic Group C streptococci), but any food may be contaminated by an infected handler. Human infections may produce severe sore throat, sometimes scarlet fever, or rheumatic fever. Treatment is usually with penicillin or erythromycin.

IV. Bacteria not Conclusively Proved to be Pathogenic in 'Normal' People

A. *Aeromonas hydrophila* and Related Species

The genus *Aeromonas* comprises several species of Gram-negative, rod-shaped bacteria, typically 0.3–1.0 μm × 1.0–3.5 μm, but coccobacillary and filamentous forms also occur. They are facultative and motile by polar flagella. Aeromonads are widespread in the environment, particularly in water. They are often found in the human intestines (normal and diarrheal). They are a reported cause of diarrhea in humans (especially young children and most often during the summer), and several modes of pathogenesis have been described. Several outbreaks of foodborne diarrhea have been recorded in various countries, but documentation has tended to be weak. Even the causation of diarrhea has been questioned, in that Koch's postulates were not fulfilled.

B. *Enterococcus* spp.

The enterococci are Gram-positive cocci that had been called Group D streptococci before being assigned to a genus of their own. They are common in the human intestines and are used as indicators of fecal contamination of water. They have been associated with urinary tract infections, wound infections, and bacteremia in the severely disabled. They have repeatedly been proposed as causes of diarrhea in humans, but have failed to cause illness in human volunteers. Transmission via food and water is proposed, but unproven.

C. *Plesiomonas shigelloides*

These bacteria are straight, Gram-negative rods, 0.8–1.0 μm × 3.0 μm, occurring singly, or in pairs or short chains, and motile by polar flagella. They are common in aquatic environments worldwide and are found in humans with watery diarrhea or with septicemia, often accompanied by meningitis. Their role in causing diarrhea has not been proved, but two waterborne outbreaks have been reported in Japan. Transmission via food has been suspected for some time, but not conclusively proven.

D. *Pseudomonas aeruginosa*

The pseudomonads are straight or slightly curved Gram-negative rods, 0.5–1.0 μm × 1–5 μm, aerobic, non-spore forming, and motile by one or more polar flagella. *P. aeruginosa* is widespread in the environment, especially in water. It is an opportunistic pathogen, causing a variety of infections, most of which are superficial. It is alleged to cause gastroenteritis in humans if ingested in large numbers. Transmission via food and ingested water is proposed, but unproven.

V. Concept of Emerging Foodborne Pathogens

The concept of emerging pathogens is so widely accepted that a journal is published especially to describe them. Although agents featured in *Emerging Infectious Diseases* may be foodborne, many are not. There most certainly are emerging foodborne pathogens, if one looks at the recorded causes of foodborne disease over time. In 1952, all recorded foodborne disease in the US was attributed to infections with *Salmonella typhi* or other salmonellae and *Shigella* spp., or to intoxications from *Clostridium botulinum* or *Staphylococcus aureus* (i.e. only four or five pathogens). By 1982, 17 microbial causes of foodborne disease were reported in the US. People with long careers in food safety can often recall the year in which a new agent was first described. All the same, not every agent implicated as a cause of foodborne disease has been incriminated after longer investigation. The advent of 'molecular fingerprinting' methods has helped in at least two ways. Sometimes, apparently identical agents have been isolated from humans and from food or water during an outbreak investigation, but found to have different fingerprints. Alternately, the same methods have permitted incrimination of other microbes in foodborne outbreaks when the rest of the evidence was not convincing. It must also be recognized that foodborne disease is usually suspected in instances of acute onset after relatively brief incubation periods and with gastroenteritis as the predominant symptom. Chronic illnesses, those affecting parts of the body other than the digestive tract, and those causing sporadic illnesses rather than obvious outbreaks are much harder to recognize.

A. 'New' Agents

Some pathogens that are now widely recognized problems in food were not known in, say, 1980. The first recorded outbreaks of *E. coli* O157:H7 hemorrhagic colitis occurred in 1982. Now, this agent is the cause of destruction of large quantities of food in the US, owing to the 'zero-tolerance' policy of the US Department of Agriculture. It seems clear, given the worldwide distribution of *E. coli* O157:H7, that it had been in circulation for many years, but somehow had escaped notice. It is also noteworthy that the types of *E. coli* that were known to cause human disease (and sometimes been transmitted via food or water) before 1982 were essentially human-specific, whereas *E. coli* O157:H7 is a zoonosis.

The largest outbreak in the US (in terms of number of people affected) in 1982 was caused by the Norwalk virus – an agent hardly known as a cause of foodborne disease until then. Now, the US Centers for Disease Control and Prevention (CDC) says that 67% of foodborne illnesses in the US are caused by the Norwalk virus and its close relatives.

B. Agents not Previously Recognized as Foodborne

The emetic form of *Bacillus cereus* food poisoning was attributed to *Staphylococcus aureus* at one time, on the basis of clinical picture, even though *B. cereus* was isolated from the rice vehicle and *S. aureus* or its toxin was not. Now gastroenteritis, with a probable rice vehicle, is usually assumed to be due to *B. cereus*.

The agent of Johne's disease in cattle, *Mycobacterium avium* subspecies *paratuberculosis* (MAP), has been proposed to be the cause of Crohn's disease in humans. Crohn's disease is a degeneration of the bowel, from which MAP has been isolated with varying frequency in a number of published studies. If MAP causes Crohn's disease, it may or may not be transmitted via food. Some studies have reported that the agent can withstand pasteurization in milk, but other reports disagree.

Aflatoxin has probably occurred in improperly stored grains since before grains were domesticated. Because the toxin usually causes illness only after chronic exposure in humans, aflatoxin was not recognized as a foodborne threat until its presence in peanuts killed turkeys in the early 1960s. Eventually, it was found to be a serious threat to human health in grains and pulses and to pass into the milk of cattle fed contaminated grain.

C. Agents in Food, not Previously Recognized as Pathogenic

Listeria monocytogenes was thought to be rarely pathogenic for humans, and its presence in food was ignored, until a significant outbreak of listeriosis was associated with coleslaw in Canada. After an outbreak associated with soft cheese in California, *L. monocytogenes* became a greatly feared pathogen. Now, its presence in ready-to-eat foods leads to the recall and destruction of large quantities of such products in the US.

Bovine spongiform encephalopathy (BSE, sometimes called 'mad cow disease') was recognized and battled in the UK for approximately 10 years before there was any suspicion that the prion disease was transmissible to humans. By November 2001, there had been over 100 human illnesses, called variant Creutzfeldt–Jakob disease (vCJD), most of them in the UK, all of which seem to be resulting in death. Although the exact mode of transmission of BSE to humans has not yet been determined, no probable alternative to the food vehicle has been suggested. It appears that portions of cattle that were incubating or ill with BSE – most likely brain or spinal cord – were eaten and resulted in vCJD.

Table 11.1 US subpopulations especially susceptible to foodborne disease[a]

Category	Number	Year
Age >65	29 400 000	1989
Pregnant women	5 657 900	1989
Newborns	4 002 000	1989
Cancer outpatients	2 411 000	1986
Nursing home residents	1 553 000	1986
AIDS patients	135 000	1993
Organ transplant patients	110 270	1981–1989

[a] Adapted from Council for Agricultural Science and Technology (CAST) (1994, Table 3.1, p. 25).

Giardia lamblia (also called *G. duodenalis* or *G. intestinalis*), a protozoon more often transmitted via water than food, was regarded as a commensal until tourists returning to the US from Leningrad (now St Petersburg) in the then Soviet Union showed a highly distinctive form of diarrhea that was found to be caused by this agent. It is now recognized as a frequent cause of waterborne (and sometimes foodborne) disease in North America, and probably elsewhere.

D. Problems with Koch's Postulates

A problem with implicating agents such as *L. monocytogenes* or *Toxoplasma gondii* as significant foodborne pathogens is that most infections are mild or asymptomatic. There are, however, increasing numbers of people in affluent countries whose resistance is subnormal (Table 11.1). In the cases of *L. monocytogenes* or *T. gondii*, the most extreme effects occur during pregnancy, resulting in stillbirths or birth defects. The great variation in human susceptibilities has also complicated the fulfillment of Koch's postulates. It is now recognized that some people have conditions that put them at special risk of (*inter alia*) serious consequences of certain low-grade, foodborne pathogens. At the same time, increasingly strong restrictions have been placed on the use of human subjects in research. This makes it essentially impossible to conduct human volunteer studies to incriminate low-grade, human-specific pathogens that affect only people listed in Table 11.1. Similar problems occur in trying to determine dose–responses of both infectious agents and intoxicants in compromised populations.

E. 'Old' Agents, Newly Named

Campylobacter is a genus that was formerly part of the genus *Vibrio*. Even more recently distinguished from *Campylobacter* are the genera *Arcobacter* and *Helicobacter*, which may sometimes be foodborne. Although some of the species in these genera may be entirely newly discovered, others have been recorded previously under other names, whereby comparing their former and present roles in disease causation becomes difficult. The further development of new taxonomic criteria ensures that this problem will continue.

VI. Summary

In addition to 'emerging' foodborne pathogens, some seem to be disappearing, at least in affluent countries. Measures such as machine milking and pasteurization of milk have largely ended the continuing problem of infectious agents from milkers being transmitted to those who drank the milk. Some agents are still transmitted via raw milk; these are largely zoonoses. Some, such as *Brucella abortus*, *B. melitensis*, and *Mycobacterium bovis*, have been significantly reduced by on-farm measures. Some agents that occur in foods may threaten only especially vulnerable populations (see Table 11.1), and some alleged pathogens may be virtually harmless.

Bibliography

Chin, J. D. (ed.) (2000). 'Control of Communicable Diseases Manual', 17th ed. American Public Health Association, Washington, DC.

Claridge, J. E. and Spiegel, C. A. (1995). *Corynebacterium* and miscellaneous gram-positive rods, *Erysipelothrix*, and *Gardnerella. In* 'Manual of Clinical Microbiology', 6th ed. (P. R. Murray, E. J. Baron, M. A. Pfaller, F. C. Tenover and R. H. Yolken, eds), pp. 357–378. American Society for Microbiology Press, Washington, DC.

Council for Agricultural Science and Technology (CAST) (1994). Foodborne Pathogens: Risks and Consequences. Task Force Report No. 122, CAST, Ames, IA.

Facklam, R. R. and Sahm, D. F. (1995). *Enterococcus. In* 'Manual of Clinical Microbiology', 6th ed. (P. R. Murray, E. J. Baron, M. A. Pfaller, F. C. Tenover and R. H. Yolken, eds), pp. 308–314. American Society for Microbiology Press, Washington, DC.

Gilligan, P. H. (1995). *Pseudomonas* and *Burkholderia. In* 'Manual of Clinical Microbiology', 6th ed. (P. R. Murray, E. J. Baron, M. A. Pfaller, F. C. Tenover and R. H. Yolken, eds), pp. 509–519. American Society for Microbiology Press, Washington, DC.

International Commission on Microbiological Specifications for Foods (1996). 'Microorganisms in Foods. 5. Characterization of Microbial Pathogens'. Blackie Academic & Professional, London.

Janda, J. M., Abbott, S. L. and Carnahan, A. M. (1995). *Aeromonas* and *Plesiomonas. In* 'Manual of Clinical Microbiology', 6th ed. (P. R. Murray, E. J. Baron, M. A. Pfaller, F. C. Tenover and R. H. Yolken, eds), pp. 477–482. American Society for Microbiology Press, Washington, DC.

Johnson, E. A. (1990). Infrequent microbial infections. 'Foodborne Diseases' (D. O. Cliver, ed.), pp. 259–273. Academic Press, San Diego, CA.

Kirov, S. M. (2001). *Aeromonas* and *Plesiomonas* species. *In* 'Food Microbiology: Fundamentals and Frontiers', 2nd ed. (M. P. Doyle, L. R. Beuchat and T. J. Montville, eds), pp. 301–327. American Society for Microbiology Press, Washington, DC.

Mead, P. S., Slutsker, L., Dietz, V., McCaig, L. F., Bresee, J. S., Shapiro, C., Griffin, P. M. and Tauxe, R. V. (1999). Food-related illness and death in the United States. *Emerg. Infect. Dis.* **5**, 607–625.

Moyer, N. P. and Holcomb, L. A. (1995). *Brucella. In* 'Manual of Clinical Microbiology', 6th ed. (P. R. Murray, E. J. Baron, M. A. Pfaller, F. C. Tenover and R. H. Yolken, eds), pp. 549–555. American Society for Microbiology Press, Washington, DC.

Nolte, F. S. and Metchock, B. (1995). *Mycobacteria. In* 'Manual of Clinical Microbiology', 6th ed. (P. R. Murray, E. J. Baron, M. A. Pfaller, F. C. Tenover and R. H. Yolken, eds), pp. 400–437. American Society for Microbiology Press, Washington, DC.

Olsen, S. J., MacKinnon, L. C., Goulding, J. S. and Slutsker, L. (2000). Surveillance for foodborne disease outbreaks – United States, 1993–1997. *Morbid. Mortal. Weekly Rep. Surveill. Summ.* **49**(SS01), 1–62.

Olson, J. G. and McDade, J. E. (1995). *Rickettsia* and *Coxiella. In* 'Manual of Clinical Microbiology', 6th ed. (P. R. Murray, E. J. Baron, M. A. Pfaller, F. C. Tenover and R. H. Yolken, eds), pp. 678–685. American Society for Microbiology Press, Washington, DC.

Palumbo, S., Abeyta, C., Stelma, G., Jr, Wesley, I. W., Wei, C., Koberger, J. A., Franklin, S. K., Schroeder-Tucker, L. and Murano, E. A. (2001). *Aeromonas*, *Arcobacter*, and *Plesiomonas. In* 'Compendium of Methods for the Microbiological Examination of Foods', 4th ed. (F. P. Downes and K. Ito, eds), pp. 283–300. American Public Health Association, Washington, DC.

Viruses

12

Dean O. Cliver and Suzanne M. Matsui

I. Introduction

A. History of Food Virology

Some diseases now known to be viral were recorded as foodborne before the nature of viruses was understood. In particular, outbreaks of poliomyelitis associated with drinking raw milk were recorded before the poliomyelitis viruses (polioviruses) had been isolated. The polioviruses only infect humans and other primates and are shed in feces; the problem of milk transmission had been solved by improved sanitation and almost universal pasteurization of milk before vaccines became available and eradicated poliomyelitis in the developed world.

Similarly, an outbreak of hepatitis A was observed to be associated with eating raw oysters before the agent had been isolated or vaccines developed. Now, the hepatitis A virus is also known to be essentially human-specific and shed in feces. Because people still eat raw oysters, the disease continues to be transmitted in this way, although stepped-up surveillance of oyster-growing waters for fecal contamination has improved the situation to some degree.

Recent years have seen a growing awareness that viruses play a significant role as causes of foodborne disease. A major event was the estimate, published by the US Centers for Disease Control and Prevention (CDC) that 67% of foodborne illnesses in

Table 12.1 Classification of most foodborne viruses		
Particle diameter (nm), nucleic acid strands	**Nucleic acid type**	
	RNA	**DNA**
25–35, single	Astroviruses	Parvoviruses[a]
	Caliciviruses	
	Picornaviruses	
	Hepatitis E virus	
70–90, double	Reoviruses	Adenoviruses
	Rotaviruses	

[a] Rarely or never foodborne.

the US are caused by Norwalk-like viruses. Hepatitis A virus and rotaviruses were also said to be frequent causes of foodborne disease.

B. Special Common Properties of Foodborne Viruses

Viruses are highly evolved parasites that multiply (replicate) only within suitable living host cells. This means that viruses cannot replicate in food or in the environment. The particulate forms in which the viruses are transmitted are too small to be visible even with microscopes that can view bacteria, so their appearance remained a mystery until the electron microscope was invented. Many important features of virus structure are still under investigation. A virus particle may comprise either RNA or DNA that is single- or double-stranded, a coat of protein, and sometimes an outer envelope of lipid-containing material. Most of the foodborne viruses contain RNA (usually single-stranded) and lack an envelope outside their protein coats (Table 12.1). In addition to the classification criteria shown in the table, each virus group (genus) has its own characteristic genetic organization, which is to say the order in which the genetic elements are arranged along the RNA strand.

Viruses are produced by host cells, rather than replicating independently. The viral particle must encounter an appropriate receptor on the plasma membrane of a potential host cell in order to begin the infectious cycle. None of the RNA viruses transmitted via food is known to include DNA in its replicative cycle, so no reverse transcription is required. On the other hand, synthesis of RNA on an RNA template requires enzymes that the host cell does not possess, so the virus genome must specify an RNA-dependent RNA polymerase. Steps in the replicative cycle of the picornaviruses, which have been relatively well studied, are summarized in Table 12.2.

Viruses transmitted via foods are produced in the human body and shed in feces or, more rarely, in vomitus. They infect after being ingested, presumably first in the cells of the intestinal lining. Resistance to acidity such as that of the stomach is a general property of foodborne viruses. Molluscan shellfish from fecally contaminated growing waters are frequent vehicles of viruses; it appears that the shellfish concentrate the virus efficiently during their filter-feeding activities and yield it only slowly if subjected to depuration.

Table 12.2	Steps in picornavirus replication

Stage	Events
1. Adsorption	Randomly diffusing virus particle encounters a *receptor* with complementary charge array on a cell's plasma membrane and attaches.
2. Engulfment	The host cell's plasma membrane wraps itself completely around the viral particle, whereby the viral particle is now inside the cell's cytoplasm.
3. Uncoating	The coat of protein is removed, thus liberating the single strand of RNA into the cell's cytoplasm.
4. Translation	The viral RNA is translated by the cell's enzymes and ribosomes into a large peptide that comprises some polypeptides that will make up the coat of the progeny virus and others that are essential to viral replication inside the cell; a first event is the cutting, by a proteolytic portion of the large peptide, of the large peptide into functional units. Later translation produces many more copies of the viral coat peptides than of the enzymes that participate in replication in the cell.
5. Transcription	One of the functional units that derives from the large peptide is an enzyme that causes RNA-dependent RNA synthesis; this leads to production of complementary ('minus') strands on the original viral RNA template, and eventually production of 'plus' strands – for incorporation into progeny virus – on the minus strand templates.
6. Maturation	As viral components accumulate in the cell's cytoplasm, they assemble themselves spontaneously into progeny viral particles comprising one molecule of single-stranded RNA (*ca* 7500 bases) coated by 60 copies of each of three or four different polypeptides. The three-dimensional array of positive and negative charges on the surface of the viral particle determines which receptors it can attach to, and provides the antigenic specificity to which the host's immune system responds.
7. Release	Progeny virus leaves the cell, often in small, membrane-bound clusters. The process may be prolonged – there is not one cataclysmic burst as is common in bacteria that have been infected by bacteriophages.
8. Cytopathology	Depending on the virus, the host cell may survive indefinitely, and even continue to produce progeny virus. However, some viruses cause the host cell to die promptly, in certain instances by blocking RNA synthesis on a DNA template, which the host cell requires to produce its own enzymes, etc.
9. Transmission	Virus liberated from host cells passes down the digestive tract and is shed in feces. If the virus is ingested by another person, perhaps with food or water, the cycle may begin again.

II. Hepatitis A and E Viruses

A. Agent of Hepatitis A

The hepatitis A virus (HAV) is a picornavirus, so the particle comprises a coat of protein around a single strand of plus-sense RNA and has a diameter of *ca* 28 nm, with no distinctive surface features. The copy number for coat protein polypeptides seems to be 60, as in other picornaviruses; however, the existence of the smallest or fourth polypeptide that is typically present in picornavirus coats is debated. For whatever reason, the HAV particle is exceptionally resistant to inactivation outside the body, especially by heat and drying. A result is that HAV in food can withstand processes that would inactivate other viruses, and indeed most vegetative cells of bacterial pathogens. Only one serotype of HAV is known worldwide; this has greatly facilitated the development of vaccines that could, diligently applied, lead to eradication of the disease.

B. Hepatitis A Disease

Infection begins with ingestion of the virus, possibly but not necessarily with food or water. It is reasonable to suppose that the first site of infection is the intestinal lining, but this has yet to be demonstrated. Somehow, virus reaches the liver and infects parenchymal cells there. Replication of HAV does not kill the liver cells; some of the progeny virus drains with bile into the intestine and is eventually shed with feces.

In time, the viral antigen present in the body evokes an immune response. Antibody of the IgM class is produced and cytotoxic T cells begin to destroy the infected cells in the liver. Because the liver is a vital organ, destruction of liver cells seriously disrupts the body's functions. Symptoms include fever, loss of appetite, nausea, and abdominal discomfort, often followed after a few days by jaundice. Infections in children under 5 years old tend to be mild or asymptomatic. Even adults sometimes have asymptomatic infections or mild illness without jaundice.

The virus circulates fairly briefly in the bloodstream, so that hepatitis A is less frequently transmitted by blood transfusions than other viral hepatitides. Fecal shedding typically begins 10–14 days before the onset of symptoms, and high levels of virus are shed during the incubation period. Illness, if any, begins 15–50 days (with an average of about 4 weeks) after the virus is ingested. Symptoms may persist for days or weeks, but death is rare. Some victims experience permanent loss of some liver functions. The immune response is durable and normally lasts for life. Immunity begins with production of IgM-class antibody, which declines after a few months. Long-term immunity is based upon IgG-class antibody, so only detection of IgM-class antibody is of diagnostic significance, indicating current or recent infection with HAV.

C. Transmission of Hepatitis A

As has been stated, HAV is shed exclusively in feces. Nonhuman primates are of no consequence in most situations, so the source is human feces shed during illness or, more likely, during the incubation period or an inapparent infection. However, outbreaks have certainly occurred from food handled by persons who were visibly jaundiced. Most feces produced by infected persons are disposed of via water-carriage toilets, so that the virus is consistently present in the sewage of all but the smallest communities. If this sewage is not properly treated and disinfected before discharge, it represents a significant source of virus contamination of drinking water and of shellfish in estuaries.

Direct, person-to-person fecal–oral transmission of HAV is perhaps most common. Fecal contamination also occurs when food is handled by an infected person without proper hand washing. Because this can happen at any point until the food is eaten, there is no way to protect against it absolutely. That is, proper cooking and other precautions can be undone by the unwashed hands of an HAV-infected person. Still, vehicles in outbreaks from this source are more often salads and sandwiches than thoroughly cooked foods.

D. Noteworthy Outbreaks

- People who ate raw clams in Shanghai, China, during 1988 year-end festivities experienced an outbreak of hepatitis A, apparently due to sewage contamination of the shellfish beds. Nearly 300 000 people were affected. An outbreak of gastroenteritis, which may have been viral, preceded the onsets of hepatitis.

- Over 200 people in a number of states acquired hepatitis A from frozen strawberries in 1997. The strawberries had been grown in Mexico and processed and frozen in San Diego, California; where and how the contamination occurred could not be determined. The outbreak drew special attention because many of those ill were children who had been served the strawberries as part of the US Department of Agriculture school lunch program.

- A young man who worked at a fast-food establishment in Northern California was responsible for setting up the salad bar each day. While he was incubating a case of hepatitis A, he evidently contaminated the salad materials on several days, whereby 36 diners contracted hepatitis A from eating at the restaurant on different occasions.

E. Hepatitis E Virus

It appears that only imported cases of hepatitis E occur in North America; however, the disease is very important in less affluent areas of the world and is often transmitted via drinking water, but perhaps not via food. The virus evidently bears some resemblance to the human caliciviruses (single-stranded RNA, diameter 32–35 nm) but has been removed from that taxon. A closely related virus occurs in swine in North America, but is not known to infect humans. Hepatitis E generally resembles hepatitis A, with three significant exceptions: the incubation period is slightly longer, with a range of 15–62 days and a mean that varies from 26 to 42 days on different occasions; the target age group are generally young adults, rather than children; and infections can be lethal, especially in pregnant women.

III. Norwalk-like Gastroenteritis Viruses

A. Agents

The original Norwalk virus was derived from a gastroenteritis outbreak at a school in Norwalk, Ohio, in 1968. Since then, many similar outbreaks have been recorded, and many related agents described. The viruses have been assigned to the calicivirus family and are collectively called Norwalk-like viruses (NLV) or small round structured viruses (SRSV). Viral particles comprise a single molecule of positive-strand RNA coated with many copies of a single capsid protein. The diameter is *ca* 27–28 nm, and the surface has a defined structure by electron microscopy (EM). A great many strains of NLV, constituting at least two genogroups, are known. There is a genogroup of human caliciviruses, the Sapporo-like viruses (SLV), which cause gastroenteritis in humans. They are somewhat larger (*ca* 35 nm in diameter) and display the characteristic indentations that were perceived by some as cup-shaped (calici from Greek *kalyx*, a cup). The human caliciviruses are essentially human-specific; no *in-vitro* cell culture host or animal model has yet been identified.

B. Disease

The disease is marked by a 12–48-hour incubation period (mean 36 hours), which is relatively short for a virus. Symptoms consist principally of nausea, vomiting (more prevalent in children), and diarrhea (more prevalent in adults), although various aches and low-grade fever may also occur. The illness is usually over in 12–60 hours, but fecal shedding of the virus sometimes continues for more than a week. During illness, intact viral particles may be found both in the diarrheal stool and in the vomitus. Frequently, the amount of virus shed is under the limit of detection by direct EM, and immune EM may be needed to identify the viral particles. Infection leads to production of antibody against the virus, but immunity apparently does not last much more than a year. In some studies, people with antibody against a NLV were more susceptible than those without antibody.

C. Transmission

Only small numbers of NLVs are required to cause infection (less than 100 particles). They can pass through groups (e.g. children in daycare centers) very rapidly. These may be the only foodborne viruses shed in vomitus; fecal shedding, again, occurs during illness and sometimes continues for a week or more. Although the virus is evidently not as durable in the environment as HAV, predominant vehicles are, similarly, mollusks and uncooked, ready-to-eat foods. Because these viruses are very contagious and immunity is not durable, attack rates in common-source outbreaks tend to be high. The NLVs often produce secondary cases – those who have been infected by eating contaminated food pass their infections to other people who have not eaten the food.

D. Noteworthy Outbreaks

- In 1982, a baker's assistant who worked at a large bakery in the vicinity of Minneapolis and St Paul, Minnesota, prepared a large batch of butter-cream frosting while experiencing diarrhea. The frosting was served on various pastries and transmitted Norwalk gastroenteritis to at least 3000 patrons. Although the individual illnesses were brief, large numbers of teachers and hospital staff were affected, which significantly disrupted activities at local schools and hospitals.
- A North Yorkshire, UK, cook vomited into a sink and cleaned it out with a sanitizing agent that was used in the restaurant where he worked. The sink was used to prepare potato salad the next day, and at least 47 people who ate the salad at a wedding reception contracted viral gastroenteritis.
- The Duke University (North Carolina) football team ate a box luncheon and then flew to Florida State University (Tallahassee, Florida) for a game the next day. Of those who ate the box lunch, at least 43 (62%) were stricken with Norwalk viral gastroenteritis – many during the game. Another 11 people among the Duke party, and 11 members of the Florida State team, became ill the following day as secondary cases. The two teams had no contact off the playing field and had shared no food or beverages.

- In 1993, a multistate outbreak of NLV gastroenteritis was traced to oysters harvested from a bed near where an ill oyster harvester had disposed of contaminated raw sewage. Despite a large dilution factor, the harvested oysters contained enough NLV to transmit disease. Sequence analysis of the virus found in infected individuals linked the outbreaks with the contaminated oysters.

IV. Other Gastroenteritis Viruses

A. Agents

Two further groups of viruses that cause food-associated gastroenteritis are the astroviruses and the rotaviruses (Table 12.1). The astroviruses are about the size of the NLVs and picornaviruses, but are distinguished by surface features that resemble five- or six-pointed stars. Like the other small foodborne viruses, they contain positive-sense single-stranded RNA. Human and most animal astroviruses cause acute gastroenteritis in their respective hosts, but avian astroviruses have been observed to cause both intestinal and extraintestinal infection. All eight known serotypes of human astrovirus have been adapted to growth in cell culture.

The rotaviruses contain double-stranded RNA that is divided into eleven segments. The coat protein comprises a double layer of 70–75 nm diameter, the outermost of which must be removed by proteolysis before the viral particle can enter a host cell. The particle is perceived as looking like a wheel (rota-) in electron micrographs. Many serotypes, assigned to several groups, are known. The majority of human illnesses appear to be caused by members of serogroup A, but common-source outbreaks may more often involve members of other serogroups. There are human and animal strains of rotavirus; human and animal reassortant viruses (containing certain RNA segments from one parent virus and other RNA segments from the other parent virus) may cause human infection and contribute to the diversity of rotavirus infection. Human rotaviruses found in clinical specimens have been difficult to propagate in cell culture.

B. Diseases

The gastroenteritis caused by astroviruses differs from that of the NLVs in having a somewhat longer incubation (up to 4 days) and duration (sometimes a week) and in having vomiting as a less prominent feature. Astroviruses also appear to cause a milder illness with less dehydration than seen with rotavirus. Serological surveys in the UK have shown that infections usually occur in the first few years of life. Immunity is probably relatively durable, although outbreaks have been reported in populations expected to be immune (see Section IV.D).

Rotaviruses frequently affect young children throughout the world. In developing countries, they are a significant cause of child mortality. The illness involves vomiting, fever, and watery diarrhea beginning after 24–72 hours of incubation and lasting

4–6 days, on average. Shedding is said seldom to continue beyond the eighth day of infection, except in the immune impaired. A recent study from Japan indicates that symptomatic rotavirus infection occurs commonly among healthy adults, challenging the presumption of durable immunity.

C. Transmission

Both the astroviruses and the rotaviruses are apparently transmitted by a fecal–oral route, with vehicles such as food and water occasionally involved. Although person-to-person transmission may predominate, vehicular outbreaks have been recorded on various occasions.

D. Noteworthy Outbreaks

- Contaminated food from a common supplier apparently transmitted astrovirus to >4700 students and teachers at ten primary and four junior high schools in Katano City, Osaka, Japan. Among those suffering from gastroenteritis, astrovirus was found in the stools of some, confirmed by solid-phase immune electron microscopy and isolation in cell culture. Some of these same patients showed antibody production against a reference astrovirus strain.
- Eating delicatessen sandwiches in a university cafeteria in the District of Columbia, evidently led to rotavirus gastroenteritis in at least 108 students in March and April of 2000. Two cooks at the dining hall were ill, but their illnesses occurred during the outbreak, rather than before it.

V. Other Viruses and Food

A. Tick-borne Encephalitis Virus

The tick-borne encephalitis virus is very much an exception among viruses transmitted via food. The virus is enveloped with a lipid-containing layer and contains RNA. A tick biological vector transmits the virus to dairy animals – principally goats, given the terrain, but also sheep and cows, in Slovakia and perhaps contiguous areas. Virus shed in the milk of these animals will infect humans, if they drink the milk unpasteurized, and cause a serious encephalitis. Fortunately, outbreaks have now become very rare.

B. Other Enteric Viruses that May be Foodborne

Several other kinds of human viruses are transmitted by the fecal–oral route (though they are not primarily associated with gastroenteric syndromes) and might thus be transmitted via food or water. The three types of viruses that caused poliomyelitis (polioviruses), which has now been eradicated in much of the world by vaccination, are members of a larger group of human *enteroviruses*. The enteroviruses are members of the picornavirus family and include, in addition to the polioviruses, other groups called coxsackieviruses A and B and echoviruses, as well as enterovirus types with numbers after their names. These cause a variety of illnesses, including meningitis and encephalitis; members of the various groups have been implicated in rare outbreaks of foodborne illness.

The reoviruses were discovered before the rotaviruses and resemble them on a slightly larger scale (Table 12.1). Although they certainly infect humans and are transmitted by a fecal–oral cycle, they are not significantly associated with human illness, enteric or other. The adenoviruses are the size of reoviruses, but contain double-stranded DNA and show a classic icosahedral shape. Most adenoviruses are associated with respiratory infections, but serotypes 40 and 41 cause gastroenteritis in humans and are apparently transmitted by the fecal–oral route. Food-associated adenovirus outbreaks are apparently unknown.

C. Bovine Spongiform Encephalopathy Prions

Clearly, prions are not viruses; however, this seemed the most appropriate place in this book to mention the disease transmitted from cattle that have bovine spongiform encephalopathy (BSE) to humans, as variant Creutzfeldt–Jakob disease (vCJD). The discussion will be kept as brief and simple as possible because, although the pathology in animals has been reasonably well studied, transmission from cattle to humans has only been inferred, and a great deal remains to be learned about how this took place.

Prions are fairly small polypeptides that occur in the brain and spinal cord of many animal species, possibly with no essential function. Each species has its own amino acid sequence (often more than one is possible per species), which leads to a normally folded configuration. Some also have an abnormally folded configuration that makes them resistant to proteases, whereby they may accumulate in specific parts of the brain as deposits that can be seen microscopically and lead to the formation of microscopic voids. These voids make the brain look like a sponge, hence the name 'spongiform encephalopathy'. This degeneration of the brain is accompanied by certain neurological abnormalities that culminate in death. The exact patterns and symptoms differ among humans and animal species, and some species are susceptible to more than one spongiform encephalopathy or prion disease. These diseases are often transmissible, in that an abnormally folded prion, on contact with a normal prion, can evidently transmit its abnormal configuration. Although transmissible spongiform encephalopathies (TSEs) have been known for some time, most of these have not been transmissible across species. The species barrier probably depends on the extent to which the amino acid sequence of the possible recipient's prions leads to folding that resembles that of the donor's prions.

When an outbreak of BSE occurred in the UK, beginning in *ca* 1986, it was viewed as an animal health problem, and control was undertaken by preventing the feeding of ruminant-derived meat-and-bone meal (MBM – thought to contain prions from the brains of BSE cattle, or even from sheep with scrapie, another TSE) to cattle. However, as the BSE outbreak was showing signs of coming under control in the UK, an outbreak of feline spongiform encephalopathy (FSE) was recognized in domestic and zoo cat species; this was followed by recognition that humans were also affected and were showing vCJD, a variant of another spongiform encephalopathy that had been recognized earlier and was only transmissible by certain tissue transplants. Meanwhile, the feeding of MBM from the UK in other European countries had led to the spread of BSE to these countries as well.

The BSE prions apparently do not occur in voluntary muscle nor in milk; they are probably absent from most cattle tissues other than brain and spinal cord until quite late in the incubation period, when they can be detected in the distal ileum segment of the small intestine. The prions are apparently heat-resistant to the extent that no cooking process normally applied to food will inactivate them, so control to date has been based on excluding bovine brain and spinal cord from the human food chain. Cattle at least 30 months of age at slaughter, in much of Europe, must now have their brains tested for BSE before the carcass can be released as food. More than 100 vCJD cases had occurred in the UK by November 2001, plus perhaps four in France and one or two elsewhere. Although the source of the infection has apparently been removed, present ignorance of the incubation period of vCJD makes it impossible to predict how many people will ultimately be affected. Meanwhile, the impact on animal agriculture worldwide, but particularly in Europe, has been enormous. Central nervous system tissue is prohibited in beef in many countries other than the US, whether or not BSE is thought to be present.

VI. Detection and Monitoring

Given that viruses are significant causes of foodborne disease, it is reasonable to aspire to methods of detecting them in foods. This presents special problems because viruses are likely to be present at low levels (yet still a threat to consumer health) and heterogeneously distributed in food, and cannot be enriched from a sample, as bacterial pathogens often are. Furthermore, the most important of the foodborne viruses, HAV and the NLVs, do not have laboratory hosts that are useful in detecting these agents in foods. Only general principles will be addressed here, as space does not permit detailed instructions in detecting foodborne viruses. As will be explained, detection methods are available and improving, but they are far from routine or standard. Most often, outbreaks of foodborne virus disease have been confirmed by diagnosing the infection among the victims, rather than demonstrating the virus in food.

Because of the difficulty of detecting foodborne virus, these methods are not suited to routine surveillance of food for virus contamination. Detection of foodborne virus is a reasonable undertaking in the course of an outbreak investigation, but may also be appropriate in prevalence surveys and in studies of processes to eliminate viruses from foods. Most of these applications pose the difficult problem of differentiating infectious from inactivated virus when no laboratory host is available in which to do infectivity testing.

A. Sampling and Sample Processing

Samples are selected for testing, either because of a general perception of hazard or because a food history has implicated these particular foods in an outbreak that is under investigation. Because of the long incubation period of hepatitis A (average 4 weeks), foods representative of what was eaten that led to a hepatitis outbreak may well be unavailable for testing by the time the etiology is established. Otherwise, the most common problem with samples from outbreak investigations is that they have been exhausted in bacteriological testing or allowed to deteriorate badly before virological testing was considered.

The size of the sample may be governed by what is available, but also by the test method that is to be used. The choices, all largely adaptations of methods that have been used to diagnose viral diseases, include transmission electron microscopy, cell culture infectivity, serologic reactions, and reactions with the viral nucleic acid (so-called molecular methods) or combinations of these. It is a very different matter preparing a sample for nucleic acid-based detection than for detection by cell culture infectivity – possible volumes of inoculum are much greater for cell culture methods than for most others. For example, the large magnification factors attained with the electron microscope greatly limit the volume that can be examined. Since no enrichment of virus can be done, it is important that samples be extracted carefully (to separate virus as completely from food components as possible) and then concentrated with minimum loss of virus. Samples in various studies have usually ranged from 10 to 100 g, largely on the basis of expediency.

Several means of liquefaction of solid food samples are available. If the virus is inside the food (e.g. mollusks), the soft tissue may be homogenized or the digestive tract may be first dissected out and then extracted. If virus is likely to be located on the surface of the food, one tries to dislodge the virus and leave most of the food solids behind. Foods contaminated in handling are likely to have only superficial virus, but few samples of this kind become available after an outbreak. In any case, superficial contaminants can be dislodged by mechanical pummeling or shaking, leaving most of the food solids intact.

Food solids can be removed from the sample suspension by centrifugation or filtration, perhaps aided by additives that promote coagulation of the solids. Virus in the extract can be concentrated by chemical precipitation, by ultracentrifugation or ultrafiltration, or by capture with homologous antibody – possibly attached to paramagnetic beads. The intention is to reduce the volume of suspension to be tested, with minimal loss of virus. Some food components may be concentrated along with the virus and need to be further treated or counteracted with specific additives, again depending on the detection method to be employed. Removal of the food's bacterial or fungal flora may also be required, though cell cultures, if used in testing, can sometimes be protected from these contaminants by simply adding antibiotics to the medium.

B. Available Detection Methods

The original detection methods for foodborne and waterborne viruses were generally based on the infectivity of the viruses for cell cultures. Eventually, it became clear that the viruses most often transmitted by vehicles, based on the epidemiological record, were not detectable by available cell culture types; but the quest for susceptible cell lines continues.

Electron microscopy has been an important tool in diagnosis and characterization of viral diseases. In addition to lack of sensitivity for testing food and water samples, it also suffers from the limitation that a great variety of viruses look just alike. This can sometimes be remedied by using a known antiserum to collect or agglutinate the viruses on the electron microscope grid.

Serological methods of virus detection depend on capture of the virus by antibody, followed by magnification by electron microscopy, by amplification by enzyme immunoassay, or by nucleic acid analysis. Serological methods, of themselves, have tended not to be sufficiently sensitive for use in detecting foodborne viruses; most serological methods used to diagnose viral diseases are based on detection of the ill person's antiviral antibody (often of the IgM class) with viral antigen prepared in advance. Food testing offers no counterpart of this approach.

Nucleic acid-based methods began with labeled probes that had specific nucleotide sequences that could couple with viral nucleic acid. These were highly specific, depending on the length of the probe, but not very sensitive. The polymerase chain reaction (PCR) offered greater sensitivity but, given that foodborne viruses usually contain RNA rather than DNA, required that viral RNA first be reverse-transcribed (RT) to permit application of the method. Several adaptations and combinations of RT-PCR with other methods are now available, and offer both great sensitivity and specificity. Remaining problems with nucleic-acid methods are their susceptibility to interference by food components with both RT and PCR, and the inability of the methods to distinguish between infectious and inactivated virus. Viral RNA, if unprotected, is very vulnerable to environmental ribonuclease; on the other hand, the RNA will withstand quite high temperatures, so that the viral RNA is not disrupted by temperatures as high as 95° or 99°C, which will inactivate viruses almost instantaneously. Several ingenious techniques are under investigation to avoid false-positive test results with inactivated viruses.

C. Prospects for Monitoring via Indicators

Since viruses transmissible via foods are typically shed with feces, it is reasonable to try to circumvent the problems of detecting foodborne viruses by instead seeking indicators of fecal contamination. Indicators explored for this purpose have included bacteria, human viruses, and bacteriophages. Since foodborne viruses are human-specific, the indicator should show that fecal contamination of human origin is present, but this is seldom possible. Bacterial indicators range from coliforms, fecal (thermotolerant) coliforms, and enterococci to *Escherichia coli* or *Bacteroides fragilis*. Their specificity as fecal indicators varies, and they are consistently lacking in specificity for human feces. However, the greatest problem with bacteria as indicators is that they do not die at the rates of virus inactivation, and they may even multiply in the environment (including food), so their correlation with the presence of virus has been found to be very poor. Human enteroviruses, such as the vaccine polioviruses, are often present in community sewage and are specific indicators of human fecal contamination, but they present their own detection problems and are unlikely to be present in feces from a single individual who contaminated a food in handling. The bacteriophages considered in this indicator role include those that infect *E. coli* (coliphages), either via receptors on the cell wall (somatic coliphages) or on the F-pilus (and happen incidentally to contain single-stranded RNA – FRNA coliphages, as do most foodborne viruses), as well as phages that infect *B. fragilis*. All bacteriophages share the advantages of rapid readout times and relatively inexpensive detection methods, but none has really proven well correlated with the presence of human enteric viruses in the foods context. Studies on this problem are continuing as well, in the hope of finding a test method that could be applied routinely in monitoring mollusks, if not other foods.

VII. Prevention

A. Preventing Contamination of Foods

Since viruses transmitted via foods are shed almost exclusively in human feces (with the minor exception of the NLVs shed in human vomitus), preventing contamination is simply a matter of keeping human feces out of food. Of course, one is obliged to note that a great many nonviral foodborne diseases could be prevented on the same basis, but are not. Where contamination via water (e.g. of shellfish) is concerned, the challenge begins with the fact that our basic standard of community hygiene calls for water-carriage toilets. Every time a toilet is flushed to dispose of feces, a considerable volume of water becomes contaminated; when this sewage reaches the treatment plant, it is a challenge to remove or inactivate any virus that may be present before the wastewater is discharged to be used for another purpose or to enter an estuary in which mollusks are growing. Food contamination via feces-soiled hands is preventable by hand-washing, but the necessary cleaning does not always take place, owing to carelessness, cultural blocks, or unavailability of water for this purpose. The most extreme case is outbreaks of mollusk-associated viral disease where the waters were contaminated by direct fecal discharge from the harvesters. The mode of fecal contamination in the recent outbreak involving frozen strawberries has never been demonstrated. Vaccination can be very effective. Poliomyelitis, having been eradicated in the Americas, is no longer foodborne there. Eventually, the hepatitis A vaccines may be used to prevent infections of food workers or to eradicate the disease in the entire population.

B. Inactivation of Foodborne Viruses

Viruses cannot multiply in food, like some other pathogens. Therefore, the virus can only persist or be inactivated before the food is eaten. Virus on food surfaces or in water is accessible to chemical inactivation by strong oxidizing agents (e.g. chlorine) or by ultraviolet light. HAV withstands drying on surfaces reasonably well, but many other viruses are inactivated by drying. Virus within a food ordinarily can only be killed by heat: although ionizing radiation penetrates food well, the target size of viruses is so small that relatively large and costly irradiation doses are required. Viruses will gradually lose infectivity at any temperature above freezing, but inactivation of viruses within the shelf-life of most foods requires elevated temperatures. The heat resistance of most viruses is little greater than those of many nonsporing bacterial pathogens transmissible via foods. However, HAV is again an exception, in that as much as 1% of the virus may withstand pasteurization (72°C, 15 seconds) in milk. HAV contamination of milk is relatively unlikely, if milking is done by machine, but the observation shows that HAV will withstand thermal treatments that will kill most potentially foodborne vegetative bacterial cells.

VIII. Summary

Viruses are frequent causes of foodborne disease in the US, and probably in other countries as well, although data are harder to find. Essentially all of the foodborne viruses are shed in human feces and infect perorally. The great majority of these are RNA viruses, often small and containing single-stranded RNA. They infect via the intestinal lining, but some are transported to the liver (and occasionally other organs) before causing disease. Virus particles are submicroscopic in size and cannot multiply in foods. They have been known to contaminate foods at levels that caused thousands of illnesses, but most foodborne virus transmission probably results from contamination of small quantities of food, eaten by small numbers of people, by feces on unwashed human hands. It is often possible to detect viruses in food samples, but the difficulty is such that one must have a compelling reason for conducting the test. Washing hands and cooking foods are probably the two measures that contribute the most to preventing foodborne viral disease.

Bibliography

Cromeans, T., Nainan, O. V., Fields, H. A., Favorov, M. O. and Margolis, H. S. (1994). Hepatitis A and E Viruses. *In* 'Foodborne Disease Handbook: Vol. 2. Diseases Caused by Viruses, Parasites, and Fungi' (Y. H. Hui, J. R. Gorham, K. D. Murrell and D. O. Cliver, eds), pp. 1–56. Marcel Dekker, New York.

Cubitt, W. D. (1996). Historical background and classification of caliciviruses and astroviruses. *Arch. Virol.* (Suppl) **12**, 225–235.

Green, K. Y., Ando, T., Balayan, M. S., Berke, T., Clarke, I. N., Estes, M. K., *et al.* (2000). Taxonomy of the caliciviruses. *J. Infect. Dis.* **181**(Suppl. 2), S322–S330.

Greenberg, H. B. and Matsui, S. M. (1992). Astroviruses and caliciviruses: emerging enteric pathogens. *Infect. Agents Dis.* **1**, 71–91.

Grešíková, M. (1994). Tickborne encephalitis. *In* 'Foodborne Disease Handbook: Vol. 2. Diseases Caused by Viruses, Parasites, and Fungi' (Y. H. Hui, J. R. Gorham, K. D. Murrell and D. O. Cliver, eds), pp. 113–135. Marcel Dekker, New York.

Sattar, S. A., Springthorpe, V. S. and Ansari, S. A. (1994). Rotavirus. *In* 'Foodborne Disease Handbook: Vol. 2. Diseases Caused by Viruses, Parasites, and Fungi' (Y. H. Hui, J. R. Gorham, K. D. Murrell and D. O. Cliver, eds), pp. 81–111. Marcel Dekker, New York.

Smith, J. L. (2001). A review of hepatitis E virus. *J. Food Prot.* **64**, 572–586.

Parasites

13

J. P. Dubey, K. Darwin Murrell, and John H. Cross

I. Introduction

The chapter is divided into three main sections on the basis of the classes of foods that most often serve as vehicles for the parasites. Included are meatborne parasites, fishborne parasites, and other parasites disseminated in fecally contaminated food and water.

II. Meatborne Parasites

A. *Toxoplasma gondii*

Toxoplasma gondii is an intracellular coccidian protozoan parasite. Felids, including the domestic cat (*Felis catus*) and wild Felidae, are definitive hosts, and various warm-blooded animals including humans are intermediate hosts. As with other coccidian parasites, oocysts are excreted only by the definitive hosts, cats. Unsporulated noninfective oocysts (10×11 μm) are shed in feces of infected cats. Oocysts sporulate outdoors in cat feces within 1 or more days, depending on environmental conditions. Sporulated oocysts can survive in soil and elsewhere in the environment for many months, and they are resistant to freezing and drying. Each sporulated oocyst contains two sporocysts, in each of which there are four banana-shaped sporozoites (8×2 μm). Sporulated oocysts are infectious to virtually all warm-blooded hosts, including human beings.

After ingestion of sporulated oocysts in food or water, sporozoites are released from oocysts in the gut lumen and become tachyzoites. Tachyzoites ($6 \times 2\,\mu m$) multiply in many cells of the body by division into two zoites. Within 3–4 days, tachyzoites may become encysted in tissues and are called bradyzoites. Bradyzoites ($7 \times 2\,\mu m$) are enclosed in a thin, elastic wall, and the entire structure is called a tissue cyst. Tissue cysts are formed in many locations, but particularly in the central nervous system (CNS), striated and smooth muscles, and many edible organs. They are often elongated (up to $100\,\mu m$) in muscles and round (up to $70\,\mu m$) in the CNS. Tissue cysts may persist for the life of the host.

All hosts, including the definitive felid hosts, can become infected by ingesting tissue cysts. In the intermediate host, bradyzoites become tachyzoites within 18 hours of infection, and the tachyzoite–bradyzoite cycle is repeated. However, in the definitive host, the cat, bradyzoites give rise to a conventional coccidian cycle in the epithelium of the small intestine. This coccidian cycle consists of an asexual cycle (schizonts) followed by the sexual cycle (gamonts). The male parasite (microgamete) fertilizes the female (macrogamont) parasite, giving rise to an oocyst. The entire asexual and sexual cycle can be completed in the feline intestine within 3 days of ingestion of tissue cysts.

The ingestion of tissue cysts or oocysts can cause parasitemia in the mother and lead to infection of the fetus. Congenital *T. gondii* infections are frequent in humans, sheep, and goats.

Clinical signs and symptoms vary with hosts, mode of infection, immune status, and the organ parasitized. Congenitally acquired toxoplasmosis is generally more severe than is postnatally acquired toxoplasmosis. Chorioretinitis, hydrocephalus, mental retardation, and jaundice may all be seen together in severely affected infants, but chorioretinitis is the most common sequela of congenital infection in children.

Most infections in immunocompetent human beings are asymptomatic; however, toxoplasmosis can be fatal in immunocompromised people, such as those with acquired immunodeficiency syndrome (AIDS) or receiving immunotherapy for tumors or in connection with organ transplants. In the immunocompetent person, *T. gondii* rarely causes serious illness. Flu-like symptoms accompanied by lymphadenopathy often occur. Fever, headache, fatigue, muscle and joint pains, a maculopapular rash, nausea, and abdominal pain and loss of vision can occur.

Toxoplasma gondii is a major cause of abortion in sheep and goats, and causes mortality in many other species of animals. In ewes infected during pregnancy, lambs may be mummified, macerated, aborted, or stillborn or may be born weak and die within a week of birth. Severe congenital toxoplasmosis has been reported in cats, dogs, and pigs, but not in cattle or horses. Clinical toxoplasmosis also occurs in adult animals including dogs and cats but has not been documented in cattle or horses. Toxoplasmosis is a leading cause of mortality in Australian marsupials, especially in zoos.

Diagnosis is based on biologic, serologic, or histologic methods or by combinations of them. *T. gondii* can be isolated from patients by inoculation of laboratory animals (generally mice) and tissue cultures with appropriate material. Finding of *T. gondii* antibodies aids diagnosis. There are numerous serologic tests for detecting humoral antibodies. Diagnosis can also be made by finding *T. gondii* in host tissue removed by biopsy or at necropsy.

Sulfadiazine and pyrimethamine (Daraprim) are two drugs widely used for treatment of human toxoplasmosis. These drugs act on multiplying tachyzoites and have little or no effect on tissue cysts. Spiramycin is used in some countries as a prophylactic to minimize transmission of the parasite from mother to the fetus.

To prevent human infection, hands should be washed thoroughly with soap and water after handling meat. All cutting boards, sink tops, knives, and other materials that come in contact with uncooked meat should be washed with soap and water because the stages of *T. gondii* that occur in meat are killed by water. The meat of any animal should be cooked to 66°C before being eaten by humans or animals, and tasting meat while cooking it or while seasoning homemade sausages should be avoided. Pregnant women definitely should avoid contact with cat feces or litter, soil, and raw meat. Pet cats should be fed only dry, canned, or cooked food. Cat litter should be emptied every day, preferably not by a pregnant woman. Gloves should be worn while gardening. Vegetables should be washed thoroughly before eating because of the risk of contamination with cat feces.

Because most cats become infected by eating infected tissues, cats should never be fed uncooked meat, viscera, or bones, and efforts should be made to keep cats indoors to prevent hunting. Trash cans should be covered to prevent scavenging. Freezing meat overnight in a domestic freezer (-8 to -12°C) can kill most *T. gondii* tissue cysts. Cats should be spayed to control the feline population on farms. Dead animals should be removed promptly to prevent scavenging by cats.

B. *Sarcocystis* Species

Sarcocystis spp. are coccidian parasites with an obligatory prey–predator two-host cycle. The asexual cycle develops only in the intermediate host, which in nature is often a prey animal. Sexual stages develop only in a carnivorous definitive host. The definitive host becomes infected by eating sarcocysts containing infective zoites (bradyzoites). The bradyzoites transform into male and female gamonts in the small intestine, and oocysts are produced after fertilization. The oocysts sporulate in the lamina propria and contain two sporocysts, each with four sporozoites. The oocyst wall is thin and often breaks so that both sporocysts and oocysts are excreted in feces, usually 1 week after ingestion of sarcocysts. The asexual cycle occurs initially in vascular endothelium, later in cells in the bloodstream, and finally in muscles. Sarcocysts mature in about 1–2 months and become infectious for the carnivore host.

There are two known species of *Sarcocystis* for which humans serve as the definitive host, *S. hominis* and *S. suihominis*. Humans also serve as accidental intermediate hosts for several unidentified species of *Sarcocystis*. Symptoms in persons with intestinal sarcocystosis are different from those in persons with muscular sarcocystosis and vary with the species of *Sarcocystis* causing the infection.

Infection with *S. hominis* is acquired by ingesting uncooked beef containing *S. hominis* sarcocysts. *S. hominis* is only mildly pathogenic for humans. Symptoms include diarrhea and stomach ache. Infection with *S. suihominis* is acquired by eating undercooked pork. It is more pathogenic than *S. hominis*. It can cause nausea, vomiting, stomach ache, diarrhea, and dyspnea within 24 hours of ingestion of uncooked pork from naturally or experimentally infected pigs. Sporocysts were shed 11–13 days after ingestion of the infected pork. The diagnosis of intestinal sarcocystosis is easily made by fecal examination. As said earlier sporocysts or oocysts are shed fully sporulated in feces. It is not possible to distinguish one species of *Sarcocystis* from another by examination of sporocysts.

Sarcocysts have been found in striated muscles of human beings, mostly as incidental findings. The clinical significance of sarcocysts and the life cycles of sarcocysts of humans are unknown. The antemortem diagnosis of muscular sarcocystosis can only be made by histological examination of muscle collected by biopsy. The finding of immature sarcocysts with metrocytes suggests recently acquired infection. The finding of mature sarcocysts only indicates past infection.

There is no treatment known for *Sarcocystis* infection of humans. There is no vaccine to protect livestock or humans against sarcocystosis. Uncooked meat should never be eaten by humans.

C. *Trichinella* Species

Trichinella spiralis is a nematode parasite. It is the most frequent agent of human trichinellosis and was most frequently derived from domestic pork. The other species of *Trichinella* (*T. nativa*, *T. nelsoni*, *T. britovi*, *T. pseudospiralis*, *T. murrelli*, *T. papue*) are morphologically similar and can infect humans. All species of *Trichinella* have a direct life cycle with complete development in a single host. The host capsule surrounding the infective larva is a modified striated muscle structure called a nurse cell, which is digested away in the stomach when the infected muscle is ingested by the next host. The free larvae (L_1) then move into the upper small intestine and invade the columnar epithelial cells. Within 30 hours, the larvae undergo four molts to reach the mature adult stages, the males and females. After mating, the female begins shedding live newborn larvae (NBL), about 5 days postinfection; the NBL are early developmental forms of the L_1 stage. Adult worms may persist in the intestine of humans for many weeks. The NBL migrate throughout the body, via the blood and lymph circulatory system. Although the NBL may attempt to invade many different tissues, they only develop further in striated skeletal muscle cells. Here the NBL continue to grow and develop during the first 2 weeks of intracellular life until they reach the fully developed L_1 infective stage. The life cycle is completed when the host's infected muscle is ingested by a suitable host.

The ingestion of 500 or more larvae by a human is sufficient to cause clinical disease. In heavy infections, illness is reflected in gastrointestinal signs such as nausea, abdominal pain, and diarrhea. Coinciding with muscle invasion by NBL is acute muscular pain, facial edema, fever, and eosinophilia. Cardiomyopathy is not uncommon; it results from unsuccessful invasion of larvae in cardiac muscle.

Diagnosis of trichinellosis may be made by either a direct or an indirect demonstration of infection. Larvae that have reached the musculature within the first 3 weeks of infection are more readily detected by compression and histologic techniques; otherwise, digestion of muscle tissue is the preferred method. Circulating antibodies can be detected even in lightly infected patients 3–4 weeks after infection and as early as 2 weeks in heavily infected individuals. A variety of serologic tests can be used, but the enzyme-linked immunosorbent assay (ELISA) has proved most useful.

To prevent *Trichinella* infections in pigs the following should be practiced: (1) strict adherence to garbage feeding regulations, particularly cooking requirements for waste material (100°C for 30 minutes); (2) stringent rodent control; (3) preventing exposure of pigs to dead animal carcasses, including pigs; (4) prompt and proper disposal of dead pig and other animal carcasses (e.g. burial, incineration, or rendering); and (5) construction of effective barriers between pigs, wild animals, and even domestic pets.

Meat inspection has proved to be a very successful strategy for controlling trichinellosis, but is not used for this purpose in the US. Inspection currently is performed by either of two direct methods: microscopical examination (trichinoscope) or muscle sample digestion. The trichinoscope method is expensive, labor intensive, and is not standardized worldwide. The practical limit of sensitivity is about three trichina larvae per gram of diaphragm pillar muscle. The digestion method is rapidly supplanting the trichinoscope procedure in most countries where inspection for trichinae is practiced. This method involves the artificial digestion (pepsin–HCl) of diaphragm tissue pooled into batches in order to reduce the number of samples and time required for examination. More recently, immunological tests have been developed for use in abattoir testing but are not yet practiced widely.

Other strategies for reducing consumer risk are: (1) cooking meat to an internal temperature of 60°C for at least 1 minute; (2) freezing meat to −15°C for 20 days, −23°C for 10 days, or −30°C for 6 days if the meat is less than 15 cm thick. Because game is an important source of infection for humans and pigs, all such meat should be considered suspect and should only be eaten after thorough cooking.

D. *Taenia* Species

Taenia are tapeworms, of which there are two main species, *Taenia saginata* ('beef tapeworm'), and *T. solium* ('pork tapeworm') that infect humans. The adult tapeworm stage of *T. saginata* and *T. solium* resides in the human small intestine and is composed of a chain (strobila) of segments (proglottids), which contain both male and female reproductive systems. As the segments mature and fill with eggs, they detach and pass out of the anus, either free or in the fecal bolus. The life span of an adult tapeworm may be as long as 30–40 years. The number of eggs shed from a host per day may be very high (500 000–1 000 000), which leads to high environmental contamination. These eggs contain an infective stage (oncosphere), which matures in the environment. Symptoms vary in their intensity, and may include nervousness, insomnia, anorexia, loss of weight, abdominal pain, and digestive disturbances.

Biological and epidemiological studies in Southeast Asia have demonstrated the presence of a subspecies of *T. saginata asiatica* ('Taiwan Taenia'), which has pigs rather than cattle as an intermediate host.

When *T. saginata* eggs are ingested by cattle, the oncosphere stage is released in the intestine, and it penetrates the gut and migrates throughout the body via the circulatory system. Oncospheres that invade skeletal muscle or heart muscle develop to the cysticercus stage, a fluid-filled cyst or small bladder. When beef that is either raw or inadequately cooked is eaten by humans, the larval cyst is freed and it attaches to the intestinal wall by means of a small head (scolex) with suckers. Over the span of a couple of months, the tapeworm develops and begins to shed eggs, completing the life cycle.

The development of *T. solium* is similar in its intermediate host (pigs or humans) except that the cysticerci are distributed throughout the liver, brain, CNS (neurocysticercosis), skeletal muscle and the myocardium. Neurocysticercosis is increasingly recognized as a serious public health problem, especially in developing countries. Although it is uncommon in the US, Canada, and Western Europe, its prevalence in Latin America, China, and Africa appears to be relatively high.

Feedlot outbreaks of bovine cysticercosis are usually traced to human carriers who have handled cattle or feed with contaminated hands; from the passage out of the anus of egg-bearing segments (proglottids) into the environment while working in the lot; or by indiscriminate defecation in and around feed-storage facilities. Sewage sludge and effluent should be regarded as high risk factors in the epidemiology of this zoonosis, especially when used on agricultural lands. Swine and bovine cysticercosis are also significant economic problems in certain regions owing to condemnation of infected carcasses at slaughter.

Most cases of human cysticercosis (*T. solium*) are asymptomatic and are not recognized by either the individual or a physician. Symptomatic infections may be characterized as either disseminated, ocular, or neurological. Disseminated infections may localize in the viscera, muscles, connective tissue and bone; subcutaneous cysticerci may present a nodular appearance. These localizations are often asymptomatic, but may produce pain and muscular weakness. CNS involvement may include the invasion of the cerebral parachynus subarachnoid space, ventricles, and spinal cord. The larval cysts may persist for years. Symptoms of infection may include partial paralysis, dementia, encephalitis, headache, ocular diseases, meningitis, epileptic seizures, and stroke.

Detection of the adult worm (intestinal) infection is based primarily on identification of proglottids or eggs in the feces. Recently, a species-specific DNA probe for *T. solium* and *T. saginata* eggs has been developed for use in a dot blot assay. Serological diagnosis of humans with *T. solium* cysticercus infection is now quite reliable with the introduction of an enzyme-linked immunotransfer blot (EITB) using purified worm glycoprotein antigen; this test has proven to be 98% sensitive in parasitologically proven cases, and is 100% specific.

To prevent taeniasis, consumers should cook their meat to at least 60°C. On-farm management must ensure that animals are protected from ingesting feed or water contaminated with human feces. The use of sewage sludge and effluents for agricultural purposes should receive more careful scrutiny.

III. Fishborne Parasites

A. *Capillaria philippiensis*

Capillaria philippiensis is a trichurid nematode. It is small: males are 1.5–3.9 mm and females 2.3–5.3 mm. The worms reside in the small intestine, and eggs produced by the females pass in the feces and reach fresh water. The eggs survive in water in the laboratory for months. The larva develops into the infective stage in 3 weeks and, when the eggs are eaten by freshwater fish, they hatch in the fish intestine. When the fish is eaten by a definitive host, the larvae develop into adults in 2 weeks, and the first generation of female worms produce larvae. These larvae mature in 2 weeks, and the females produce eggs that pass in the feces. There are always a few adult female worms that remain in the intestinal tract and continue to produce larvae that develop into adults. This is a process of autoinfection that maintains the infection and increases the population. Autoinfection is an integral part of the life cycle.

Man is presently the only confirmed definitive host, but fish-eating birds are probably the natural hosts. Many species of small freshwater fish in the Philippines and Thailand are intermediate hosts. Human infections are reported mostly from the Phillippines and Thailand, with sporadic reports from Japan, Korea, Taiwan, Egypt, India, Iran, Italy, and Spain. As the numbers of parasites increase in the human host, the symptoms of diarrhea, abdominal pain, and borborygmus become more severe. The loss of potassium, protein, and other essential elements leads to weight loss cachexia. If treatment is not begun in time, the patient will die. Treatment is replacement of potassium, an antidiarrheal drug, and an anthelminthic. The recommended drugs are mebendazole 400 mg/day in divided doses for 20 days or albendazole 400 mg/day for 10 days. The method of diagnosis is detection of *C. philippiensis* eggs, larvae, and adults in the feces.

B. *Gnathostoma* Species

Gnathostomes are nematodes, and several species of *Gnathostoma* larvae are acquired from eating a variety of animals. *Gnathostoma* larvae are short and stumpy with a globose head armed with rows of hooklets. The body also has spines half way down the body. The larvae measure 11–25 mm for males and 25–54 mm for females. Adult worms are found in tumors in the stomach wall of fish-eating mammals. Eggs produced by female worms pass in animal feces, reach water and embryonate. The larva hatches from the egg and is eaten by a freshwater copepod, where it develops into the second stage. When the infected copepod is eaten by a bird, fish, frog, turtle, or mammal, the larva enters the tissue and develops into the third stage. When the second intermediate host is eaten by a *paratenic* host, the larva enters the tissue but does not develop further. It will, however, migrate in the tissue and cause disease. When the second intermediate host or paratenic host is eaten by the final host, the larva will penetrate the intestinal wall, migrate to the liver and other organs, and eventually back to the peritoneal cavity, where it penetrates the stomach wall and provokes a tumor. Cats and dogs are definitive hosts for *G. spinigerum*, and pigs are the definitive host for *G. hispidium*. Intermediate hosts are fish, frogs, snakes, chickens, ducks, rats, etc. The major source of infection in Thailand is a snake-headed fish, *Ophicephalus* sp., which is eaten raw or fermented. In Japan, raw fish is eaten as shashimi and causes infection and, in Mexico, fish prepared as ceviche is a source of infection.

Humans are unnatural hosts and the parasite never matures in humans. The worm migrates through any organ. In the subcutaneous tissue it develops migratory tracts, causing necrosis and hemorrhage. Toxic products cause transient swellings, pain, and edema. The larvae may enter the eyes causing retinal damage and, in the CNS, invasion leads to encephalitis, myelitis, radiculitis, and subarachnoid hemorrhage.

The diagnosis can be made on the basis of the symptoms and a history of eating uncooked fish. Serologic tests are available. A definitive diagnosis is made only when the larvae are recovered from surgical specimens, urine, or vaginal discharge. Larvae may emerge from the subcutaneous tissue following long-term treatment with albendazole.

C. *Anisakis simplex*

Anisakiasis is a nematode larval infection. *Anisakis simplex* and *Pseudoterranova decipiens* are the species most commonly involved. These are parasites of marine mammals found in the stomachs of cetaceans and pinnepeds. Eggs are passed in the feces and remain on the ocean floor until a larva develops. The larva hatches from the egg and is eaten by a small crustacean (euphausid). The third-stage larva develops in the crustacean and, when it is eaten by squid or marine fish, the larva penetrates the gut and passes into the peritoneal cavity or into musculature. When eaten by the mammalian definitive host, the larva is released from the fish or squid tissue, and enters the animal's stomach.

Whales, dolphins, and porpoises are definitive hosts for *A. simplex*, and seals, sea lions, and walruses are the hosts for *P. decipiens*. Mackerel, herring, cod, salmon, and squid are second-intermediate hosts for *A. simplex*, and cod, halibut, flatfish, and red snapper are the second-intermediate host for *P. decipiens*.

The larvae are the stage of the parasite that is pathogenic to humans. The larvae are creamy white to yellowish brown in color, 20–50 mm long and 0.3–1.22 mm wide.

When humans eat infected marine fish and squid raw, prepared as shashimi, sushi or ceviche, the larvae penetrate the tissue of the gastrointestinal tract and cause an eosinophlic granuloma. *A. simplex* has been reported in the stomachs of over 1000 Japanese. *P. decipiens* is reported from humans in Northern Japan and along the California coast. In these areas, the sea lion populations have increased to high levels along with the intermediate host. *P. decipiens* infection usually causes 'tickle throat'.

The diagnosis is made by the recovery of the parasite surgically following the occurrence of an acute abdominal pain. However, the larvae are now being recovered by the use of fiberoptic gastroscopy. Immunodiagnostic methods have been used, but are not readily available or completely reliable.

D. *Clonorchis sinensis*

The Chinese liver fluke is widespread in Asia, being reported from China, Japan, Korea, Taiwan, and Vietnam. The adult worms measure 10–25 mm long and 3–5 mm wide, are lanceolate, flat and pinkish in color. The hermaphroditic adult worms reside in the distal tributaries of the bile passages. Eggs produced by the worms pass down the bile ducts to the intestines and are passed with the feces. The eggs must reach fresh water where they are eaten by the first intermediate snail host (*Parafossaurulus*, *Bythnia*, and *Alocinma* spp.). The larval miracidium is released from the egg and enters the snail tissue to go through polyembryony producing sporocysts, rediae, and cercariae. The cercariae leave the snail and search for cyprinid fish second-intermediate hosts. The cercariae enter the skin of the fish and encyst as metacercariae. When the fish is eaten raw, the metacercariae excyst in the intestine of the definitive host and migrate through the Ampulla of Vater to the bile radicles where they mature into adults.

The parasites in the biliary tract provoke hyperplasia of the epithelium, leading to fibrous development of the ducts. Fever and chills may develop, and the liver may become large and tender. Long-term infections may lead to carcinoma of the bile ducts.

Humans and animals acquire the infection by eating raw, pickled, or smoked carp fish; *Pseudorasbora* or *Ctenopharyngodon* spp. are the most common sources of infection. There are more than 113 species of freshwater fish recorded as second-intermediate hosts.

E. *Opisthorchis viverrini*

Opisthorchis viverrini is a liver fluke found in Thailand and surrounding countries in Southeast Asia, and *O. feliensis* is reported from Eastern Europe and Siberia. The life cycles of these flukes are similar to that of the Chinese liver fluke, but the intermediate hosts are different. *Bithynia* spp. are snail hosts for *O. viverrini*, and *Codiella* spp. for *O. feliensis*. There are many fish intermediate hosts for these species, but the important ones for *O. viverrini* are *Puntius* spp., and *Abranis* spp. for *O. feliensis*. Infections are acquired by eating improperly cooked, preserved, or fermented fish. Diseases associated with these species are similar to those caused by *C. sinensis*. The diagnosis of the liver fluke infection is by detection of eggs in feces or duodenal aspirates, or by immunological methods.

F. Other Heterophyid Flukes

Metagonimus yokogawi and *Heterophyes heterophyes* are two species of flukes acquired by eating raw freshwater fish. They are tiny intestinal parasites that pass eggs in the feces. The eggs will hatch in the water, releasing a miracidium. *Pirenella* spp. of snails are the first intermediate host for *H. heterophyes*, and *Semisulcospira* spp. are snail hosts for *M. yokogawi*. Cercariae produced by polyembryony in the snail search for a fish second-intermediate host; *Mugil* spp. for *H. heterophyes* and for *M. yokogawi*. *H. heterophyes* is endemic in Egypt and China, and *M. yokogawi* in Japan and Korea.

Both flukes are very small, 1–2 mm long and 0.3–0.7 mm wide. Attachment of the fluke may cause small ulcers in the mucosa leading to abdominal pain, diarrhea, and lethargy. The severity depends on the number of worms involved. There are reports of eggs of these worms in ectopic locations such as the brain and heart. Infections are often reported in sports fishermen who eat the fish raw shortly after catching them in the cool mountain streams in Northern Japan. The diagnosis is based on finding eggs in the feces.

G. Echinostomes

Echinostomes are fluke parasites that are acquired by eating a variety of aquatic animal life, especially freshwater fish in Asia. They are generally parasites of other animals, and humans are accidental hosts. The flukes reside in the intestines, and eggs pass in the feces and reach water. The miracidium hatches from the egg and enters a susceptible snail, *Gyraulus* spp.; cercariae produced may enter other snails or other aquatic animal life: *Echinostoma lindoensis*, *E. ilocanum*, and *E. misyanum* are endemic in certain parts of Indonesia, the Philippines, and Malaysia. The parasites have a characteristic collar of spines around the oral sucker, are spindle shaped and measure 4–7 mm by 1.0–1.35 mm. Little disease is associated with infection except for an occasional colic and diarrhea. The diagnosis is based on the presence of eggs in the feces.

H. *Diphyllobothrium latum*

Diphyllobothrium latum is a tapeworm. There are several species of fish tapeworms, but the most important is *D. latum*, which is widespread in the temperate and subarctic regions of the Northern Hemisphere. Fish-eating mammals are definitive hosts. The worm resides in the small intestine and produces eggs, which pass in the feces into water. A ciliated larva (coricidium) develops in the egg and, when released from the egg, swims in the water until ingested by a copepod. A procercoid larva develops in the copepod and, when eaten by a fish second-intermediate host, the larva develops into a plerocercoid larva or sparganum in the fish tissue. When the fish is eaten uncooked, the plerocercoid larva attaches to the intestinal mucosa and develops into an adult tapeworm. The worm is the largest parasite to infect humans and ranges in length from 2 to 15 m with a maximum width of the gravid proglottids of 20 mm. The scolex is 2 mm long and 1 mm wide, and has a dorsal ventral groove or bothrium. The mature proglottid has a rosette-shaped uterus that fills with eggs.

There is little disease associated with fish tapeworm infections. The worms may compete for vitamin B_{12} with the host and cause megaloblastic anemia. This occurs more often in Finland, where patients experience fatigue, weakness, the desire to eat salt, diarrhea, epigastric pain, and fever. The diagnosis is made by finding eggs in the stools.

A variety of fish serve as the second-intermediate host, including pike, perch, turbot, salmon, and trout. Infection occurs if fish is poorly cooked, pickled, or smoked. Japanese eat the fish as shashimi or sushi. Definitive hosts other than dogs, include foxes, bears, mink, seals, and sea lions.

At times humans may acquire infections with the spargana of *Spirometra* spp. This is a diphyllobothroid tapeworm of felines and canines. The plerocercoid larva is a larval migrans that causes transient migratory swelling.

IV. Parasites Disseminated in Fecally Contaminated Food and Water

A. *Isospora belli*

Isospora belli is a coccidian protozoan that causes intestinal coccidiosis in humans. *Isospora belli* oocysts are elongate, ellipsoidal and are $20 \times 19 \mu$m. Sporulated oocysts contain two ellipsoidal sporocysts without a Stieda body. Each sporocyst is $9–14 \times 7–12 \mu$m and contains four crescent-shaped sporozoites and a residual body. Sporulation occurs within 5 days, both within the host and in the external environment. Thus, both unsporulated and sporulated oocysts may be shed in feces.

Infection occurs by the ingestion of food contaminated by oocysts. Merogony and gametogony occur in the upper small intestinal epithelial cells, from the level of the crypts to the tips of the villi. In AIDS patients, the parasite may be disseminated to extraintestinal organs including mesenteric and mediastinal lymph nodes, liver, and spleen. Single zoites surrounded by a capsule (cyst wall) have a prominent refractile or crystalloid body indicating that the encysted organisms are sporozoites. Organisms with a cyst wall are found only in extraintestinal organs.

Isospora belli can cause severe symptoms with an acute onset, particularly in AIDS patients. Infection has been reported to cause fever, malaise, cholecystitis, persistent diarrhea, weight loss, steatorrhea, and even death.

Diagnosis can be established by finding characteristic bell-shaped oocysts in the feces or coccidian stages in intestinal biopsy material. Affected intestinal portions may have a flat mucosa similar to that which occurs in sprue. The stools during infection are fatty and at times very watery. Sulfonamides are considered effective against coccidiosis.

B. *Cyclospora cayetanensis*

Cyclospora cayetanensis is an intestinal coccidian protozoan of humans. *Cyclospora cayetanensis* oocysts are approximately 8μm in diameter and they contain two ovoid $4 \times 6 \mu$m sporocysts. Each sporocyst has two sporozoites. Unsporulated oocysts are excreted in feces. Sporulation occurs outside the body. Other stages in the life cycle are not known. Outbreaks of cyclosporosis in humans have been associated with ingestion of fruits, salads, and herbs contaminated with oocysts.

Both immunocompetent and immunosuppressed patients of all ages may have diarrhea, fever, fatigue, and abdominal cramps. Infection has been reported from several countries.

Diagnosis can be made by fecal examination. *Cyclospora* oocysts are approximately 8 μm, remarkably uniform, and contain a sporont (inner mass) that occupies most of the oocyst. They autofluoresce and are acid-fast and need to be distinguished from *Cryptosporidium* oocysts. Unlike cryptosporidial oocysts, *C. cayetanensis* oocysts have a much thicker oocyst wall and their contents are more granular than are those of cryptosporidial oocysts.

Treatment with sulfamethoxazole and trimethopium is considered effective in relieving symptoms. Irradiation is one proposed method of killing coccidian oocysts on fruits and vegetables.

C. *Cryptosporidium parvum*

Cryptosporidium parvum is a coccidian protozoan parasite. It is a main cause of intestinal, parasitic disease in humans. Its oocysts are approximately 5 μm in diameter and contain four sporozoites. After ingestion of food or water contaminated with sporulated oocysts, sporozoites excyst and penetrate the surface of the microvillus border of a host cell. Asexual development leads to first- and second-generation meronts. Merozoites released from second-generation meronts form male (micro) and female (macro) gamonts. After fertilization, oocysts are formed. Oocysts sporulate in the host and sporulated oocysts are excreted in feces. As many as 79 species of mammals are considered a host for *C. parvum*; in addition to those capable of interspecies transmission, there is now thought to be a human-specific genotype of *C. parvum* and even new species (e.g. *C. canis*) that were once thought to be *C. parvum*.

Cryptosporidial infection is common but disease is rare in immunocompetent humans. Infections in immunosuppressed patients (such as those with AIDS) can be life threatening, and even fatal. The most characteristic symptom is profuse watery diarrhea, cramps, abdominal pain, vomiting, and low-grade fever. Rarely, cryptosporidiosis may involve extraintestinal organs, including gall bladder, lungs, eyes, and vagina.

Diagnosis is made by fecal examination, including direct examination of fecal floats, staining of fecal smears using acid-fast and other stains, and staining with anticryptosporidial antibodies. There is no specific therapy. Removal of oocysts during filtration and sedimentation of municipal water is an engineering problem. Boiling of water kills all coccidian oocysts.

D. *Giardia lamblia*

Giardia lamblia is a flagellated protozoan parasite with bilateral symmetry. Trophozoites and cysts are the two stages in its simple fecal–oral cycle. Trophozoites are pear-shaped, 10–20 μm long, and 5–15 μm wide. They contain two nuclei and eight (four lateral, two ventral, and two caudal) flagella. The dorsal surface is convex, the ventral surface is usually concave and has a sucking disc at the broader end. Cysts are round to ellipsoidal and are 8–19 μm by 11–14 μm in size. There are four nuclei and no flagella. Trophozoites attach to the intestinal epithelium by the adhesive disc. Trophozoites divide into two by longitudinal binary fission. Cysts and trophozoites may be passed in feces. Trophozoites do not survive for long periods outside the host, whereas cysts can survive in the external environment.

Humans become infected by ingesting food or water contaminated with cysts. The exact mechanism of pathogenesis is unknown because the parasite is extracellular. Giardiasis can cause severe intestinal disorders resulting in diarrhea, nausea, steatorrhea, and weight loss.

Diagnosis is made by fecal examination. To find trophozoites, smears of freshly passed stool should be made in isotonic saline (not water). Flotation in zinc sulfate is useful in detecting cysts. The discharge of *Giardia* in the stool is not always regular; therefore, multiple stool examinations may be necessary. Immunodiagnostic methods including an ELISA and fluorescent antibody test to detect *Giardia* in stools have been described.

Quinacrine and metronidazole are two drugs recommended for the treatment of giardiasis. Because of the potential for wild and domestic animals as reservoirs of infection, good personal hygiene is necessary to prevent infection with *Giardia*. Boiling will kill *Giardia* in water.

E. Soil Transmitted Nematodes and Visceral Larval Migrans

There are several nematodes transmitted via ingestion of contaminated soil. Nematodes (e.g. *Trichuris*, *Ascaris*) that complete full development in humans will not be discussed further. Visceral larval migrans (VLM) usually results from the migration of nematode larvae in the accidental human host. Although several nematodes may cause VLM, the most common cause of VLM in humans are (1) *Toxocara canis* and (2) *Bayliascaris procyonis*. Dogs and other canids are the definitive hosts for *T. canis*, whereas raccoons are the definitive hosts for *B. procyonis*.

Adult worms live in the stomach and small intestine and pass unembryonated eggs in feces. First-stage larvae develop inside the egg and undergo a molt to form second-stage larvae. *Toxocara* and *Bayliascaris* eggs are sticky and survive in the environment for many years. Humans become infected by ingesting embryonated eggs in contaminated water or food, or ingestion of contaminated soil. In accidental human hosts, second-stage larvae can migrate extensively in many organs but do not mature. Even one or a few larvae in the eye and brain can cause extensive damage, and symptoms vary depending on the organs involved. Other nematodes (e.g. *Toxocara cati* from cats, *Ascaris suum* from pigs; *Ancylostoma* from dogs and cats) can also cause VLM or cutaneous larval migrans.

In the definitive host (dog), *T. canis* larvae return to the gut and mature into adult worms and they can lay a large number of eggs for a long period. *T. canis* is transmitted transplacentally and via milk to pups. Three-week-old pups can shed *T. canis* eggs in feces, and these are a health hazard for humans, especially children. *B. procyonis* is only transmitted fecally, and eggs in feces deposited by raccoons (e.g. in chimneys, garages, and near hiking trails) are a source of infection, especially for children.

Diagnosis of VLM in humans is difficult because of imprecise symptoms. Eosinophilia can arouse suspicion of helminth infection. Serologic tests can aid diagnosis. Mebendazole and albendazole have been used to treat VLM in humans. Dogs should be dewormed regularly to reduce environmental contamination by eggs.

V. Summary

This chapter has reviewed information on diseases associated with foodborne parasites, including etiologic agent, life cycle, symptoms, diagnosis, treatment, prevention, and control. Parasites covered are species of the genera *Anisakis*, *Bayliascaris*, *Capillaria*, *Clonorchis*, *Cryptosporidium*, *Cyclospora*, *Diphyllobothrium*, *Giardia*, *Gnathostoma*, *Isospora*, *Opisthorchis*, *Sarcocystis*, *Taenia*, *Toxocara*, *Toxoplasma*, *Trichinella*, and others.

Bibliography

Ash, L. R. and Orihel, T. C. (1984). 'Atlas of Human Parasitology', 2nd ed. American Society of Clinical Pathologists Press, Chicago.

Chin, J. D. (ed.) (2000). 'Control of Communicable Diseases Manual', 17th ed. American Public Health Association, Washington, DC.

Fayer, R., Gamble, H. R., Lichtenfels, J. R. and Bier, J. W. (2001). Waterborne and foodborne parasites. *In* 'Compendium of Methods for the Microbiological Examination of Foods', 4th ed. (F. P. Downes and K. Ito, eds), pp. 429–438. American Public Health Association, Washington, DC.

Garcia, L. S. and Bruckner, D. A. (1997). 'Diagnostic Medical Parasitology', 3rd ed. American Society for Microbiology Press, Washington, DC.

Hayunga, E. G. (2001). Helminths acquired from finfish, shellfish, and other food sources. *In* 'Food Microbiology: Fundamentals and Frontiers', 2nd ed. (M. P. Doyle, L. R. Beuchat and T. J. Montville, eds), pp. 533–547. American Society for Microbiology Press, Washington, DC.

Kim, C. W. and Gamble, H. R. (2001). Helminths in meat. *In* 'Food Microbiology: Fundamentals and Frontiers', 2nd ed. (M. P. Doyle, L. R. Beuchat and T. J. Montville, eds), pp. 515–531. American Society for Microbiology Press, Washington, DC.

Ortega, Y. R. (2001). Protozoan parasites. *In* 'Food Microbiology: Fundamentals and Frontiers', 2nd ed. (M. P. Doyle, L. R. Beuchat and T. J. Montville, eds), pp. 549–564. American Society for Microbiology Press, Washington, DC.

Various authors (1995). Parasitology. *In* 'Manual of Clinical Microbiology', 6th ed. (P. R. Murray, E. J. Baron, M. A. Pfaller, F. C. Tenover and R. H. Yolken, eds), pp. 1141–1256. American Society for Microbiology Press, Washington, DC.

Xiao, L., Morgan, U. M., Fayer, R., Thompson, R. C. A. and Lal, A. A. (2000). *Cryptosporidium* systematics and implication for public health. *Parasitol. Today* **15**, 287–292.

Intoxications

PART

III

Naturally Occurring Toxicants in Foods

Steve L. Taylor and Susan L. Hefle

14

I. Introduction

Since all chemicals are toxic at some dose, every chemical in foods, including all of the additive chemicals and all of the naturally occurring constituents, could be considered a foodborne toxicant. While this approach is philosophically correct, it is clearly impractical to view food constituents in this manner. In foods that are normal constituents of our diets, the component chemicals, both natural and synthetic, are usually present at levels that are too low to produce harmful effects when the foods are eaten in reasonable amounts and prepared in a usual manner. Some exceptions exist to this statement, since atypical consumers with allergies to food constituents can experience adverse reactions to foods that are normally present in our diets. Also, this statement ignores any possible role of food components in cancer, heart disease, or hypertension, although the purported associations with these chronic diseases are controversial (see Chapter 21).

Thus, the degree of hazard associated with a chemical is more important than the chemical's toxicity. Toxicity is an intrinsic property of all chemicals. Hazard is the capacity of a substance to produce injury under the circumstances of exposure. Hazard takes into account the dose and frequency of exposure as well as the relative toxicity of the particular chemical. This chapter will focus on the naturally occurring chemicals in foods and their degrees of hazard.

Foodborne Diseases 2nd Edn
ISBN: 0-12-176559-8

II. Natural Sources of Toxicants in Foods

Two general classes of naturally occurring toxicants occur in foods: naturally occurring constituents and the contaminants that are produced in foods by natural processes. The naturally occurring constituents of foods can be considered both normal and unavoidable. The naturally occurring contaminants are not always present and can be avoided, if the contamination is avoided.

A. Naturally Occurring Constituents of Food

Naturally occurring constituents are present in foods of animal, plant, or fungal origin. Most are not hazardous under normal circumstances of exposure. Foods of animal origin include meat, milk, eggs, fish, mollusks, and crustacea. The most hazardous chemicals occurring in this category are found in poisonous seafoods, although many of these seafood toxicants are naturally occurring contaminants rather than normal, unavoidable constituents. Foods of plant origin include vegetables, fruits, grains, seeds, nuts, spices, and beverages. Most plant-derived chemicals are not particularly hazardous, but a few that are potentially hazardous will be discussed. The major food of fungal origin is mushrooms, many of which are hazardous.

B. Naturally Occurring Contaminants of Food

Foods can be contaminated naturally by bacteria, mold, algae, and insects. These biological contaminants can produce chemicals that remain in the food even after the biological source has been removed or destroyed. Contaminants from bacteria, molds, and algae are discussed in detail in other sections and chapters. Insects can also produce and secrete toxic chemicals into foodstuffs. Thus, prevention of insect infestation of foods may be more than merely an aesthetic consideration.

III. The Preoccupation with Natural Foods

In the past several decades, much attention has been focused on manmade toxicants, such as additives and pesticide residues, in foods. As a result, some consumers have accepted the notion that foods containing no added chemicals are safer than foods that contain added chemicals. It is impossible to demonstrate that natural is synonymous with safer, yet consumers are demanding such products.

A. Definitions of 'Natural'

No legal definition exists for the word *natural* in US food laws, so there is no legal limitation on the use of this term. The part of the definition in Webster's Dictionary that is most relevant to foods states: 'planted or growing by itself; not cultivated or introduced artificially, e.g. grass; existing or produced by nature; consisting of objects so existing or produced; not artificial'. Most consumers would likely define natural as being anything that is not artificial – another term that defies definition. To the food industry, natural can mean additive free, preservative free, or antioxidant free. It can also refer to foods derived from plant or animal materials, processed agricultural commodities, or raw, unprocessed agricultural commodities.

B. Concerns over Additives in Foods

Why are consumers concerned about chemicals added to foods, and are their concerns justified? The concerns likely arose from notable government regulatory actions against certain food additives and incidental contaminants (pesticides), which contrasts with an absence of action against naturally occurring chemicals in foods. Also, the health foods industry promoted natural products as being beneficial to health and has been able to convince some consumers that natural products and supplements are valuable components of the diet. More recently, some naturally occurring food components have been labeled as 'nutraceuticals', chemicals that have beneficial health effects when added to foods. This has also fostered the notion that natural foods are safer. The reasons for consumer concerns seem clear, but are these concerns justified?

1. Regulatory emphasis on additives

The US food laws assume natural foods to be safe, while most additives – including all newly developed ones – must be proven to be safe. The US Food and Drug Administration (FDA) has rarely taken any action to limit the availability of naturally occurring foods because of the presence of toxic or carcinogenic chemicals of natural origin.

One exception was the FDA's decision in the 1970s to ban sassafras bark from sassafras tea and root beer, based upon the demonstrated carcinogenicity of safrole, a naturally occurring component. Meanwhile, the FDA permits up to 20 ppb total aflatoxins, which are potent naturally occurring carcinogens, in foods such as peanuts.

In contrast to the general lack of regulatory activity on naturally occurring toxicants, the FDA has acted on numerous food additives because of concerns about their safety. The provisions of the *Delaney Clause* have been used to ban cyclamates (a noncaloric sweetener) and FD&C Red #2 (a food dye) because of evidence of carcinogenic activity in animal feeding trials. Saccharin, another noncaloric sweetener, is allowed at present, but warning statements are required on product labels alerting consumers that saccharin can cause cancer in laboratory animals. The FDA has required the specific declaration of FD&C Yellow #5 (tartrazine) and sulfites on food labels, and has banned the use of sulfites on fresh fruits and vegetables because of concerns that these additives can trigger adverse reactions in some sensitive individuals. At the same time, the US Environmental Protection Agency (EPA) has banned a number of pesticides, such as DDT and ethylene dibromide (EDB), because of their potential carcinogenicity. With all of these regulatory restrictions occurring, it is little wonder that consumers have developed a suspicious attitude toward chemicals added to foods. Also, even when FDA approves a new food additive, such as aspartame (another noncaloric sweetener) or Olestra (a fat substitute), one or more consumer-interest groups have sought to raise public concerns about the safety of these new additives.

2. Perspective on diet and health

The widespread use of additives and pesticides coupled with the regulatory actions just described has created concerns over chronic diseases such as cancer. However, health developments in the US in recent decades reveal that an epidemic of chronic disease is unlikely. Life expectancy is increasing, the death rate from heart disease is dropping, the death rate from stroke is declining, and fewer people under the age of 49 are dying of cancer. Meanwhile, the use of most chemicals in foods continued, leading to obvious questions about whether concerns over synthetic chemicals in foods are justified.

C. Hazards Associated with Naturally Occurring Substances

While consumer concerns about the safety of chemicals added to foods are understandable even if not justified, consumers' lack of concern about naturally occurring chemicals in foods may not be justified.

1. Naturally occurring acute toxins in foods

Certain foods naturally contain some rather well-known toxic substances, such as mercury and arsenic in seafoods, and cyanogenic glycosides, sugar derivatives that release cyanide on contact with stomach acid, in lima beans. Table 14.1 lists the estimated per capita intakes of certain acute toxins among US consumers.

While these chemicals are certainly toxic, several factors allow us to consume these foods with little chance of adverse reactions. First, the normal level of exposure to these toxins is usually below the threshold for toxicity. Second, the form of the chemicals may lessen their toxicity. For example, mercury residues in fish may be bound to selenium, a form that is far less toxic than other forms of mercury. Third, these toxins,

Table 14.1 Examples of naturally occurring acute toxins in foods

Chemical	Source	Estimated yearly per capita consumption (US) (mg)
Arsenic	Shellfish, seafood	50–332
Cyanide	Lima beans, cassava	40
Mercury	Tuna, swordfish	1.8–3.6
Myristicin	Nutmeg	44
Solanine	Potatoes	10 000

with the possible exception of mercury, do not accumulate in the body. Therefore, small doses from one meal are metabolized and excreted before another dose is consumed.

2. Toxicity of food nutrients

Since all chemicals are toxic, even the basic nutrients in foods are slightly toxic though not hazardous in most cases in the levels found in typical, nonsupplemented consumer diets. Laboratory animals can be killed by feeding them glucose, sucrose, or salt, but only at rather high doses. However, some of the micronutrients, such as vitamin A and selenium, are hazardous if consumed in amounts only a few times greater than our requirements for these nutrients. With vitamin A, acute toxicity has been noted in adults ingesting 2–5 million international units (IU); the recommended dietary allowance for vitamin A is 4000 IU for adult females and 5000 IU for adult males.

3. Chronic toxicity of naturally occurring substances

Numerous examples exist of naturally occurring carcinogens. Some of the mycotoxins, such as aflatoxin B_1 and sterigmatocystin, which can be natural contaminants of many foods, are carcinogenic (see Chapter 19). Some of the naturally occurring toxicants in plants, including the pyrrolizidine alkaloids and safrole, are carcinogenic. Many of the essential oils contain carcinogenic substances such as safrole. Even very common spices such as pepper have demonstrated carcinogenic activity. Aflatoxin B_1, which is allowed in foods at levels below 20 ppb, is a much more potent carcinogen than some of the additive substances that have been banned.

4. Hazards associated with alternatives to processed foods

As discussed previously, some consumers have become quite suspicious of chemicals added to their foods. Some of these consumers seek alternatives, such as foraging for food in the wild, or purchasing foods and remedies from health food stores. Health food stores have become so successful that many such products are now offered through more typical grocery stores and the food industry is adding purportedly beneficial chemicals, the nutraceuticals, to a variety of food products.

a. Foraging in the forest

Foraging for foods in the wild can be hazardous for anyone not expert in the identification of edible species. The harvesting of plants and mushrooms in the wild is fraught with risks because toxic species are rather common and identification can be difficult. Mushroom poisoning occurs exclusively among individuals consuming mushrooms harvested in the wild. In the period of 1988–1992, the Centers for Disease Control and Prevention (CDC) reported five outbreaks of mushroom poisoning in the US involving 18 cases. The ingestion of plants harvested in the wild also occasionally results in adverse reactions, often among campers and backpackers. Individuals harvesting their own herbs for the preparation of herbal teas are also victims on occasion. An elderly couple in the State of Washington in 1977 harvested what they thought were comfrey leaves for the preparation of herbal tea. They mistakenly harvested fox-glove, a very toxic plant, and both individuals died as a result.

b. Health food stores and their products

Health or natural food stores often have a limited array of products, which deters many consumers from purchasing all of their foods from these outlets. Most of the items sold are or are purported to be natural and often organic. Some items are promoted as health remedies. Because the natural chemicals present in these products are not legally required to be evaluated for safety, hazardous products may be more likely to be encountered in health foods stores than in typical supermarkets.

Identity of ingredients   It is very difficult to verify the identity of dried leaves in some herbal preparations.

Hazards associated with ingredients   A few health foods and remedies contain acutely toxic substances. For example, apricot kernels contain amygdalin, also known as vitamin B_{17} or laetrile. Amygdalin is purported to have anticarcinogenic properties, but in fact is a glycoside that releases HCN on contact with stomach acid. Cyanide is reasonably toxic and at least one death has been attributed to ingestion of apricot kernels. Other fruit pits also contain cyanogenic glycosides and are hazardous to ingest in large quantities.

Many foods, including some health foods, contain substances that may pose a significant carcinogenic hazard, although there is no direct evidence that they do. Comfrey tea, for example, is widely recommended by health food stores for its general healing properties. It contains several pyrrolizidine alkaloids known to induce liver damage and tumors in laboratory animals. These alkaloids are potent carcinogens, and many consumers of comfrey tea ingest the product frequently.

Large doses of ginseng, a popular health food, have been reported to cause secondary female characteristics in men, including the loss of body hair and breast enlargement. Effects of high intakes of ginseng may also mimic corticosteroid intoxication, including high blood pressure, nervousness, and insomnia. The estrogenic effects are definitely due to chemicals present in ginseng. Other toxic effects may be due to other plants frequently sold as ginseng.

Some health food products can elicit allergic reactions in sensitive consumers. Chamomile tea can induce allergic reactions in individuals sensitive to ragweed (common in the US), and people with pollen allergies may experience reactions after ingesting bee pollen.

Method of preparation   The method of preparation of health foods can be important in avoiding certain associated hazards. A few herbal teas require very careful filtration to remove potentially toxic components. Remedies containing valerian root are frequently marketed as tranquilizers, but active ingredients are easily inactivated by heat, acid, or alkali, so the commercial product is often worthless.

Dose of hazardous ingredients   The dose of the toxic constituents is always an important determinant of the hazard associated with a product. Because health foods are often harvested in the wild after growing under uncontrolled conditions, concentrations of toxic substances can be unpredictable. Tansy, an occasional ingredient of various remedies sold in health food stores, contains thujone, a convulsant and hallucinogen. The level of thujone varies widely in tansy, so consumers can never be certain how much of this toxic substance they are ingesting with the remedy.

Economic fraud   Fraudulent practices, which are not hazardous themselves, are common in health and natural food stores. The most common is the marketing of herbal remedies that have no actual therapeutic value. Applicable active ingredients are either absent or negligible in wild lettuce sold as a soporific, black cohosh for menstrual problems, red raspberry leaves (in herbal teas) as stimulants and smooth muscle relaxants, and horsetail for the treatment of kidney and bladder ailments as well as tuberculosis. Federal law prohibits unfounded health claims on labels, so these claims are usually made in literature that accompanies the products, which is also available at health food stores.

Conventional medical therapy   Another risk associated with health food remedies is the avoidance of conventional medical therapy. Often consumers will use these products in an attempt to cure illnesses that would be much better treated by established medical means. As a consequence, the illness may worsen before suitable medical treatment is sought. Unnecessary complications and even death may occur.

IV. Intoxications from Naturally Occurring Toxicants

Many naturally occurring, potentially hazardous chemicals exist in foods. Ingestion of these toxicants does not always cause illness, but intoxications can result under certain circumstances of exposure as indicated in Table 14.2. Chronic illnesses, such as cancer, are difficult to correlate with foods or specific foodborne toxicants, since many confounding variables can affect the onset and course of the disease. Thus, many of the examples discussed here will involve acute intoxications.

A. Statistics on Chemical Etiology of Foodborne Disease

Chemical etiology accounted for 4.7–7.3% of all foodborne disease outbreaks reported in the US from 1988 to 1992 (Table 14.3). Staphylococcal food poisoning and botulism could be classified as chemical intoxications, but the CDC record these in the category of bacterial infections. Foodborne disease, in general, is under-reported; however, outbreaks of chemical etiology probably would constitute 5–10% of the total number of outbreaks, even with complete reporting.

The agents involved in the foodborne disease outbreaks of chemical etiology in the period of 1988 to 1992 are listed in Table 14.4. Naturally occurring toxicants account

Table 14.2 Intoxications from consumption of naturally occurring foodborne toxicants

Abnormal though natural contaminants that adversely affect *normal* consumers eating *normal* amounts of the food
- Algal toxins in seafood
- Staphylococcal enterotoxins in various foods
- Botulinal toxins in various foods
- Mycotoxins in various foods
- Insect- and mite-derived toxins in foods

Unusual 'foods' that adversely affect *normal* consumers eating *normal* amounts of this 'food'
- Poisonous mushrooms
- Poisonous plants such as foxglove and *Senecio*
- Poisonous fish such as pufferfish

Normal constituents of food that can cause illness if ingested by *normal* consumers in *abnormal* amounts
- Cyanogenic glycosides in lima beans, cassava, and fruit pits
- Phytoestrogens in ginseng

Normal components of foods that, consumed in *normal* amounts, are harmful to *abnormal* consumers
- Food allergies
- Lactose intolerance
- Celiac disease

Normal foods processed or prepared in an *unusual* manner and consumed in *normal* amounts by *normal* consumers
- Lectins in underprocessed kidney beans
- Trypsin inhibitors in underprocessed legumes
- Heavy metals in acidic beverages stored improperly

Normal foods eaten by *normal* consumers in *normal* amounts over an extended period of time (**alleged**)
- Cholesterol in atherosclerosis
- Saturated fats in atherosclerosis
- Dietary fat in cancer
- Caloric intake in cancer
- Sodium in hypertension

Table 14.3 Reported incidence of foodborne disease outbreaks of chemical etiology, 1988–1992

Year	Total number of outbreaks	Outbreaks of chemical etiology	
		Number	Percentage
1988	451	29	6.4
1989	505	37	7.3
1990	532	27	5.1
1991	528	31	5.9
1992	407	19	4.7

for the majority of the outbreaks. Fish and shellfish toxins are the most prevalent causes of chemical intoxications. Ciguatera, paralytic shellfish, and tetrodotoxin poisoning will be discussed in a separate chapter. Scombroid fish poisoning is actually a microbial intoxication and is discussed in Chapter 15. Mushroom poisoning is also a rather frequent form of foodborne chemical intoxication, and will be discussed in this chapter. Among the agents listed under miscellaneous chemicals in Table 14.4 are several naturally occurring substances including alkaloids from various plants. Some

Table 14.4 Etiologic agents in reported incidents of chemical food poisoning, 1988–1992	
Agent	**Outbreaks (cases)**
Ciguatoxin	42 (176)
Heavy metals	3 (26)
Mushrooms	5 (18)
Paralytic shellfish poison	5 (65)
Scombroid poisoning	76 (514)
Miscellaneous chemicals	12 (128)

others listed are not naturally occurring. Most of the heavy metal intoxications are due to packaging errors. Many of the miscellaneous chemicals are detergents, sanitizers, and other processing aids that are hazardous if used improperly.

B. Naturally Occurring Contaminants

As indicated in Table 14.2, many examples exist of naturally occurring contaminants in foods. Separate chapters in this book are devoted to bacterial toxins (Chapters 16–18), mycotoxins (Chapter 19), and fish and shellfish toxins arising from algae (Chapter 15). Considered together, these naturally occurring contaminants are a rather frequent cause of foodborne disease. Since the most common of these naturally occurring toxicants are covered elsewhere in the book, the only examples remaining to discuss would be some of the miscellaneous microbial contaminants in foods and the insect-derived contaminants.

1. Miscellaneous microbial toxins in foods

As noted, separate chapters in this book are devoted to the principal intoxications resulting from bacteria growing on foods: staphylococcal food poisoning, botulism, and *Bacillus cereus* intoxications. Histamine poisoning arising principally from bacteria growing on certain species of fish (although several cheese-related outbreaks have also occurred) is discussed in Chapter 15. The remaining microbial toxins have only caused human illness in a few isolated circumstances.

a. Bongkrek food poisoning

Bongkrek food poisoning was once common in Indonesia but has not occurred in the US. The symptoms of this foodborne disease include hypoglycemia, severe spasms, convulsions, and death.

This illness is associated with the consumption of bongkrek, which are flat white cakes made from pressed coconut and fermented with the fungus, *Rhizopus oryzae*. The cakes are wrapped in banana leaves during the fermentation. The toxin known as bongkrek acid (Fig. 14.1) is produced by *Pseudomonas cocovenenans*, which overgrows the fungus under some conditions. Bongkrek acid is a heat-stable, unsaturated fatty acid. The bacterial overgrowth can be prevented by use of oxalis leaves rather than

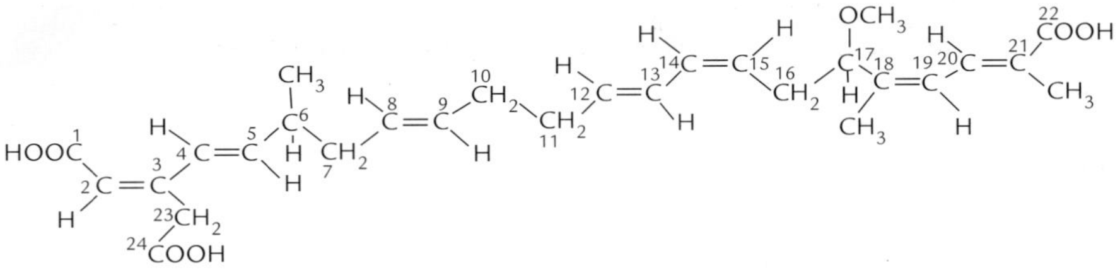

Figure 14.1 Structure of bongkrek acid.

banana leaves. The oxalis leaves contribute to a rapid drop in pH to about 5.5, which prevents the growth of *P. cocovenenans*. The substitution of oxalis leaves has largely eliminated bongkrek food poisoning.

b. Fermented corn flour poisoning

Between 1961 and 1979, 327 known cases of intoxication with 101 deaths were reported in the People's Republic of China associated with the ingestion of fermented corn flour. The disease had a short onset period and was characterized by abdominal discomfort, mild diarrhea, and vomiting. In severe cases, the disease progressed to jaundice, coma, delirium, oliguria, hematuria, rigidity of the extremities, urinary retention, toxic shock, and death.

The toxin responsible for this illness is produced by *Flavobacterium farinofermentans*. The toxin was identified as bongkrek acid (Fig. 14.1). This foodborne illness can be prevented by properly controlling the fermentation of the corn flour by ensuring that the pH of the fermentation is sufficiently high. The fermented corn meal produced in the affected regions of China was made under conditions of neutral pH. Under alkaline conditions, the growth of *F. farinofermentans* would be discouraged, and the toxin is unstable.

c. Intoxications caused by nitrate-reducing bacteria

The reduction of nitrate produces nitrite, which can be acutely toxic to humans. The symptoms of nitrite intoxication are methemoglobinemia, nausea, vomiting, headache, weakness, shortness of breath, cyanosis, collapse, and occasionally death. Most cases of nitrite intoxication involve inadvertent addition of meat-curing salts containing nitrite to other foods; these episodes do not involve bacterial conversion of nitrate to nitrite.

Nitrate occurs in plants naturally but levels can be increased as the result of application of nitrogen-based fertilizers. Some plants will accumulate more nitrate/nitrite in their edible tissues than others; but considerable variability is observed depending upon species, variety, plant part, stage of maturity, and environmental conditions such as drought, harvest temperatures, nutrient deficiencies, insect damage, use of herbicides and/or insecticides, and fertilizer application. Some plants possess the enzyme nitrate reductase, which will convert nitrate to nitrite, but in most plants, the majority of the nitrate/nitrite remains in the nitrate form, which is less toxic.

Bacteria can also possess nitrate reductase and convert the nitrate in plant tissues to nitrite. Storage of spinach and carrot juice at elevated temperatures that encourage bacterial growth has contributed to cases of nitrite intoxication involving bacterial formation of nitrite. Many common bacteria possess nitrate reductase, including pseudomonads, Enterobacteriaceae, staphylococci, and *Clostridium perfringens*. Proper storage of nitrate-containing plants at refrigerator temperatures will prevent bacterial nitrite formation.

2. Insect- and mite-derived toxins in foods

Insect infestations are rather frequent in the food supply and are considered by most consumers to be aesthetically undesirable. However, insects can produce chemicals, some of which are potentially hazardous, that remain in the foods even if the insects are removed. The common flour beetles, *Tribolium confusum* and *Tribolium castaneum*, produce *o*-benzoquinones. These *o*-benzoquinones are putative carcinogens for laboratory animals, although their carcinogenicity for humans remains unknown. Curiously, the *o*-benzoquinones are mutagenic to the flour beetles themselves when insect contamination of the flour becomes high. No cases of human illness have been reported from the ingestion of foods contaminated with flour beetles. However, several cases of human illness have been reported from the ingestion of foods contaminated with dust mites. Dust mites are common inhalant allergens. Dust mites can contaminate foods, sometimes in high numbers if food storage conditions are poor. Apparently, sufficient levels of dust mite allergens are produced in the food to elicit allergic symptoms in susceptible consumers when the contaminated food is ingested. Since the dust mites are microscopic, the consumer would not be aware of the contamination.

3. Naturally occurring constituents eaten in normal amounts

Some plants and a few animals contain hazardous levels of various toxicants. These plants and animals should not be eaten but are consumed intentionally or accidentally on occasion, resulting in foodborne chemical intoxications.

a. Animals

Most species of animals do not contain hazardous levels of toxicants. The only good examples are poisonous fish. Puffer fishes and the toxins they contain, which may be of algal origin, are described in the separate chapter on seafood toxicants.

b. Plants

Many plant species contain hazardous levels of toxic constituents. Some of these plants, such as hemlock and nightshade, are classical agents used in early times to poison enemies. In modern times, intoxications from the ingestion of poisonous plants have resulted primarily from misidentification of plants by individuals harvesting their own foods in the wild. As a group, the most hazardous constituents of plants are the alkaloids, although there are many alkaloids and they vary widely in toxicity. Alkaloids is a term used collectively to describe a large and diverse group of chemicals that have alkali-like properties and at least one nitrogen atom in a heterocyclic ring structure. Alkaloids are present in a large number of plant species, and humans are routinely exposed to alkaloids whenever they consume plant foods of several types. Fortunately,

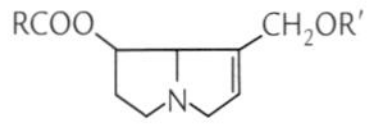

Figure 14.2 The general structure of the pyrrolizidine alkaloids.

the levels of alkaloids in these edible plants are usually insufficient to elicit acute symptoms, although the chronic toxicity of some of these alkaloids at lower levels in the diet remains unknown. The pyrrolizidine alkaloids and the principal alkaloids of potatoes, solanine and chaconine, will be discussed as examples, although a thorough review of naturally occurring constituents of plants is far beyond the scope of this book.

Pyrrolizidine alkaloids The pyrrolizidine alkaloids are a group of chemicals that contain a single nitrogen in a pyrrolizidine ring (Fig. 14.2). Pyrrolizidine alkaloids are found in many plant species distributed throughout the world and are known to cause both human and livestock intoxications. The contamination of food grains with seeds of pyrrolizidine alkaloid-containing plants in Afghanistan and India in the 1970s caused hundreds to become ill and several dozen to die. In cases from central Asia, Africa, and Central and South America, cases of pyrrolizidine alkaloid poisoning often result from the ingestion of plants harvested in the wild that contain hazardous levels of these compounds. In the US and Europe, intoxications have occurred with products purchased from retail outlets, as in the contamination of herbal teas.

In one such example, a retail herbal tea was contaminated with *Senecio longilobis* (common name: thread-leaf groundsel), a well-known poisonous plant. This herbal tea, called gordolobo yerba, was sold to the Mexican–American population of Arizona and promoted as a cure for colic, viral infections, and nasal congestion in infants. The number of infants and others who ingested the hazardous tea is not known, but six infants died. This tea contained 1.5% of dry weight of pyrrolizidine alkaloids, and one of the deceased infants was estimated to have consumed 66 mg of the alkaloids over a 4-day period. Many herbal teas commonly contain lower levels of pyrrolizidine alkaloids that are not typically hazardous on an acute basis. For example, comfrey (*Symphytum officinale*) typically contains a total alkaloid level of 0.003–0.02% including a pyrrolizidine alkaloid, symphytine, which is apparently insufficient to elicit acute illness. However, several acute intoxications have occurred with comfrey tea containing elevated levels of pyrrolizidine alkaloids.

The acute symptoms of pyrrolizidine alkaloid intoxication include ascites (accumulation of fluid in the abdominal cavity), enlarged liver with veno-occlusive liver damage, abdominal pain, nausea, vomiting, headache, apathy, and diarrhea. The liver damage is irreversible and can result in death.

Pyrrolizidine alkaloids can also cause chronic intoxications if ingested in smaller quantities over extended periods of time. The effects of pyrrolizidine alkaloids on the liver are cumulative with irreversible liver damage occurring in small increments over months or even years. Cirrhosis and liver cancer are the principal manifestations of chronic intoxication with pyrrolizidine alkaloids. Several of the pyrrolizidine alkaloids are well documented carcinogens in laboratory animals. The lifelong ingestion of herbal products containing low levels of potentially carcinogenic pyrrolizidine alkaloids presents unknown carcinogenic risks.

Solanine and chaconine Solanine and chaconine (Fig. 14.3) are structurally similar steroidal glycoalkaloids found in potatoes, usually at doses that are not hazardous. Potatoes (*Solanum tuberosum*) are one of a number of species in the nightshade family

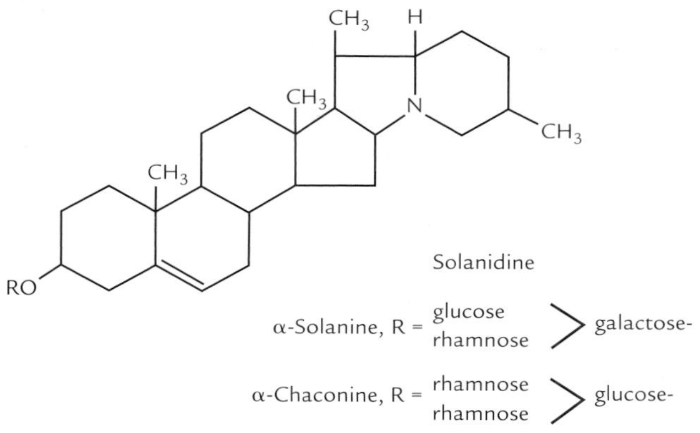

Figure 14.3   The structures of solanine and chaconine.

that also includes tobacco (*Nicotiana tabacum*) and deadly nightshade (*Atropa belladonna*). Other members of the nightshade family include tomatoes and eggplants, both of which also contain structurally related steroidal glycoalkaloids in concentrations that are not often hazardous. Potatoes contain variable amounts of solanine and chaconine in about equal proportions.

Solanine and chaconine are relatively potent inhibitors of acetyl cholinesterase, an enzyme necessary for proper functioning of the nerve synapse. Severe cases of intoxication, including deaths, with solanine and chaconine result from central nervous system depression. Solanine and chaconine also disrupt the membranes of red blood cells and other cellular membranes resulting in gastrointestinal and hemorrhagic symptoms. The typical symptoms of potato glycoalkaloid poisoning include both gastrointestinal (abdominal pain, nausea, diarrhea, and vomiting) and neurological (apathy, drowsiness, mental confusion, dyspnea, a rapid and weak pulse, low blood pressure, and, in severe cases, coma and death) symptoms. Chronic symptoms have not been noted.

Potatoes, as typically consumed, do not contain sufficient quantities of these glycoalkaloids to elicit acute symptoms. Potato tubers usually contain only 20–130 mg/kg tuber of total glycoalkaloids, and there are no reports of intoxication from ingestion of such potato tubers. Elevated levels of glycoalkaloids are found in green potato tubers, potato sprouts, potato leaves, and the tubers of some wild potato species. Human intoxications have occurred from the consumption of potato components having glycoalkaloid contents of 300–800 mg/kg tuber. The toxic oral dose of these glycoalkaloids for humans is estimated at 2–5 mg/kg body weight. With potato tubers containing glycoalkaloids at 200 mg/kg tuber, a level considered to be the upper safe limit for commercial potaotoes, an average adult would need to consume four or five 170-g tubers to become ill. Thus, human cases are almost always associated with the ingestion of green tubers that may contain much elevated levels of the glycoalkaloids.

c. Fungi

Many species of poisonous mushrooms exist and may fool even expert mushroom hunters occasionally. While most of the intoxications occur among mushroom hunters

Table 14.5 Some poisonous mushrooms and their toxins

Species	Common names	Known toxins (Group)
Amanita phalloides	Death cap	Amatoxin (Group I)
		Phallotoxin (Group I)
Amanita muscarina	Fly agaric	Muscarine (Group III)
		Bufotenine (Group VI)
		Amatoxin (Group I)
		Phallotoxin (Group I)
		Ibotenic acid (Group V)
		Muscimol (Group V)
Clitocybe dealbata	—	Muscarine (Group III)
Coprinus atramentarius	—	Coprine (Group IV)
Gyromitra esculenta	—	Gyromitrin (Group II)
Psilocybe mexicana	Mexican mushroom, magic mushrooms, shrooms	Psilocybin (Group VI)
		Psilocin (Group VI)

harvesting from the wild, increasingly wild-harvested mushrooms are available from retail outlets, which increases the hazard to more typical consumers. Table 14.5 provides some examples of known poisonous mushrooms and their toxins. The toxins are grouped into seven categories, depending upon their structures and the symptoms produced.

The amatoxins of Group I are definitely the most hazardous of the mushroom toxins. *Amanita phalloides* contains 2–3 mg of amatoxins per gram of dry tissue. A single mushroom can kill an adult human. The amatoxins are cyclic octapeptides, while the related phallotoxins are cyclic heptapeptides. The phallotoxins are much less toxic than the amatoxins, perhaps due to poor absorption from the gastrointestinal (GI) tract.

Symptoms of amatoxin poisoning begin within 6–24 hours of ingestion of the mushrooms. The first stage involves primarily the GI tract with abdominal pain, nausea, vomiting, diarrhea, and hyperglycemia. A short period of remission usually follows. The third and often fatal stage of the intoxication involves severe liver and kidney dysfunction. The symptoms experienced in this third stage include abdominal pain, jaundice, renal failure, hypoglycemia, convulsions, coma, and death. Death results from hypoglycemic shock, usually between the fourth and seventh days after the onset of symptoms. Recovery may require 2 weeks with intensive medical intervention.

The Group II toxins are hydrazines, such as gyromitrin. A bloated feeling, nausea, vomiting, watery or bloody diarrhea, abdominal pain, muscle cramps, faintness, and loss of motor coordination typically occur 6–12 hours after consumption of *Gyromitra esculenta* mushrooms. In rare cases, the illness can progress to convulsions, coma, and death.

The Group III toxins, characterized by muscarine, affect the autonomic nervous system. Within a few minutes to a few hours of consumption of mushrooms containing these toxins, the patient will experience perspiration, salivation, and lacrimation (PSL) syndrome, blurred vision, abdominal cramps, watery diarrhea, constriction of the pupils, hypotension, and a slowed pulse. Death does not usually occur when these are the only toxins in the poisonous mushrooms (e.g. *Clitocybe dealbata*). Fly agaric also contains Group I toxins, so a fatal combination of symptoms may occur.

Coprine is the classic example of a Group IV toxin. It causes symptoms only in conjunction with alcohol. Symptoms typically begin about 30 minutes after consuming alcohol and may occur for as long as 5 days after mushroom ingestion. The symptoms include flushing of the face and neck, distension of the veins in the neck, swelling and tingling of the hands, metallic taste, tachycardia, and hypotension, progressing to nausea and vomiting. These also are the result of actions on the autonomic nervous system.

The Group V and VI toxins primarily act on the central nervous system, producing hallucinations. The Group V toxins, isoxazoles including ibotenic acid and muscimol, cause dizziness, incoordination, staggering, muscular jerking and spasms, hyperkinetic activity, a coma-like sleep, and hallucinations within 30 minutes to 2 hours of ingestion. The Group VI toxins, including psilocybin and psilocin, are indoles. Symptoms begin about 30–60 minutes after ingestion and include pleasant or apprehensive mood, unmotivated laughter and hilarity, compulsive movements, muscle weakness, drowsiness, hallucinations, and finally sleep. Recovery is spontaneous. Mexican mushrooms have been used as recreational drugs for their hallucinogenic effects. However, the level of hallucinogens in these mushrooms is widely variable, so prolonged and severe side effects can be experienced. Death has occurred in small children who accidentally ate *Psilocybe* mushrooms. More commonly, patients experience persistent sequelae and are admitted to mental institutions.

Group VII toxins are not represented in Table 14.5. However, it is known that many mushroom species will cause nausea, vomiting, diarrhea, and abdominal pain within a few minutes to a few hours of ingestion. The toxins responsible have not yet been identified.

Another class of mushroom toxins not captured in this classical grouping is orellanine and orelline. These toxins are found in various *Cortinarius* species especially *C. orellanus*. The orellanus toxins are nephrotoxins. Symptoms may not appear until 36 hours up to 14 days after ingestion of the mushrooms. Symptoms include nausea, paresthesias, anorexia, GI disturbances, headache, an intense burning thirst, oliguria, and eventually renal failure. Renal function may return to normal slowly or result in the need for prolonged intermittent dialysis. Fatality rates once approached 15% but, with dialysis, few fatalities are reported. Intoxications with orellanus toxins have occurred primarily in Europe, although a few cases have occurred in the US.

4. Naturally occurring constituents eaten in abnormally large amounts

Since the dose of a toxic chemical is the major determinant of hazard, some foodborne intoxications occur only if unusually large amounts of the toxin-containing food are consumed. The cyanogenic glycosides in cassava and lima bean are examples.

Many plants contain cyanogenic glycosides, such as linamarin in lima beans (Fig. 14.4). Cyanide can be released from these compounds by enzymes present in plant tissues during storage and processing of the food, or by stomach acid after the food has been ingested. The amount of cyanide present in plants varies with the species, variety, and part of the plant (Table 14.6). The commercial lima beans (white American) contain far less cyanide than some of the wild varieties.

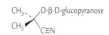

Figure 14.4 Structure of linamarin.

Table 14.6 Maximum cyanide yields of various plants

Plant	HCN yield (mg/100 g wet weight)
Bitter almonds	250
Bitter cassava	
Dried root cortex	245
Whole root	53
Fresh root bark	89
Fresh stem bark	133
Leaves	104
Sorghum – whole plant, immature	250
Lima bean varieties	
Java, colored	312
Puerto Rica, black	300
Burma, white	210
Arizona, colored	17
America, white	10

Adapted from Montgomery (1980).

Cyanide binds to heme proteins in the mitochondria and to hemoglobin in the blood, inhibiting cellular respiration. Cyanide prevents oxygen binding to hemoglobin, causing 'cyanosis', where the skin and mucous membranes display a bluish coloration. The symptoms of cyanide intoxication include a rapid onset of peripheral numbness and dizziness, mental confusion, stupor, cyanosis, twitching, convulsions, coma, and death. The lethal oral dose for humans is 0.5–3.5 mg/kg body weight. Cyanide is rapidly absorbed from the GI tract.

Reasonable quantities of lima beans or cassava contain doses of cyanide insufficient to produce symptoms. Lima bean varieties used in the US have low HCN yields – about 10 mg/100 g on a wet-weight basis. Assuming a lethal dose of cyanide is 0.5 mg/kg, a 70-kg adult would be unlikely to ingest 35 mg of cyanide or 350 g of lima beans. Cyanide is excreted from the body and does not accumulate. Cassava has been associated with some cases of cyanide intoxication in parts of Africa and South America where people have little else to eat on occasion. Deaths have not been reported. Adverse reactions including deaths have occurred from the ingestion of fruit pits, because the level of cyanide found in these sources is much higher than that found in lima beans or cassava.

5. Naturally occurring constituents eaten by abnormal consumers

Some naturally occurring constituents of foods are hazardous only for consumers with enhanced sensitivities, such as those individuals with food allergies or metabolic food disorders discussed in Chapter 1 on foodborne disease processes.

6. Naturally occurring constituents from foods processed or prepared in an unusual manner

Humans have learned through the years how to detoxify many of the hazardous substances found in foods. For example, in the raw state, soybeans contain trypsin inhibitors, lectins, amylase inhibitors, saponins, various antivitamins, and other potentially hazardous factors. The most important of these, the trypsin inhibitors and lectins, are inactivated by heating the soybeans. Fermentation can also destroy certain of these toxic factors, so fermented soybean products, such as soy sauce and tofu, are not hazardous. All other legumes contain similar toxicants. Many other foods contain toxicants that are removed or inactivated during processing. Food technologists should be aware of the naturally occurring toxicants in their raw materials and make certain that those toxicants do not appear in hazardous amounts in new food products.

An example is the presence of lectins in undercooked or raw kidney beans. Lectins bind to sugar residues on the surfaces of cell membranes, causing hemolysis of red blood cells and intestinal damage. Lectins from kidney beans cause nausea, abdominal pain, vomiting, and bloody diarrhea. The lectins are inactivated by thorough cooking. Problems have occurred among recent immigrants to England, who did not understand the importance of thorough cooking of kidney beans, a staple of the British diet. Consumers who soaked the raw beans and ate them with little or no cooking had a prompt onset of GI symptoms.

V. Summary

Clearly, a great many substances that occur naturally in foods are toxic. The distinction between natural and artificial has nothing to do with safety. Whereas intense government scrutiny is devoted to synthetic substances in food, there are very few regulatory limits on the incidence and levels of natural toxicants. Safety from these natural substances depends largely on proper processing and preparation of food, and on the inherently low levels of some of the toxicants. Sometimes, however, inappropriate substances are ingested and are found to be naturally deadly.

Bibliography

Committee on Food Protection, National Academy of Sciences (1973). 'Toxicants Occurring Naturally in Foods', 2nd ed. National Academy of Sciences, Washington, DC.

Erban, A. M., Rodriguez, J. L., McCullough, J. and Ownby, D. R. (1993). Anaphylaxis after ingestion of beignets contaminated with *Dermatophagoides farinae*. *J. Allergy Clin. Immunol.* **92**, 846–849.

Hu, W. J., Zhang, G. S., Chu, F. S., Meng, H. D. and Meng, Z. H. (1984). Purification and partial characterization of flavotoxin A. *Appl. Environ. Microbiol.* **48**, 690–693.

Jaffe, W. G. and Seidl, D. S. (1992). Toxicology of plant lectins. *In* 'Handbook of Natural Toxins, Vol. 7, Food Poisoning' (A. T. Tu, ed.), pp. 263–290. Marcel Dekker, New York.

Keating, J. P., Lell, M. E., Straus, A. W., Zarkowsky, H. and Smith, G. E. (1973). Infantile methemoglobinemia caused by carrot juice. *N. Engl. J. Med.* **288**, 824–826.

Liener, I.E. (ed.) (1980). 'Toxic Constituents of Plant Foodstuffs', 2nd ed. Academic Press, New York.

Marais, J. P. (1997). Nitrate and oxalates. *In* 'Handbook of Plant and Fungal Toxicants', (J. P. F. D'Mello, ed.), pp. 205–218. CRC Press, Boca Raton, FL.

Montgomery, R. D. (1980). Cyanogens. *In* 'Toxic Constituents of Plant Foodstuffs' (I. E. Liener, ed.), pp. 143–160. Academic Press, New York.

Noah, N. D., Bender, A. E., Reaidi, G. B. and Gilbert, R. J. (1980). Food poisoning from raw red kidney beans. *Br. Med. J.* **6234**, 236–237.

Sinden, S. L. and Deahl, K. L. (1994). Alkaloids. *In* 'Foodborne Disease Handbook, Vol. 3, Diseases Caused by Hazardous Substances' (Y. H. Hui, J. R. Gorham, K. D. Murrell and D. O. Cliver, eds), pp. 227–259. Marcel Dekker, New York.

Spoerke, D. G., Jr (1994). Mushrooms: epidemiology and medical management. *In* 'Foodborne Disease Handbook, Vol. 3, Diseases Caused by Hazardous Substances' (Y. H. Hui, J. R. Gorham, K. D. Murrell and D. O. Cliver, eds), pp. 433–462. Marcel Dekker, New York.

Tyler, V. E. (1993). 'The Honest Herbal – A Sensible Guide to the Use of Herbals and Related Remedies', 3rd ed. George F. Stickley, Philadelphia, PA.

Van Veen, A. G. (1967). The Bongkrek toxins. *In* 'Biochemistry of Some Foodborne Microbial Toxins' (R. I. Mateles and G. N. Wogan, eds), pp. 43–50. MIT Press, Cambridge, MA.

Wirtz, R. A., Taylor, S. L. and Semey, H. G. (1978). Concentrations of substituted *p*-benzoquinones and 1-pentadecene in the flour beetles *Tribolium consfusum* J. Du Val and *Tribolium castaneum* (Herbst). *Comp. Biochem. Physiol.* **61B**, 25–28.

Seafood Toxins

15

Eric A. Johnson [1] and Edward J. Schantz

I. Introduction

Seafood consumption has increased in many countries during the past two decades. In the US, seafood consumption increased nearly 21% during the 1980s. Finfish and shellfish have an abundance of certain nutrients, and can contribute to a healthy and delicious diet for humans. Currently, people throughout the world receive about 6% of their total protein and 16% of their animal protein from fish.

Although seafoods are nutritious and healthful, they have also served as vehicles for a variety of foodborne illnesses. They have been associated with the transmission of bacterial and viral gastroenteritis, and have also supported outbreaks of bacterial intoxications, including staphylococcal poisoning and botulism. These topics are addressed in other chapters of this book. Seafoods are unique in also transmitting

[1] Corresponding author.

Foodborne Diseases 2nd Edn
ISBN: 0-12-176559-8

diseases mediated by nonprotein, heat-stable, low-molecular-weight toxins produced mainly by microalgae and bacteria. Seafood toxins cause a variety of illnesses of humans and animals in many areas of the world. Several of these illnesses, such as paralytic shellfish poisoning, puffer fish poisoning and neurotoxic shellfish poisoning, have been known for centuries; whereas others, such as amnesic shellfish poisoning and diarrhetic shellfish poisoning, were recognized more recently. Contaminated shellfish, including mussels, clams, cockles, oysters, scallops, and other varieties feeding on toxic microalgae, are the main vehicles of seafood intoxications. For some toxin-mediated illnesses, a single clam or mussel contains enough poison to kill a human, but without noticeable health or organoleptic effects to the shellfish. The normal appearance and taste of toxic seafoods presents much difficulty for control of these illnesses.

In addition to shellfish, certain finfish species that have been spoiled by bacteria (scombroid poisoning) or that have fed on toxic algae (ciguatera) also can cause human foodborne illness. Cyanobacteria (blue–green algae) and certain eukaryotic algae, such as *Pfiesteria*, have caused devastating waterborne poisonings of animals in recent years and also sporadic human poisonings. It is possible that they may also cause outbreaks of human foodborne illness. With increased interstate and international transport of seafoods, nearly all human populations that consume finfish and shellfish are at risk of seafood poisonings. Evidence indicates that seafood poisonings are increasing in frequency and newer ones are emerging. In the US, Canada, and certain other countries, governmental monitoring of coastal waters for toxic algae and toxins in seafoods, and cautionary warnings have reduced the risk of consumption of poisonous seafood.

II. Overview of the Causes of Seafood Intoxications

The main recognized human intoxications from fish and shellfish include amnesic shellfish poisoning (ASP), ciguatera fish poisoning (CFP), diarrhetic shellfish poisoning (DSP), neurotoxic shellfish poisoning (NSP), paralytic shellfish poisoning (PSP), puffer fish poisoning (PFP), scombroid fish intoxication (SFP), and certain other rare intoxications. Except for scombroid toxin (histamine), which is produced by bacterial spoilage of improperly refrigerated fish, most of the toxins are produced by marine unicellular algae or phytoplankton. Of the more than 5000 known species of phytoplankton, about 60–80 are known to produce harmful toxins. Less than 25 toxic species were recognized only a decade ago. Occasionally, the algae grow to large numbers and form 'blooms' that are visible as patches near the water surface. 'Red tide' is a common name for a harmful algal bloom in which the red pigments of the algal species give the ocean patch its characteristic color. Blooms vary in color depending on the alga and appear as red, brown, or green. Early records and folklore indicate that toxic algal blooms have occurred for hundreds of years, but their actual incidence and the associated causative algae were not accurately identified until relatively recently.

Toxic algal blooms are reportedly increasing worldwide in frequency, magnitude, and geographical extent. They have a marked adverse effect on fisheries, aquaculture, and human health. The ecological factors contributing to this increased occurrence of toxic algal blooms are not completely understood, but recognized contributing factors include increased availability of nutrients through pollution and oceanic currents and upwelling. Changes in climate have also been proposed to contribute to blooms.

Microalgal toxins are produced as secondary metabolites and are transferred through the food web, where they accumulate in shellfish and finfish. Since the toxins accumulate in the seafood through the food chain, the toxic seafoods usually appear unspoiled and seemingly harmless. Most of the toxins are tasteless and colorless at poisonous levels. This mode of toxin contamination combined with the heat stability of the toxins presents considerable difficulty in prevention of seafood intoxications. Currently, prevention of seafood intoxications depends mainly on surveillance and detection of toxins in the commodities at the point of harvest.

III. Incidence and Economic Costs of Seafood Intoxications

Globally, about 60 000 seafood intoxications and at least 100 deaths are reported each year. Like most other foodborne illnesses, this is certainly an underestimation by at least 10–50-fold, since many cases of seafood intoxications are mild, are not reported, or are misdiagnosed. The most common intoxications worldwide are PSP, SFP, and CFP. Globally, PSP is probably the most widespread geographically of seafood intoxications, while most seafood intoxications are clustered in geographic locations near to the area of harvest.

Since 1978, recognized seafood intoxications in the US included ASP, CFP, NSP, PSP, and SFP. Incidents of DSP affecting humans have not been confirmed in the US, but a large outbreak occurred in Canada. Only CFP, PSP, and SFP are reported in *Morbidity and Mortality Weekly Report* by the Centers for Disease Control and Prevention (CDC), in the category of 'chemical poisonings'. In the US, CFP and SFP are responsible for more than 80% of the seafood intoxications. In the latest published statistics from the CDC for chemical poisonings, this category accounted for ~17% of the outbreaks and ~1% of the total foodborne illness cases for the period 1992–1997. The incidence of chemical poisonings during the period 1992–1997 was lower than during the previous reporting period of 1973–1987, during which chemical etiology accounted for ~25% of the outbreaks and ~4% of the cases. This may be due to the change of outbreak definition, which is now defined as two or more cases of a similar illness resulting from ingestion of a common food. Before 1992, a single case of a seafood intoxication was reported as an outbreak. Foodborne illness outbreaks occurring on cruise ships were also not included in the CDC tabulations. In the US, seafood ranked third on the list of products that transmitted foodborne disease between 1983 and 1992. An estimated 2970 cases of foodborne illness were caused by seafood toxins in Canada during that period, compared to 260 reported cases in the US.

In the US during 1993–1997, 60 outbreaks including 205 cases were due to ciguatoxin fish poisoning; 69 outbreaks including 297 cases were due to SFP; and one outbreak including three cases was due to shellfish poisoning (presumably PSP), as reported by the CDC. Unlike earlier reporting periods, there were no deaths due to seafood-associated intoxications. In the earlier (1988–1992) reporting period for foodborne disease, 42 outbreaks and 176 cases were due to ciguatoxin, 76 outbreaks and 514 cases with one death due to SFP, and five outbreaks and 65 cases with two deaths were due to PSP.

The costs of seafood intoxications have been estimated in Canada. For the estimated 150 PSP cases each year in Canada, the total cost was estimated to be $539 000, with the cost per case about $3600. The PSP monitoring program in 1988 cost $3 000 000. It was estimated that 57 DSP cases occur annually costing about $112 000, or $2000 per case; 190 SFP cases occur yearly costing over $300 000 per year, or $1600 per case; and 133 ciguatera cases occur annually costing $640 000 or about $4800 per case. The total annual cost associated with the estimated 550 illnesses associated with seafood toxins was estimated to be about $1 603 000 and the product monitoring costs were $3.2 million. In the US the incidence of seafood toxin illnesses was estimated to be 58 260 cases at a cost of $158 million dollars.

IV. Amnesic Shellfish Poisoning (Domoic Acid)

Amnesic shellfish poisoning is a life-threatening shellfish intoxication. It results from eating shellfish contaminated with domoic acid, which is produced by diatoms in the species *Pseudonitzschia* (Table 15.1). It is a newly recognized seafood intoxication that was described in 1987 among persons who ate poisonous blue mussels from Prince Edward Island, Canada. Until this outbreak, the diatom *Pseudonitzschia* was not thought to produce toxins poisonous to humans or animals. Most of the persons in the Canadian outbreak experienced gastroenteritis including vomiting (75%), diarrhea (42%), abdominal cramps (49%), while some older persons with underlying chronic diseases developed neurological symptoms including memory loss, confusion, disorientation, seizure, coma, or cranial nerve palsies within 48 hours. Certain patients experienced short-term memory loss for at least 5 years, and some patients were unable to recognize family members or perform simple tasks. Memory loss was more common in elderly patients greater than 70 years of age than in the young. Evidence suggests that persons with impaired renal function may be at greater risk of domoic acid neurotoxicity because of impaired ability to inactivate and excrete the toxin. Of the 107 persons affected in the Canadian outbreak, three patients died within 3 weeks of eating the mussels. Currently there is no medical treatment for ASP other than supportive care. Medications have been administered to control seizures and potentially reduce the extent of brain lesions.

Investigators were unable to find infectious levels of pathogenic bacteria or viruses, or toxic levels of substances such as heavy metals or organophosphorous pesticides in the poisonous Canadian mussels. Using the mouse assay designed to detect saxitoxin

Table 15.1 Seafood intoxications

Toxic syndrome	Source of toxin	Geographic areas	Foods affected	Major toxin	Onset time; duration of illness	Major symptoms	Treatment	Prevention
Amnesic shellfish poisoning (ASP)	Diatom: *Pseudonitzschia* spp.	NE Canada; rare or affects animals in NW US, Europe, Japan, Australia, and New Zealand	Mussels, clams, crabs, scallops, anchovies	Domoic acid	Hours; months to years	n, v, d, p, r	Supportive (respiratory)	Seafood surveillance; quarantine of seafood. Rapid reporting
Ciguatera	Toxic dinoflagellates: *Gambierdiscus toxicus*, *Prorocentrum* spp., *Ostreopsis* spp., *Coolia monotis*, *Thecadinium* spp., *Amphidinium carterae*	Tropical areas around the world; in US mainly near Florida	Edible tropical fish; commonly barracuda, kahala, snapper, grouper	Ciguatoxin	Hours; months to years	n, v, d, t, p	Supportive	Seafood surveillance; quarantine of seafood, region. Rapid reporting
Diarrhetic shellfish poisoning (DSP)	Toxic dinoflagellates: *Dinophysis* spp., *Prorocentrum* spp.	Mainly in Europe, Japan. Rare cases in Chile, Southeast Asia, and New Zealand	Mussels, clams, scallops	Okadaic acid	Hours; days	d, n, v	Supportive	Seafood/water surveillance; quarantine of seafood/region. Rapid reporting
Neurotoxic shellfish poisoning (NSP)	*Gymnodinium breve*; possibly other species	Gulf of Mexico, South Atlantic Blight; New Zealand	Bay scallops, clams, oysters, quahogs, cohinas	Brevetoxin	30 minutes to a few hours; few hours	n, v, d, b, t, p	Supportive	Seafood/water surveillance; quarantine of seafood/region. Rapid reporting

(continued)

Table 15.1 (*continued*)

Toxic syndrome	Source of toxin	Geographic areas	Foods affected	Major toxin	Onset time; duration of illness	Major symptoms	Treatment	Prevention
Paralytic shellfish poisoning (PSP)	Toxic dinoflagellates: *Alexandrium* spp., *Gymnodinium catenatum*, *Pyrodinium bahamense*	Worldwide in coastal regions	Mussels, clams, bay scallops, quahogs, some finfish	Saxitoxin	5–30 minutes; occasionally few hours–few days	n, n, d, p, r	Supportive (respiratory)	Seafood surveillance; quarantine of seafood/region. Rapid reporting
Puffer fish poisoning (PFP)	Puffer fish, poison in liver, gonads, and roe. Possibly produced by bacteria	Pacific regions near Japan and China; rare in US	Puffer fish or globefish	Tetrodotoxin	Similar to PSP	n, v, d, p, r, bp	Supportive (respiratory)	Regulated food source, preparation. Rapid reporting
Scombroid fish poisoning	Bacterial decomposition of fish held at elevated temperatures (>5°C)	Worldwide	Various fish; common in mahimahi, tuna, bluefish, mackerel, skipjack	Histamine	10–90 minutes; hours	a, d, h, v	Supportive; antihistamine	Regulated food handling; keep temperature <5°C

Symptoms: a, allergic-like; b, bronchoconstriction; bp, decrease in blood pressure; d, diarrhea; n, nausea; p, paresthesias; r, respiratory distress; t, reversal of temperature sensation; v, vomiting.

(Association of Official Analytical Chemists, 1997), it was found that mussel extracts caused death of the mice, usually within 15–45 minutes of intraperitoneal injection. However, the symptoms were distinct from those of PSP, with a unique scratching syndrome of the shoulders and hind leg, followed by convulsions and death. Upon examination of the mussels, it was found that the digestive glands of poisonous animals contained green phytoplankton. Investigators were rapidly able to purify domoic acid, and showed that it caused similar unique symptoms in mice. The toxic mussels contained up to 900 mg of domoic acid per kg of tissue. Although several algae can produce domoic acid, it was found that a bloom of the pennate diatom *Nitzschia pungens* f. *multiseries* that was occurring at the time of the outbreak contained domoic acid. Isolates of this *Nitzschia* strain produced domoic acid in culture, demonstrating that the diatom was the causative agent and not just a vehicle for the toxin. Domoic acid was produced as a secondary metabolite in axenic cultures of the alga *N. pungens* f. *multiseries*. Evidence indicated that domoic acid was produced by a limited number of species in the genera *Nitzschia*, *Digenea*, *Vidalia*, *Amansia*, and *Chondriaarmata*. Presently, species of the genus *Pseudonitzschia* are regarded as the principal agents of domoic acid production.

Domoic acid is water soluble, with a molecular weight of 311 Da. It contains a glutamate-like moiety and is an analog of kainic acid. Kainic acid binds to certain receptors in the central nervous system and stimulates the release of glutamate in the manner of excitotoxins. Evidence suggests that domoic acid affects calcium transport and stimulates a calcium-dependent process that regulates release of glutamate from presynaptic nerve endings. Domoic acid has been shown to be excitotoxic in a number of animal models including rodents and primates, and induces seizures at high doses. In rodents, domoic acid produces a loss of neurons in various brain regions, particularly the hippocampus. On autopsy, brain tissue from the victims of the 1987 Canadian outbreak had lesions in several regions including the hippocampus, amygdala, thalamus, and cerebral cortex. The median lethal dose, administered intraperitoneally to mice, was estimated to be 3.6 mg domoic acid per kg body weight. Human intoxication in ASP cases occurred after ingestion of an estimated 1–5 mg of domoic acid per kg of body weight. Domoic acid has also been shown to cause deaths among various marine animals, including birds and sea mammals such as sea lions and humpback whales.

Initially the intraperitoneal mouse assay for PSP detection was used to survey seafood for domoic acid. However, this assay is not consistently sensitive enough to detect the toxin at the Canadian regulatory level of 20 μg/g tissue. A nondestructive extraction method, combined with reverse-phase chromatographic separation and ultraviolet detection at 242 nm, has been employed and was adopted as an official first action by the Association of Official Analytical Chemists (AOAC) in 1990. The detection limit of this method is about 1 μg domoic acid per gram of tissue. Other biochemical methods have also been investigated for sensitive and accurate detection of domoic acid. The current Canadian and UK guideline for the limit of PSP in seafood is 0.8 mg/kg of edible meat.

Although toxic *Pseudonitzschia* species occur in oceans worldwide, only two outbreaks of ASP affecting humans or animals have been reported. The first was the Canadian outbreak in 1987, followed by a large outbreak in seabirds in September 1991 in Monterey, California. High levels of domoic acid were found in *Pseudonitzschia australis* harvested from the area. Since anchovies are a major food source for seabirds in the area, it is possible that the intoxication could be transmitted in herbivorous finfish such as anchovies. Outbreaks of domoic acid poisoning have also caused mortality in sea lions and humpback whales. The ASP outbreak affected 24 people who consumed razor clams and became ill with gastrointestinal symptoms; two people developed mild neurological symptoms. Surveys have detected razor clams and Dungeness crabs in Washington and Oregon that contained domoic acid.

Monitoring of phytoplankton blooms for domoic acid in response to the 1987 outbreak has contributed to prevention of ASP. It has been recommended that shellfish from suspect regions be tested and those that contain $\geqslant 20\,\text{mg/kg}$ should not be harvested for human consumption. If shellfish are found with levels above $5\,\text{mg/kg}$ but below $20\,\text{mg/kg}$, the harvest area is monitored closely. Domoic acid is permitted in the US in bivalve shellfish and cooked crab viscera at levels of 20 or $30\,\text{mg/kg}$ tissue, respectively. In 1988, the US started the National Shellfish Sanitation Program, which is a cooperative program to reduce risks from consumption of toxic shellfish. The program includes contingency plans in case of an outbreak, certification of harvesters, processors and distributors, and tracking the shipping of shellfish. The most extensive shellfish monitoring is directed at PSP, but domoic acid is also monitored to a lesser extent. As of 1998, monitoring programs in Canada and the US have found domoic acid in seafood products in Washington, California, Oregon, Alaska, Bay of Fundy, British Columbia, and Prince Edward Island. Domoic acid has also been detected at low concentrations in regions of Australia, Europe, Japan, and New Zealand. In addition to monitoring for domoic acid, depuration has been considered as a detoxification method for shellfish. However, depuration rates vary greatly, depending on the animal species and type of toxin, so depuration is not a consistent method of detoxification.

V. Ciguatera Fish Poisoning

Ciguatera fish poisoning is one of the most common seafood-associated illnesses and is caused by eating finfish, from tropical reef and island habitats, that have accumulated ciguatera toxins (CTXs) from epibenthic dinoflagellates in the food chain. Approximately 20 000 cases have been estimated to occur worldwide annually. The CDC generally reports 10–20 outbreaks per year, with 50–100 cases in the US. The outbreaks usually occur in Hawaii, Puerto Rico, the Virgin Islands, and Florida, but can occur in other regions to which the fish are shipped.

As many as 400 species of fish have been implicated in ciguatera poisoning in the Caribbean and Pacific regions (Table 15.1); those most commonly involved are amberjack, snapper, grouper, barracuda, goatfish, and reef fish in the family Carrangidae. Various dinoflagellates are known to produce CTXs, as shown in Table 15.1.

Ciguatera poisoning can involve gastrointestinal, neurological, and cardiovascular symptoms (Table 15.1). Gastrointestinal symptoms include diarrhea, abdominal pain, nausea and vomiting, and these usually begin a few hours after ingestion of the fish and last for only a few hours. Neurological symptoms usually begin 12–18 hours after consumption and vary in severity. Neurological signs include: reversal of temperature sensation (e.g. ice cream tastes hot; hot coffee tastes cold); muscle aches; dizziness; tingling and numbness of lips, tongue, and digits; metallic taste; dryness of mouth; anxiety; sweating; dilated eyes; blurred vision; and temporary blindness. Paralysis and death have been documented, but these are rare. There is considerable variation of symptoms and recovery time in individual patients. Recovery may require weeks, months, or even years; the chronic effects of CFP have not been elucidated. Intravenous administration of mannitol can help to relieve acute symptoms, and amitryptiline or tocainide have been suggested for treatment of chronic symptoms. The most common treatment is supportive with attention to respiratory and cardiovascular functions. The fatality rate for CFP overall is <1%, but has ranged from 0% to 12% in various fish outbreaks.

Several CTXs have been isolated from toxic fish and algae. They consist of a family of lipophilic brevetoxin-type polyether compounds; the prototype compound, gambiertoxin-4, was isolated and characterized from *Ganmbierdiscus toxicus*. The total synthesis of brevetoxin A has been accomplished. A family of CTXs has been found, since different algal species produce variant toxin structures; also, the toxins may be modified by animal or human metabolism. The structures of at least eight 'native' CTXs have been determined, and 11 CTXs formed by oxidative metabolism have been detected. The toxic mechanism of CTXs involves binding to and opening of sodium and calcium channels on excitable membranes. Like most other seafood toxins, CTXs are commonly detected by mouse bioassay. Immunoassays are available including 'dipstick' tests and commercialized kits.

Fish containing toxic levels of CTXs usually do not appear spoiled. Prevention of intoxications depends on surveillance and detection of toxins in fish and algae from endemic areas, and rapid reporting and treatment of cluster outbreaks. For the majority of US consumers, the illness is contracted from fish imported from endemic areas.

VI. Diarrhetic Shellfish Poisoning

Diarrhetic Shellfish Poisoning was reported in 1976 to occur from eating toxic mussels, scallops, or clams (Table 15.1). DSP occurs mainly in Japan and northern Europe, but also in various other regions of the world, including outbreaks in South America, South Africa, southeastern Asia, and New Zealand. DSP is caused by toxins produced by *Dinophysis* spp. DSP is not usually fatal, but shellfish may become toxic in the presence of dinoflagellates at low cell densities ($\geqslant$200 cells/ml). DSP is characterized by gastrointestinal symptoms such as severe diarrhea, nausea, vomiting, abdominal cramps, and chills that onset within 30 minutes to a few hours after eating toxic shellfish. Complete recovery usually occurs within 3 days.

Various toxins are produced by *Dinophysis* spp. including okadaic acid, pectenotoxins, and yessotoxin. Only okadaic acid has been demonstrated to cause a definitive diarrheal syndrome. Certain of the *Dinophysis* toxins appear to have mutagenic, cancer-inducing properties, hepatotoxic, or immunogenic properties; but the chronic toxic effects in humans are not known. The mouse bioassay is most commonly used to detect the presence of DSP toxins, although immunoassays and cytotoxicity-based assays have also been evaluated. Limits for diarrheal shellfish toxins (usually undetectable by mouse assay) have been proposed in Japan and certain European countries.

VII. Neurotoxic Shellfish Poisoning

Neurotoxic shellfish poisoning was observed centuries ago by Spanish explorers and Tampa Bay Indians, during certain seasons when coastal waters became red and massive fish kills occurred. Shellfish poisonings of humans were reported in the late 1800s and in 1946. NSP in the US is usually confined to regions near Florida (Table 15.1), but a bloom was spread by the Gulf Stream, leading to an outbreak in North Carolina. Outbreaks have also occurred in New Zealand.

The symptoms of NSP mimic those of ciguatera, in which gastrointestinal and neurologic symptoms predominate (Table 15.1). Symptoms begin in 30 minutes to 3 hours and include nausea, vomiting, diarrhea, bronchorestriction, and paresthesias. The illness generally subsides within 2 days. The symptoms are usually less severe than in ciguatera poisoning, and no deaths have been reported, but the illness is still debilitating. Unlike ciguatera, which can persist for weeks, NSP generally subsides within a few days. Treatment is supportive and there is no antidote. NSP blooms can become aerosolized in the surf and cause respiratory and asthma-like problems for people who breathe them on the beach. NSP appears to be rare throughout the world, with documented outbreaks mainly in the US and New Zealand. Algae related to *Gymnodinium breve* have been detected in Spain and Japan, and it is possible that intoxications could occur from shellfish harvested from these regions.

G. breve collected from toxic blooms was found to produce polyether brevetoxins with structures related to certain ciguatoxins. Brevetoxin causes openings of sodium channels on nerves and other tissues. Brevetoxin is detected by mouse bioassay or by enzyme-linked immunosorbent assay (ELISA) tests. Prevention of poisonings from shellfish depends on surveillance of waters for toxic algae, and rapid reporting. Coastal waters have been monitored for *G. breve* cell counts, and this has successfully prevented illnesses.

VIII. Paralytic Shellfish Poisoning

Paralytic shellfish poisoning is a serious and sometimes life-threatening intoxication that results from eating shellfish contaminated with saxitoxin (STX) and related toxins. PSP has a wider worldwide geographic distribution than other seafood intoxications caused by microalgal toxins (Table 15.1). PSP was first reported in 1793, after five members of Captain George Vancouver's ship crew became ill and one sailor died after eating mussels from Poison Cove in central British Columbia. PSP from toxic mussels and clams was also recognized on the Pacific Coast in the 1700s by native Americans, who connected the poisoning with red tides and associated bioluminescence. PSP occurs through ingestion of toxic bivalve molluscs (mainly mussels, clams, oysters, scallops) that have fed on toxic dinoflagellates. In the US, PSP has a wider geographical distribution than other dinoflagellate poisonings and occurs in the Pacific Northwest Coast and Alaska, and in New England from Massachusetts to Maine. Toxic algal blooms of *Alexandrium* spp. and other PSP-producing microalgal species in northern California and other cold temperate regions are seasonal, occurring mainly during the spring, and may be sustained through the summer in upwelling waters. Owing to current testing and control procedures, outbreaks are rare in commercial shellfish harvested from coastal regions. Most PSP outbreaks involve recreational collectors of bivalves, often from quarantine areas.

The symptoms of PSP generally begin within minutes after eating toxic shellfish, and initially affect the peripheral nervous system. The first signs of intoxication are a prickly feeling in the lips, tongue, and fingertips, followed by numbness in the extremities and face. The intoxication continues with an ataxic gait and muscular incoordination, followed by ascending paralysis. Death from respiratory failure may occur within 2–24 hours depending on the quantity of toxin ingested (2–4 mg is considered to be the lethal dose for a human). If one survives 24 hours the prognosis for complete recovery is good, and no chronic effects of the illness generally occur. There is no effective antidote; poisoned individuals should receive artificial respiration and supportive medical care as soon as they can be administered. Emergency treatment and first aid for victims of PSP have been described. In particular, attention should be given to cardiopulmonary resuscitation and respiratory ability, and the victim should be rapidly transported to a hospital emergency facility.

STX was the first toxin recognized in shellfish and thus has been extensively characterized. The nature of the poison responsible for PSP was elusive until it was discovered in the 1930s that culture supernatants lethal to mice were produced by phytoplankton, apparently of the genus *Gonyaulax*. A lethal substance was extracted from dinoflagellates harvested from blooms and from toxic shellfish. PSP or toxic mussel poison, now called saxitoxin, was purified and identified by Edward J. Schantz and colleagues. Toxic extracts were prepared from large quantities of harvested poisonous California mussels (*Mytilus californius*) and butter clams (*Saxidomas giganteus*) from Alaska. The toxic substances were purified and found to consist of tetrahydropurine derivatives. Good-quality crystals of STX were obtained, and the three-dimensional structure was resolved. The availability of purified toxin allowed the elucidation of the pharmacological mechanism of STX. It was demonstrated to bind selectively with high affinity to sodium channels of excitable membranes and to block the inward flux of sodium completely, in a manner very similar to tetrodotoxin. These toxins have become important neurobiological tools because of their selective and high-affinity blockade of the voltage-gated sodium channels of excitable membranes of neurons and skeletal muscle. Like most other marine microalgal toxins, PSP toxins occur as a family of related compounds and more than 20 distinct structures have been elucidated, which are referred to as saxitoxins, neosaxitoxins, or gonyautoxins. Current taxonomic understanding is that PSP is produced by various dinoflagellate species (Table 15.1), although reports have indicated that certain bacteria, including *Moraxella* spp., can produce low quantities of STX or inactive precursors and derivatives in culture.

The prevention of PSP is accomplished primarily by active monitoring of coastal algal blooms and seafoods for the presence of saxitoxin, and rapidly alerting the shellfish industry public of a health hazard from eating clams, mussels, and certain other shellfish from a designated region. Early investigators in California were instrumental in beginning, in the 1920s, a prevention program that consisted mainly of posting warning placards on the beaches with instructions not to eat clams, mussels, and certain other shellfish during the high seasons. Standardized procedures for detection of STX by mouse bioassay are currently used by governmental personnel in the US and Canada to determine if dangerous levels are present in shellfish. Industry personnel or recreational consumers of shellfish who plan to gather shellfish should contact their local, state, or national health authorities to obtain information regarding the safety of these foods.

IX. Puffer Fish Poisoning

Puffer fish poisoning has traditionally been associated with eating certain species of fish belonging to the Tetraodontiformes. These fish are commonly referred to as *fugu*, puffer fish, globefish, or swellfish, because they can inflate themselves (Table 15.1). It has been recognized for centuries that eating these fish could result in a paralytic poisoning. PFP most commonly occurs in countries that consume *fugu* as a delicacy, such as China and Japan. Puffer fish poisoning can be fatal and it has been estimated that about 1800 Japanese have died in the past 40 years by consumption of PFP-tainted and improperly prepared *fugu*. The toxicity of puffer fish varies according to its source, the variety and species of fish, and whether they are wild-caught, and grown or kept alive in aquaculture facilities.

The symptoms of PFP are similar to PSP, including initial tingling and prickling sensation of the lips, tongue, and fingers within a few minutes of eating poisonous fish. Nausea, vomiting, and gastrointestinal pain may follow in some cases. Depending on the quantity of toxin consumed, the pupillary and corneal reflexes are lost and respiratory distress follows. No antidote is currently available; treatment is supportive, with particular attention to maintaining respiration.

Puffer fish poison was first isolated in 1909 and named tetrodotoxin (TTX). The structure of TTX and derivatives was reported from Japan and the US in 1964. TTX is an amino perhydroquinazoline compound with a molecular weight of about 400, depending on the form. The chemical structure is distinct from STX, although the symptoms are analogous. It was found to be one of the most poisonous nonprotein substances known; the lethal dose is about 0.2 μg for a mouse, 4 μg for a 1 kg rabbit, and 1–4 mg for a human. Like STX, it has a highly specific action on sodium channels within excitable membranes.

TTX was long assumed to be produced by the *fugu*, but its detection in certain newts, frogs, marine snails, octopuses, squids, crabs, starfishes, and other creatures has indicated that it is formed lower in the food chain, possibly by bacteria including species of *Alteromonas*, *Vibrio*, and other bacterial genera. These bacteria produce various forms of TTXs that vary in potency, including nontoxigenic precursors or derivatives.

X. *Pfiesteria* Toxins

Toxic *Pfiesteria* heterotrophic dinoflagellates were recognized in the early 1990s to cause kills of millions to billions of finfish and shellfish in coastal waters of the mid-Atlantic and southeastern US. The predatory organism, mainly the species *P. piscicida*, has also been reported from the Mediterranean Sea, the Gulf of Mexico, and the western Atlantic. The organism exists in several life stages. The organism remains in river and coastal bottoms for years as cysts and, when induced by unknown factors in fish feces, the cysts bloom into a motile form that swarms to the upper waters and produces very potent toxins. This results in a 'feeding frenzy', after which the organism transforms to an ameba state that feeds on microorganisms and fish remains, followed by reformation of cysts, which settle in the coastal sediment to complete the cycle. Since *Pfiesteria piscicida* resembles other dinoflagellates in certain morphological stages, light microscopy is usually not discriminatory and electron microscopy is currently required for identification.

In addition to the vicious fish kills, contact with contaminated water and aerosols has been associated with serious adverse health effects in humans. People in close proximity to *Pfiesteria* cultures in at least five different laboratories, or following direct contact with waters at fish kills, have developed ill effects ranging from narcosis, eye irritation, acute burning of skin and skin lesions, stomach cramping, respiratory distress, cognitive impairment, and memory loss lasting up to several months. *Pfiesteria* toxins have not been isolated and identified. Experiments are in progress to determine if the toxic effects in humans are due to *Pfiesteria* solely or together with associated microorganisms. Currently, *Pfiesteria* should be considered as a cause of human illness from contaminated waters, as well as an occupational and laboratory hazard. There have been no definitive reports of foodborne illness, but these may just be a matter of time as well as effective investigation and diagnosis.

Recommendations have been made for closing and reopening of waters affected by *Pfiesteria* or *Pfiesteria*-like events. Closure is recommended when a significant fish kill is reported and fish are found that contain sores and lesions consistent with the toxic activity of *Pfiesteria*; or when a significant number of fish exhibit erratic behavior that cannot be attributed to other factors such as low oxygen levels in the water. Waters may be reopened for recreational and commercial activities when these signs are not apparent for 14 days.

XI. Cyanobacterial Intoxications

Unlike dinoflagellates and diatoms, which cause human food poisonings from marine finfish and shellfish, cyanobacteria (sometimes called blue–green algae) have caused animal illnesses from consumption of drinking water. The vast majority of illnesses from cyanobacteria are waterborne and affect animals. The main toxic genera of prokaryotic cyanobacteria are the filamentous species *Anabena*, *Aphanizomenom*, *Nodularia*, *Oscillatoria*, and the unicellular species *Microcystis*. Like the marine eukaryotic microalgae, they form blooms under appropriate conditions in fresh waters. Blooms usually occur in the summer and autumn during warm days, and are promoted by nutrient availability, especially nitrogen and phosphorus, which often derives from water runoffs containing fertilizers, or from livestock or human wastes. Toxic blooms occur in many lakes, ponds, and rivers throughout the world. The primary toxicoses include gastrointestinal disturbances, acute hepatotoxicosis, neurotoxicoses, respiratory distress, and allergic reactions. Although not commonly a cause of human illness, a recent waterborne outbreak caused acute liver failure in more than 100 Brazilian hemodialysis patients.

Most of the poisonings by cyanobacteria involve acute hepatoxocosis and death mediated by microcystins and nodularin, which are heat-stable small peptides. Certain cyanobacteria also produce neurotoxicoses due to the alkaloidal anatoxins and anaphatoxins. These toxins can cause death within minutes to a few hours depending on the animal species and the quantity of toxin consumed. Cyanobacteria also produce a cholinesterase inhibitor called anatoxin-a(s), which has an organophosphate structure. Certain cyanobacteria have been reported to produce PSP-like toxins, including saxitoxin and neosaxitoxin.

Cyanobacterial toxins have sporadically caused human intoxications from drinking water. In most cases, water treatment systems are adequate to remove cyanobacteria by coagulation and filtration and microcystins can be adsorbed by charcoal filters and are degraded by chlorine. However, extracellular cyanobacterial toxins may survive water treatment and are resistant to boiling. Cyanobacterial toxins in human drinking water have been documented to cause hepatic toxicity, gastroenteritis, contact dermatitis and allergic responses, and neuronal and brain damage. Some cyanobacteria, e.g. *Spirulina*, have been marketed as a health food. Although studies in rodents have shown no toxicity, it is important that *Spirulina* and other cyanobacteria food supplements be produced under hygienic conditions, and do not contain cells or toxins from toxic species.

Diagnostic procedures for human and animal illnesses include: (1) association of a bloom of a toxigenic cyanobacterial species with consumption of water; (2) the presence of characteristic symptoms in the human or animal; (3) microscopic identification of the toxic species of cyanobacteria in the suspect water; and (4) verification of the presence of toxin in the water by chemical and bioassays. Procedures for the prevention of cyanobacterial intoxications from water rely on monitoring programs and quarantine measures when toxic algae reach a certain concentration in the bloom water. Methods with increased sensitivity compared to microscopic identification are being developed to rapidly detect toxin-producing cyanobacteria. Chemicals, particularly copper sulfate, have been added to lakes to kill cyanobacteria. Reduction of agricultural runoff and animal or human fecal contamination of water will also reduce bloom formation. Obviously, if a bloom occurs, animals and humans should avoid drinking the water.

Although there are no documented outbreaks of food poisoning caused by cyanobacterial toxins, shellfish can filter cyanobacteria from water and could accumulate their toxins. Mussels (*Mytilus edulis*) fed *Microcystis* accumulated microcystins, which persisted for several days after transfer to clear water. Fresh fruits and vegetables washed with contaminated water might also acquire these toxins. Since scenarios exist by which food could transmit cyanobacterial toxins, it may be important to monitor water from suspect sources that will contact foods. The UK and the State of Oregon have set limits of 1 ppm for microcystins in drinking water and dietary supplements. It is likely that cyanobacterial poisonings of humans will persist until blooms can be prevented.

XII. Scombroid (Histamine) Fish Poisoning

Scombroid poisoning is probably the most prevalent of the seafood transmitted illnesses worldwide (Table 15.1). Scombroid poisonings have been commonly reported in Japan, Canada, the US, England, and other countries that have a high dietary intake of fish. Scombroid poisoning symptoms mimic those of an IgE-mediated food allergy with flushing of the face, neck and upper arms, nausea, vomiting, diarrhea, abdominal pain, headache, dizziness, blurred vision, faintness, itching, rash, hives, and a burning sensation in the mouth. Hypotension, tachycardia, palpitations, respiratory distress, and shock may occur in severe cases. The symptoms of scombroid illness usually occur within 10–90 minutes of eating contaminated fish. Individuals exposed to scombroid poison will usually experience only a few of these symptoms. The duration of the illness is usually less than 12 hours. Diagnosis can generally be made of scombroid poisoning by the short onset time, nonspecific yet characteristic symptoms, a history of consumption of fish, and the illness can be confirmed by detection of histamine in the spoiled fish. Scombroid poisoning has been diagnosed by measurement of plasma histamine. Corticosteroids and H_1 and H_2 antihistamines can be used to treat the symptoms.

Owing to the variety of symptoms and the resemblance to allergic responses, the illness is frequently misdiagnosed and is often confused with an allergic reaction. Many of the symptoms of allergic reactions mimic those apparent in scombroid illness, since histamine is a primary mediator of allergic disease. Normally treatment is unnecessary as the vast majority cases are mild and self-limiting, but antihistamine therapy can provide relief and rapid recovery. Hydration and electrolyte replacement may also be beneficial. Scombroid poisoning can be severe in persons with a history of allergic diseases, with pre-existing cardiac or respiratory conditions, or in people being treated with certain drugs, such as isoniazid or monoamine oxidase inhibitors. Antihistamines should be administered only under close physician care in these special situations.

Nearly all cases of scombroid poisoning have been associated with marine fish, particularly of the scombroid (dark flesh) variety, such as tuna, bonito, and mackerel. Nonscombroid fish and shellfish have also been implicated in scombroid poisoning including mahi-mahi, swordfish, salmon, dolphin, marlin, sardines, bluefish, amberjack, anchovy, and abalone. Other foods have also caused scombroid poisoning, including Swiss cheese and some fermented foods and extracts. Scombroid poisoning is caused by certain bacterial species that grow in fish stored at inappropriate, elevated temperatures; the bacteria decarboxylate histidine to histamine. Histamine is heat-stable and the toxin survives cooking. Since orally administered histamine generally does not elicit symptoms, it is believed that potentiators such as the diamines, putrescine and cadaverine, promote the illness.

Several species of bacteria produce histamine through the action of the enzyme histidine decarboxylase. Bacterial species associated with scombroid poisoning include *Morganella morganii*, *Klebsiella pneumoniae*, *Vibrio* sp., *Enterobacter aerogenes*, *Clostridium perfringens*, *Hafnia alvei*, *Lactobacillus buchneri*, and *Lactobacillus delbrueckii*. Other enteric Enterobacteriaciae, clostridia, and vibrios have been associated with scombroid poisoning, but *M. morganii* and *K. pneumoniae* are the most common species implicated. These organisms are not frequently associated with living fish, and must contaminate the fish during handling and storage. However, fish held in warm water for hours, after being caught by the long-line method, have also been implicated. Since the organisms forming scombroid toxin are not psychrophiles, temperatures above 15°C are generally required to permit adequate growth and histamine formation. Histamine production on skipjack tuna was optimal at 30°C but, once a large population of bacteria is formed, the enzyme histidine decarboxylase can stay active even at refrigerator temperatures. Most of the histamine is produced near the intestines and then diffuses into the flesh.

The standard analytical method for detection of histamine and other biogenic amines is high-performance liquid chromatography, although other methods including radioimmunoassay kits are commercially available. The generally accepted toxic level of histamine in fish is 100 mg/100 g of flesh, but the amounts actually causing illness in ingested fish have not been accurately defined. Histamine can be used to judge the freshness of certain raw fish. The US Food and Drug Administration (FDA) considers 20 mg of histamine per 100 g of flesh, or 200 ppm indicative of spoilage in tuna, and 50 mg/100 g (500 ppm) an indication of a hazard. Since other finfish and shellfish intoxications show signs resembling scombroid, the final diagnosis may depend on detection of the toxins in the foods.

The most important contributing factor to scombroid poisoning is improper refrigeration of the harvested fish, allowing bacterial proliferation. Fish should be chilled as rapidly as possible, and be brought below 15°C and preferably below 10°C within 4 hours; longer storage of fish should be maintained at 0°C (32°F) or below by using ice, brine, or mechanical refrigeration. Maintaining sanitary conditions during handling, processing, and distribution will help to prevent bacterial contamination. Histamine is heat-stable and will withstand cooking. Improved reporting of scombroid incidences to public health agencies will increase awareness of the disease and its prevention.

XIII. Other Finfish and Shellfish Toxins

Various substances have been implicated in toxic fish kills and may cause human disease. Food poisoning from eating of parrot fish has been reported in Japan, and the causative toxin was identified as palytoxin (PTX). PTXs occur in various marine organisms, such as seaweeds and crabs, and they also appear to be synthesized by microalgae. Tetramine occurs in the salivary gland of a few whelk species and has occasionally caused human poisonings. Other substances have been suggested to cause seafood-associated illnesses or toxic blooms with resulting fish kills, including unique hemagglutinins and reactive oxygen metabolites, such as superoxide anions and hydroxyl radicals. Sardine poisoning associated with high mortality ($\sim$40%) and hallucinatory fish poisoning have been described, but the causative toxins are not known.

XIV. Treatment and Prevention of Seafood Intoxications

The signs and symptoms of various finfish and shellfish poisonings as well as pharmacologic and therapeutic treatments have been published in several reviews. Therapy for most shellfish intoxications depends on rapid supportive care with particular attention given to cardiopulmonary sufficiency, respiratory distress, and shock (Table 15.1). Antidotes are not available for most of the shellfish poisonings, although certain small-molecular-weight compounds and monoclonal antibodies have been proposed to alleviate symptoms of some seafood intoxications. Scombroid poisoning can be treated with corticosteroids and H_1 and H_2 antihistamines, but physicians should refer to the patient's medication status and authoritative guidelines before administering these. Methods for determination of the type and quantity of seafood toxin ingested, to help guide treatment, are available in compilations listed in the Bibliography.

Most seafood health risks originate in the environment, primarily from harmful algal blooms, and prevention depends on control at harvest. With few exceptions, risks cannot be detected by organoleptic inspection. Surveillance, sampling, and testing by sensitive detection methods for the causative algae and toxins provide the cornerstone for prevention of seafood intoxications caused by algal toxins. In contrast, prevention of scombroid poisoning requires prompt refrigeration and maintaining the temperature of the fish near to 0°C. Descriptions of regulatory oversight and the implementation of monitoring and prevention programs are available in the literature.

Reducing the incidence of seafood intoxications will require coordinated efforts by regulatory agencies and the seafood industries. A comprehensive surveillance and identification program for the etiologic agents and toxins responsible for seafood intoxications can provide information needed to improve the safety of handling and processing. Such information will aid in identifying research needs and prevention strategies, and may help in instituting valid safety programs based on the hazard analysis critical control points (HACCP) system.

XV. Safety Precautions for Handling Toxic Seafoods and Algae

Working with toxic seafoods, toxin-producing algae in culture, and extracts or purified toxins requires care and adequate safety precautions. Protective clothing including lab coats, face and eye protection, and impervious gloves, as well as safety systems for air handling, are recommended to prevent exposure to toxins or aerosols. Chlorine can be used to kill the organisms, but spills of toxins may require additional chemical treatments. The US Army has developed procedures for chemical inactivation of various toxins. Brevetoxins, microcystins, tetrodotoxins, saxitoxins, and palytoxins can be inactivated by 30 minutes exposure to 2.5% NaOCl or more effectively by 2.5% NaOCl + 0.35 N NaOH. Algal seafood toxins are resistant to autoclaving at 121°C or 10-minute exposure to dry heat at 200°F, so chemical decontamination is usually required.

XVI. Conclusions and Perspectives

Seafoods contribute substantially to the food supply worldwide, and individual consumption of finfish and shellfish has been estimated as approximately 13 kg/year, but this varies considerably with region and personal customs. It is anticipated that worldwide consumption of seafoods will increase, owing to advances in aquaculture, and the recognized and presumed health benefits of seafoods. Unfortunately, finfish and shellfish are also important vehicles of foodborne disease.

Foodborne illnesses from finfish and shellfish are caused by infectious organisms (bacteria, viruses, parasites) or by the presence of algal or bacterial toxins. This chapter has described seafood intoxications involving nonproteinaceous, heat-stable, low-molecular-weight toxins produced by microalgae and bacteria. Scombroid fish poisoning is one of the more common seafood diseases in many parts of the world, and is caused by histamine formed by bacteria in poorly refrigerated fish. Ciguatera is also a relatively common seafood poisoning that is transmitted in certain species of fish in tropical waters that are contaminated with ciguatoxin. Other algal toxins transmitted by seafoods and causing illnesses include paralytic shellfish poisoning, puffer fish poisoning, neurotoxic shellfish poisoning, diarrhetic shellfish poisoning, and amnesic shellfish poisoning. Toxins produced by *Pfiesteria* and cyanobacteria are also recognized waterborne causes of diseases in fish and potentially could cause human foodborne illnesses. Algal blooms and associated shellfish contamination have become more common during the past two decades, and it is anticipated that seafood-associated illnesses will correspondingly increase in their incidence.

Current preventive measures mainly involve monitoring of coastal waters and seafoods for the presence of toxins. The main geographic areas of seafood intoxications have been mapped, and it is advised not to consume shellfish in these regions unless the products have been tested and deemed safe. Harmful algal blooms are spreading around the globe, and seafoods are also being shipped worldwide; the incidence of seafood-associated illnesses may show a corresponding increase. The US Centers for Disease Control and Prevention conducts surveillance and reports the incidence of ciguatera, scombroid fish poisoning, and paralytic shellfish poisoning. Other seafood intoxications are prospective risks that may warrant enhanced surveillance and reporting.

Most toxic finfish and shellfish accumulate the toxins from water, are not visibly spoiled, and cannot be distinguished from nontoxic seafoods on harvest; preventive measures rely mainly on sampling of algal blooms and foods, quantitative detection of the causative toxins, and warning the seafood industry and consumer. Scombroid fish poisoning can be prevented by adequate cooling and handling of harvested fish. Mortality for the majority of seafood diseases is generally low, and treatment is mainly supportive.

Research is needed to predict and detect the occurrence of toxic algae and toxic harvested seafoods. New technologies, such as detection of toxic species with molecular probes or by remote sensing by satellites, may lead to improvements in control of seafood intoxications from microalgal and bacterial toxins. The development of antidotes and pharmacologically specific treatments could be extremely valuable in complementing the current supportive means of alleviating seafood intoxications.

Bibliography

Ahmed, F. E. (ed.) (1991). 'Seafood Safety. Committee on Evaluation of the Safety of Fishery Products'. Food and Nutrition Board, Institute of Medicine, National Academy Press, Washington, DC.

Anderson, D. M. (2000). Harmful algal web page: http://www.redtide.whoi.edu/hab/. National Office for Marine Biotoxins and Harmful Algal Blooms. Woods Hole Oceanographic Institution, Woods Hole, MA.

Association of Official Analytical Chemists (1997). *In* 'Official Methods of Analysis of AOAC International', 16th ed., 3rd revision. AOAC International, Gaithersburg, MD.

Brown, L. R., Flavin, C. and French, H. (eds) (1999). 'State of the World'. The Worldwatch Institute. W.W. Norton & Co., New York.

Falconer, I. R. (ed.) (1993). 'Algal Toxins in Seafood and Drinking Water'. Academic Press, London.

Halstead, B. W. (1967). 'Poisonous and Venomous Marine Animals of the World'. US Government Printing Office, Washington, DC.

Jensen, G. L. and Greenlees, K. J. (1997). Public health issues in aquaculture. *Rev. Sci. Tech. Off. Int. Epiz.* **16**, 641–651.

Kao, C. Y. and Levinson, S. R. (eds) (1986). Tetrodotoxin, saxitoxin, and the molecular biology of the sodium channel. *Ann. N.Y. Acad. Sci.* **479**.

Lipp, E. K. and Rose, J. B. (1997). The role of seafood in foodborne diseases in the United States of America. *Rev. Sci. Tech. Int. Epiz.* **16**, 620–640.

Lund, B. M., Baird-Parker, T. C. and Gould, G. W. (eds) (2000). 'The Microbiological Safety and Quality of Food', Vol. 2, Aspen Publishers, Gaithersburg, MD.

McGinn, A. P. (1999). Charting a new course for the oceans. *In* 'State of the World, Millennial Edition' (L. R. Brown, C. Flavin and H. French, eds), pp. 78–95. The Worldwatch Institute, W. W. Norton & Company, New York.

Meyer, K. F., Sommer, H. and Schoenholz, P. (1928). Mussel poisoning. *J. Prev. Med.* **2**, 195–216.

Morris, J. G., Jr (1999). *Pfiesteria*, 'the cell from hell', and other toxic algal nightmares. *Clin. Infect. Dis.* **28**, 1191–1198.

Okada, K. and Niwa, M. (1998). Marine toxins implicated in food poisoning. *J. Toxicol. Toxin Rev.* **17**, 373–384.

Plumley, F. G. (1997). Marine algal toxins: biochemistry, genetics, and molecular biology. *Limnol. Oceanogr.* **42**, 1252–1264.

Price, D., Kizer, W. and Hansgen, H. K. (1991). California's paralytic shellfish prevention program, 1927–89. *J. Shellfish Res.* **10**, 119–145.

Rosen, P., Baker, F., Barkin, R., Daly, R. and Levy, R. (1988). 'Emergency Medicine'. Mosby, St Louis.

Schantz, E. J., McFarren, E. F., Schafer, M. L. and Lewis, K. H. (1958). Purified shellfish poison for bioassay standardization. *J. Off. Agric. Chem.* **41**, 160–170.

Smayda, T. J. and Shimuzu, Y. (eds) (1993). 'Toxic Phytoplankton Blooms in the Sea'. Elsevier, New York.

Sommer, H. and Meyer, K. F. (1937). Paralytic shellfish poisoning. *Arch. Pathol.* **24**, 560–598.

Taylor, S. L. (1986). Histamine food poisoning: toxicology and clinical aspects. *CRC Crit. Rev. Toxicol.* **17**, 91–128.

Todd, E. C. D. (1997). Seafood-associated diseases and control in Canada. *Rev. Sci. Tech. Int. Epiz.* **16**, 661–672.

Williams, R. A. and Zorn, R. A. (1997). Hazard analysis and critical control point systems applied to public health risks: the example of seafood. *Rev. Sci. Tech. Int. Epiz.* **16**, 349–358.

Yasumoto, T. and Murata, M. (1993). Marine toxins. *Chem. Rev.* **93**, 1897–1909.

Staphylococcal Food Poisoning

Amy C. Lee Wong and Merlin S. Bergdoll [1]

I. Introduction

Staphylococcal food poisoning is one of the most common types of foodborne disease worldwide. It is an intoxication, resulting from the ingestion of food containing one or more preformed staphylococcal enterotoxins. Symptoms usually develop quite rapidly, are of relatively short duration, and have no lasting effects. Because of the mild symptoms and rapid recovery, a doctor is seldom consulted and many cases are not reported.

In 1914, Dr M. A. Barber was the first investigator to associate staphylococcal food poisoning with a toxic substance produced by staphylococci. He had made several visits to a farm in the Philippines, and on three occasions became ill with gastroenteritis after consuming milk that had been left unrefrigerated. Barber isolated staphylococci from the milk, which apparently came from a cow's udder that had been infected with mastitis. He then inoculated some sterile milk with the staphylococci and incubated it for 8–9 hours at 36.5°C. About 2 hours after drinking some of it he became ill with cramps, faintness, nausea, and diarrhea. Two other volunteers who drank the milk also became ill with the same symptoms. Although Barber ascribed the illness to a toxin produced by the staphylococci, he did not demonstrate the presence of a toxin in culture filtrates.

Fifteen years later, in 1929, Dr Gail M. Dack rediscovered the role of staphylococci in food poisoning with his classical work on two Christmas cakes that were responsible

---

[1] Deceased.

Foodborne Diseases 2nd Edn
ISBN: 0-12-176559-8

for the illness of 11 people. These three-layer sponge cakes with cream fillings were baked possibly 1 day before delivery and eaten 2 days later. They were presumably refrigerated at the bakery but not after delivery. Dack showed that the sponge cakes were responsible for the illnesses when they were fed to human volunteers who became ill with nausea, vomiting, and diarrhea. Staphylococci were isolated from the cakes and grown in laboratory medium. The organisms were removed by centrifugation and the supernatant fluid given to human volunteers. The presence of a toxin was demonstrated when the volunteers became ill with the same symptoms as those who had eaten the cake. The toxin was called an enterotoxin because of its effect on the gastrointestinal tract, and was the first foodborne disease toxin to be so designated. In essence, this was the beginning of the research on staphylococcal food poisoning.

II. Characteristics of the Disease

A. Symptoms

The onset of symptoms is quite rapid, usually 1–6 hours after ingestion of food containing enterotoxin. The most common symptoms of staphylococcal food poisoning are nausea, retching, vomiting, abdominal cramping, and diarrhea. Vomiting is the symptom most frequently observed. In severe cases headache, fever, muscular cramping, and marked prostration may develop. Normally there is no change in blood pressure, but in severe cases dramatic decreases in blood pressure have been noted. In most cases recovery is rapid, occurring in a few hours to a day or so with no sequelae. The amounts of enterotoxin present in foods are usually small, 10–50 ng/g of food, and will not result in death if consumed. However, occasionally deaths do occur, usually in very young children or older individuals. The death rate in nursing homes is tenfold higher (0.39%) than the overall rate (0.03%). There is no effective treatment, primarily because the illness develops so rapidly and is of such a short duration. In cases where excessive vomiting and diarrhea occur, administration of fluids may be necessary to restore the electrolyte balance.

B. Emetic Dose

It is difficult to estimate the emetic dose from analysis of foods implicated in food poisoning outbreaks because in most cases the enterotoxin is not uniformly distributed in the food; in addition, it is difficult to determine how much food any one individual ate.

Nine serologically distinct staphylococcal enterotoxins have been identified (see Section III. D). Studies where purified enterotoxin was given orally to human volunteers indicated that 0.4 μg of staphylococcal enterotoxin A (SEA), B (SEB), and C (SEC) per kilogram (28 μg/70 kg man) caused vomiting and/or diarrhea in healthy adult males. The minimum dosage at which either vomiting or diarrhea was observed with all three enterotoxins was 0.05 μg/kg (3.5 μg/70 kg man). Milder symptoms were observed in two of four volunteers receiving 0.01 μg SEA/kg (0.7 μg/70 kg man). These results indicated that less than 1 μg of SEA may cause illness in susceptible individuals. Human volunteer studies have not been performed with the other six enterotoxins.

In 1985 an outbreak among school children occurred following the consumption of chocolate milk. This provided an opportunity to obtain a satisfactory estimate of the amount of enterotoxin required to cause illness. The amount of milk ingested by a majority of the children who became ill was one half-pint (246 ml). Analysis of 12 half-pint cartons revealed an average of 144 ng SEA per carton, with a range of 94–187 ng per carton.

Rhesus monkeys are the biological model used for detection of staphylococcal enterotoxins. The amount of SEA observed to produce an emetic reaction when given intragastrically was 5 μg for a 3 kg monkey, with SEB, SEC, and SEE requiring 10 μg and SED requiring 20 μg. Much smaller amounts were needed to produce an emetic reaction when the enterotoxin was given intravenously, 20 ng/kg monkey for SEA and SEC.

C. Diagnosis

Any foodborne illness with the symptoms outlined above is suspected of being staphylococcal food poisoning. The presence of large numbers of staphylococci in the suspected food may indicate poor sanitation or handling; however, it is not sufficient to incriminate that food as the cause of food poisoning. The isolated staphylococci must be shown to produce enterotoxin or enterotoxin must be detected in the suspected food. Although the latter is definitive proof of the cause, frequently the food is no longer available or an insufficient quantity of food is available for examination. Experiments with laboratory media and foods showed that enterotoxin could not be detected until the staphylococcal count reached at least 10^6 colony forming units (CFU)/g or ml. In food-poisoning outbreaks, it is not unusual to encounter counts of 10^8 CFU/g of food or even higher levels. However, large numbers of organisms may not always be present at the time of analysis. Depending on how the food has been handled, a portion of the staphylococci may have died off, leaving a small population behind.

D. Incidence

Staphylococcal food poisoning is a leading cause of foodborne illness worldwide. It has been estimated that only 1–5% of all staphylococcal food poisoning cases are reported in the US. However, the true incidence of staphylococcal food poisoning in the US, as well as in other countries, is unknown because this illness is not a reportable disease in most countries. The number of staphylococcal food-poisoning outbreaks reported to the Centers for Disease Control and Prevention (CDC) varies from year to year, with an average of 21 (16.4% of total number of confirmed bacterial foodborne outbreaks) annually for the period 1973 through 1992. The number ranged from a low of 1 in 1986 to a high of 45 in 1975, representing 1.2% and 36.6% of confirmed bacterial foodborne outbreaks in those 2 years, respectively. The size of the outbreaks also varies; the largest outbreak in the US involved 1300 people attending a picnic in Indiana.

A worldwide surveillance of foodborne disease between 1985 and 1989, which included 17 countries, showed that the highest incidence of staphylococcal foodborne diseases was 58.5 outbreaks per 10^7 population in Cuba, with Hungary, Finland,

Japan, and Israel reporting 9–15 outbreaks per 10^7 population. The largest number of outbreaks was 128 per year in Japan compared to 9.4 per year in the US. For the period 1971–1990, 14.9% and 24.6% of outbreaks in Korea and Japan, respectively, were due to staphylococcal food poisoning. In Taiwan, 169 of 555 (30.5%) of foodborne outbreaks from 1986 through 1995 were due to staphylococcal food poisoning.

III. Characteristics of the Organism

A. Classification and Characteristics

Staphylococcus was first recognized as a species in 1884 and belongs in the family Micrococcaceae. The staphylococci are Gram-positive cocci, 0.5–1.5 μm in diameter, and can occur as single cells, in pairs, or as clusters. *Staphylococcus* can be differentiated from the other three genera in the family, *Micrococcus*, *Stomatococcus*, and *Planococcus*, on the basis of the guanine plus cytosine (30–39 mol % G + C) content of the DNA, the presence of glycine in its peptidoglycan and teichoic acid in its cell wall, and the ability to grow anaerobically. Only three species of *Staphylococcus* were included in the genus in 1974. *Staphylococcus aureus*, the species involved in essentially all of the staphylococcal foodborne disease outbreaks, could be differentiated easily from the other two species, *S. epidermidis* and *S. saprophyticus*, by its ability to produce coagulase and a heat-stable endonuclease (thermonuclease; TNase) and to ferment mannitol anaerobically. In 1986, 19 species were identified, and currently at least 32 species are included in the genus.

Coagulase and TNase are two enzymes that have been used diagnostically for *S. aureus*. Coagulase is a soluble secreted enzyme that coagulates plasma. It reacts with prothrombin, normally found in plasma, to form a coagulase-thrombin complex, which in turn converts fibrinogen to insoluble fibrin, the material that constitutes the clot. TNase is a heat-stable nuclease that can degrade both DNA and RNA. It can be boiled for 30 minutes without significant loss of activity. Although most enterotoxin-producing staphylococci are coagulase- and TNase-positive, the reverse is not true.

For many years *S. aureus* was the only recognized species that produced coagulase and its identification was relatively easy because any organism that produced coagulase was automatically classified as *S. aureus*. The addition of TNase production to the characteristics of *S. aureus* aided in the identification because the production of either coagulase or TNase was adequate to classify an organism as *S. aureus*. Most of the staphylococcal species recognized to date do not produce coagulase except *S. aureus*, *Staphylococcus delphini*, *Staphylococcus hyicus*, *Staphylococcus intermedius*, and *Staphylococcus schleiferi* subspecies *coagulans*. All the coagulase-positive species produce TNase, with the exception of *S. delphini*, but only *S. aureus* can produce acid from mannitol both aerobically and anaerobically, and produce protein A and acetoin (Table 16.1). Coagulase-positive species *S. aureus* and *S. intermedius*, and the coagulase-variable species *S. hyicus* are able to produce enterotoxins. In addition, several coagulase-negative species have also been shown to produce enterotoxin. However, the degree of involvement of these species in food poisoning is unknown.

Table 16.1 Characteristics of *Staphylococcus* species

Property	Species				
	S. aureus	*S. intermedius*	*S. hyicus*	*S. delphini*	*S. epidermidis*
Pigment	+	−	−	−	−
Coagulase	+	+	±	+	−
Thermonuclease	+	+	+	−	−
Mannitol (aerobic)	+	±	−	+	−
Mannitol (anaerobic)	+	−	−	−	−
Hemolysins	+	±	−	+	±
Clumping factor	+	±	−	−	−
Acetoin production	+	−	−	−	+

B. Hosts and Reservoirs

Staphylococci are ubiquitous in nature, with humans and warm-blooded animals as the primary reservoirs. Among animals the bovine is the most important because of the involvement of staphylococci in mastitis. Certain staphylococcal species show host preferences. *S. epidermidis* is the most prevalent on human skin, while *S. hyicus* is commonly found in the nares and skin of pigs, while dogs are the preferred host for *S. intermedius*. Although animals and humans are the major reservoirs, staphylococci also can be found in the air, dust, water, and human and animal wastes.

Humans are the principal reservoir of *S. aureus*. About 30–50% of healthy individuals carry *S. aureus* in their nasal passages, throats, hair, and skin, with 40–50% of the isolates being enterotoxin producers. Prevalence is usually higher in individuals associated with hospital environments because many infections and diseases are caused by staphylococci. *S. aureus* is a common cause of boils and abscesses and more serious infections including endocarditis, osteomyelitis, enterocolitis, toxic shock syndrome, and scalded skin syndrome.

C. Growth and Survival in the Environment

Staphylococci can colonize the nasal passages, throats, and skin of humans and animals. From these sources they can be transferred to meat and other foods. A common source of contamination of dairy products is from cows' udders, particularly from animals with staphylococcal mastitis. Essentially all raw foods, especially raw meats and poultry, can be contaminated with staphylococci. They may persist on raw meats but grow very poorly, as they are poor competitors. In foods that provide a satisfactory medium, staphylococci can grow to sufficient numbers to produce enterotoxin if the foods are not refrigerated. These organisms can be transferred to equipment; if the equipment is not adequately cleansed before use, the organisms can be transferred to foods.

Heating is the most effective way to inactivate *S. aureus* in food. Heating meat to an internal temperature of 73.9–76.7°C should be sufficient to inactivate any staphylococci present. The time-temperature treatments used to pasteurize milk are adequate to destroy the organisms. The *D*-values (time required to destroy 90% of organisms) of *S. aureus* in skim milk at 60.0°C and 65.5°C are 3.44 and 0.28 minutes, respectively. *S. aureus* inoculated into frankfurters was inactivated when the frankfurters were heated to an internal temperature of 71.1°C in the smoking procedure. *S. aureus* in the interior of ham that survived the curing process was slowly inactivated during heating in a smokehouse at 48.9°C for 48 hours.

Freezing and thawing have no significant effect on the viability of *S. aureus*. However, prolonged storage at subfreezing temperatures reduces the staphylococcal number in meats. The number of *S. aureus* in raw minced beef was reduced by 91% after storage for 4 months at −22°C.

S. aureus is relatively resistant to drying. Nonfat dry milk and foods containing nonfat dry milk have been implicated in staphylococcal food-poisoning outbreaks. Staphylococci present in the milk may survive spray drying, depending on the temperature, the moisture content of the product, and the strain of *S. aureus*.

D. The Staphylococcal Enterotoxins

The agents responsible for staphylococcal food poisoning are the staphylococcal enterotoxins. Nine serologically distinct enterotoxins have been identified based on their reactions with specific antibodies. They are designated enterotoxins A (SEA), B (SEB), C (SEC), D (SED), E (SEE), G (SEG), H (SEH), I (SEI), and J (SEJ). Minor antigenic variants of SEC have been described and they are designated SEC_1, SEC_2, and SEC_3. There is no enterotoxin F (SEF) because toxic shock syndrome toxin (a major causative agent of toxic shock syndrome) was misidentified as SEF when it was first isolated. SEA is the enterotoxin most frequently associated with staphylococcal foodborne outbreaks, with SED being the second most frequent. Analysis of outbreaks indicates that unidentified enterotoxins exist.

1. Physicochemical characteristics

The enterotoxins are simple proteins that are hygroscopic and easily soluble in water and salt solutions. They have relatively low molecular weights of about 25 000–29 000. They are basic proteins, with isoelectric points of 7.0–8.6, with the exception of SEG and SEH, which have isoelectric points of 5.6 and 5.7, respectively. One characteristic of the enterotoxins is the presence of two cysteine residues in the center of the enterotoxin molecule that form a disulfide bond, forming what is referred to as the cystine loop. The only exception is SEI, which lacks the cystine loop, owing to the absence of a second cysteine residue in the center of the molecule. The size of the loop varies among the different enterotoxins. The significance of the loops is not known; however, it is assumed that they stabilize the molecular structure.

There is significant amino acid sequence homology among the enterotoxins (Table 16.2). The SECs are the most similar; antibodies to each of the SECs cross-react with each other. SEB is most similar to the SECs; some antibodies produced against SEB cross-react with the SECs. SEG shares significant amino acid identities with SEB and the SECs, respectively. SEA and SEE are also highly similar, and their antibodies

Table 16.2 Relationships among staphylococcal enterotoxins, expressed as percentages of amino acid homology

Toxin compared to	Toxin type								
	A	B	C[a]	D	E	G	H	I	J
A	=			52	82		31	28	66
B		=	63–64			39			
C			96–98			38			
D				=			31	26	52
E					=		31	27	65

[a] Comprises three closely related toxins: C1, C2, and C3.

cross-react with each other. The region directly downstream of the cystine loop contains a number of highly conserved amino acid residues among all the enterotoxins. It has been suggested that these conserved residues may be involved in their emetic activity.

2. Biological characteristics

In addition to causing staphylococcal food poisoning, the enterotoxins are involved in staphylococcal infections such as toxic shock syndrome. The enterotoxins are emetic, pyrogenic and mitogenic, suppress immunoglobulin secretion, enhance endotoxic shock, and stimulate the production of cytokines such as interleukin (IL)-1, IL-2, interferon-γ (IFN-γ), and tumor necrosis factor-α (TNF-α). Experiments with animals showed that the enterotoxins could produce enteritis, which can also occur in humans suffering from staphylococcal food poisoning. The enterotoxins also cause a hypersensitive skin reaction in some individuals. Laboratory workers have experienced sore throats, runny noses, eye irritation, and blisters on their fingers.

The enterotoxins are classified as superantigens. Superantigens are different from conventional antigens in their ability to stimulate a high percentage of T cells in the immune system. Conventional antigens are processed by antigen-presenting cells (APC), such as macrophages and B lymphocytes. The processed antigen is then presented by the major histocompatibility complex (MHC) class II molecules on the surface of the APC for recognition by multiple elements in the T-cell receptor (TCR) (Fig. 16.1). Superantigens do not require processing by APCs. They attach to the outer surface of the peptide-binding groove of the MHC class II molecule rather than in the groove as conventional antigens do. The TCR is a heterodimer composed of α- and β-chains, which include variable regions, Vα, Jα, Vβ, Dβ, and Jβ. Superantigens bind the outside of the TCR containing specific Vβ elements, independent of the antigen specificity of that receptor. Consequently, they can stimulate a large number of T cells, leading to massive release of cytokines. Conventional antigens usually activate one in 10^4–10^6 T cells, whereas superantigens may stimulate over 10% of T cells. Cytokines are highly active proteins or glycoproteins that act as signaling molecules. They can bind to and activate a variety of cells and normally play an important role in homeostatic regulatory functions. However, cytokines can also produce damage. For example, INF-γ and TNF-α can induce shock and death. It is believed that elevated cytokine

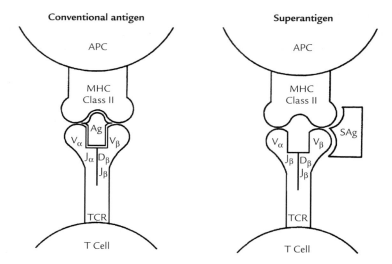

Figure 16.1 Interaction between conventional antigens (Ag) and superantigens (SAg) with antigen-presenting cells (APC) and T cells. Following processing by the APC, conventional antigen peptides are presented in the peptide-binding groove of the MHC class II molecule for recognition by multiple elements in the TCR. Superantigens interact with MHC Class II molecules (without processing) outside the peptide-binding groove. The superantigen-MHC bimolecular complex next interacts with the Vβ element of the TCR. Adapted, with permission, from L. M. Jablonski and G. A. Bohach (1997), *Staphylococcus aureus*. *In* 'Food Microbiology. Fundamentals and Frontiers' (M. P. Doyle, L. R. Beuchat, and T. J. Montville, eds.), pp. 353–375. American Society for Microbiology, Washington, DC.

levels induced by the enterotoxins or toxic shock syndrome toxin mediate some of the symptoms in staphylococcal infections such as toxic shock syndrome.

3. Mode of action

Studies in monkeys showed that the site of the emetic action is in the abdominal viscera. The vagus and sympathetic nerves in the gut are stimulated and the signal for emesis is transmitted to the vomiting center in the brain. The binding site for the enterotoxins in the intestinal tract has not been determined. SEA fed to rats passed through the intestinal tract rapidly and was removed from the circulation by the kidney within a short period of time. It has been suggested that stimulation by the enterotoxins of T-cell proliferation with concomitant massive release of cytokines, such as IL-2, cause the gastrointestinal symptoms in staphylococcal food poisoning. Cancer patients who receive IL-2 treatment experience emetic symptoms similar to those in staphylococcal food poisoning. However, studies using mutational analysis or chemical modification of the enterotoxins showed that emetic and superantigenic activities are separable, and that there is no direct correlation between these two activities. The mechanism of diarrhea caused by the enterotoxins is not known.

4. Stability of the enterotoxins

The staphylococcal enterotoxins are more stable in many respects than most proteins. In the active state, they are resistant to proteolytic enzymes, such as trypsin, chymotrypsin, rennin, and papain. Although pepsin can digest the enterotoxins at pH values of 2.0 and below, this acidic level does not exist in the stomach under normal conditions, particularly in the presence of food. This makes it possible for the enterotoxins to pass through the stomach to the intestinal tract where they stimulate emetic and diarrheal actions.

The enterotoxins are quite heat resistant. The degree of heat resistance depends on many factors, including the type of enterotoxin, purity of the preparation, amount of toxin, pH, and menstruum. Crude toxin preparations are more heat stable than purified toxins, while heat inactivation is faster in buffer than in culture media and foods. The emetic activity was not completely destroyed after crude enterotoxin preparations were boiled for 30 minutes and fed to human volunteers or injected intravenously into monkeys. However, thermal process treatments employed by the canning industry are adequate to destroy the quantity of enterotoxins usually involved in food-poisoning outbreaks. Staphylococcal food poisoning resulting from canned foods was due to inadequate processing or recontamination of improperly sealed cans after retorting. The enterotoxins are not inactivated during pasteurization. Very little loss of SEA and SED in milk or cream occurred after pasteurization at 72°C for 15 seconds. Spray-drying processes used for milk are also insufficient to inactivate the enterotoxins. Spray-dried milk has been involved in several staphylococcal food-poisoning outbreaks. In general, the heat resistance of SEA is higher than that of SEB, which is higher than that of SEC.

It has been shown that gamma irradiation processes used for pasteurization or sterilization of foods may not be sufficient to inactivate the enterotoxins. More than 2.7 and 9.7 megarad (Mrad) was required to reduce the concentration of SEB in buffer and milk, respectively, by tenfold. In lean minced beef slurries, 27–37% of SEA remained after a dose of 8 kGy. The more concentrated the beef slurry, the less the SEA was inactivated.

5. Regulation of enterotoxin production

There are at least four global regulatory systems in *S. aureus* that affect the expression of extracellular or cell-surface proteins: *agr* (accessory gene regulator), *sar* (staphylococcal accessory regulator), *xpr* (extracellular protein synthesis regulator), and *sae* (*S. aureus* exoprotein expression). The best-characterized regulon is *agr*, which affects at least 15 genes. *agr* and *sar* interact with each other and are involved in a complex system of global regulation. Expression of SEB, SEC, and SED is affected by *agr*, while SEA and SEJ expression is not. Regulation of SEE, SEG, and SEH has not been studied.

E. Conditions for Growth and Enterotoxin Production

Growth and enterotoxin production by *S. aureus* are influenced by a variety of environmental and nutritional factors including temperature, pH, water activity (a_w), inoculum size, atmospheric composition, carbon and nitrogen sources, salt levels, and competing microflora. Generally, growth is necessary for enterotoxin production, although enterotoxin production does not always accompany growth, especially in foods. The amount of enterotoxin produced is dependent on staphylococcal strain and enterotoxin type. In culture medium, SEB and SEC usually are produced in relatively large amounts (over 100 μg/ml), whereas the other enterotoxins are produced in much lower quantities, generally a few micrograms per milliliter. SED and SEH are produced in less than 1 μg/ml. SEB and SEC production are affected more by the culture conditions, whereas SEA production is more closely related to the growth of the organism.

1. Temperature

S. aureus can grow at 6.7–47.8°C, with an optimum of 37°C. Enterotoxin is produced at 10–45°C, with an optimum range of 37–40°C. Normally growth is much slower at the lower temperatures, and since enterotoxin production is related to growth, a much longer period at these temperatures would be required before enterotoxin is detectable. Enterotoxin production has been observed in a variety of foods. SEB was detected in ham after 2 weeks at 10°C, while SEA–SEE were produced in vanilla pudding over a temperature range of 10–45°C.

2. pH

Most staphylococci can grow at pH values between 4.5 and 9.3, with the optimum being 7.0–7.5. Enterotoxins are produced in a narrower pH range, between 5.15 and 9.0. The pH at which a strain can grow and produce enterotoxin depends on other cultural parameters, such as temperature, salt concentration, and atmospheric condition. Staphylococci are more resistant to salt present in foods than other organisms. They can grow in up to 20% NaCl, while enterotoxin production occurs with up to 10% NaCl. Growth and enterotoxin production are generally retarded at increased salt concentrations, and the pH range for enterotoxin production becomes narrower as the salt concentration is increased. The acid used for pH adjustment is also an influencing factor. When milk was acidified with hydrochloric acid, SEA was produced at pH levels between 4.5 and 6.5. However, when lactic acid was used, growth and enterotoxin production were observed at the higher pH levels, but not at 4.5.

3. Water activity

S. aureus can grow over a much wider a_w range than many other food-associated pathogens. Growth in some strains occurs at an a_w of 0.83; the optimum is >0.99. The minimum a_w for anaerobic growth is 0.90. The optimum a_w for enterotoxin production is 0.99, while the minimum is 0.86. The humectant used for a_w adjustment has a significant effect. For example, when NaCl was used, the minimum a_w for SEB production was 0.90–0.92. When glycerol was used, the minimum a_w was 0.98–0.99. Temperature and pH also affect the a_w at which *S. aureus* will grow and produce enterotoxin. When these parameters deviate from their optimum levels, the minimum a_w tolerated by the organisms is elevated.

4. Atmospheric conditions

Staphylococci are facultative anaerobes, but the amount and rate of growth and enterotoxin production are less under anaerobic conditions. Anaerobic production of SEB in cured meats at 22°C and 30°C has been reported. Production of SEA, SEB, and SED was observed in sausage under N_2 storage for 4 days. Staphylococcal growth and enterotoxin have also been observed in Canadian bacon and ham, turkey and hamburger sandwiches stored anaerobically.

5. Presence of other organisms

Staphylococci do not grow well in the presence of other organisms unless the initial staphylococcal population is larger than that of the other organisms present. In raw foods, one would not expect staphylococci to grow appreciably because of the presence of other microbial contaminants. An exception would be raw milk from a cow with staphylococcal mastitis. In heat-processed foods where the bacterial load is reduced significantly, any *S. aureus* present as a postprocessing contaminant would have a competitive advantage, especially if the food contains salt and has a reduced a_w.

IV. Transmission via Foods

A. Foods Involved

Eating foods that contain staphylococcal enterotoxins causes staphylococcal food poisoning. Essentially all raw foods, especially raw meats and poultry, can be contaminated with staphylococci, either by humans or animals, or both. However, in most cases, staphylococci would not grow sufficiently in many of the foods to produce enterotoxin and cooking before consumption would destroy the organism. In a majority of staphylococcal food-poisoning outbreaks, the foods were contaminated during their preparation and mishandled afterwards.

Several conditions are required for enterotoxins to be produced in foods. The food must provide a good medium for staphylococci to grow. It must be contaminated with enterotoxin-producing staphylococci and held at warm temperatures for sufficient time for the organisms to grow and produce enterotoxins.

The major vehicles for transmission of staphylococcal foodborne illness are protein-rich foods. Most meats are contaminated with staphylococci, but normally this does not present a problem because the organisms do not grow well on raw meat and are destroyed in the cooking process. In the US, baked ham is most frequently involved in outbreaks, while poultry, salads (meat, potato, etc.), and cream-filled bakery goods are responsible for many of the remaining outbreaks. The frequency of the involvement of baked ham may be due to a number of reasons. It is a common item for picnics and other large gatherings. Ham is easily contaminated by the food handler during slicing, and the high salt content of ham provides a favorable environment for the staphylococci to grow, especially if the ham is not refrigerated properly. One of the largest outbreaks involved 1300 individuals who consumed baked ham at a company picnic. Large numbers of SEA-producing *S. aureus* were isolated from the ham. Another outbreak involved 600 high school girls in Indiana and was associated with the consumption of sandwiches containing baked ham. The ham was prepared the day before and held at room temperature until the sandwiches were made. An outbreak in five elementary schools in Rhode Island involving about 100 schoolchildren was caused by improperly handled ham. The ham was prepared 2 days before the outbreak, stored in deep pans and held in a walk-in refrigerator overnight. The next morning, the ham was sliced and put back in the refrigerator. On the day of the outbreak, the sliced ham was removed from the refrigerator, heated at 64°C for 20 minutes in an oven, and delivered to the schools. The ham was then kept in warm pans for 3 hours before being consumed. SEA was detected in the ham.

A number of outbreaks have occurred from potato, egg, and meat salads. Over 1300 of about 6000 schoolchildren in 16 elementary schools in Texas developed headache, vomiting, abdominal pain, and diarrhea after eating chicken salad at lunch. On the day before the outbreak frozen chickens were cooked in a central kitchen. The chickens were then deboned, cooled to room temperature, cut into small pieces, and stored in deep pans in a walk-in refrigerator overnight. The next morning, the salad was prepared and distributed to the school cafeterias. It was stored at room temperature until served at noon. *S. aureus* was isolated from the salad and the noses of three of the food handlers.

Custard- and cream-filled bakery goods provide a suitable medium for growth of staphylococci and production of enterotoxin. Cream puffs, eclairs, and cream pies were once frequently involved in food poisoning because they were not routinely refrigerated. An outbreak resulted from the consumption of cream-filled coffee cake in Wisconsin. Chocolate eclairs were implicated in an outbreak on an airplane flight from Rio de Janeiro to New York City. About half of the 153 passengers who were served the eclairs became ill. The eclairs had been left unrefrigerated for over 12 hours before they were placed on the plane. SED-producing *S. aureus* (10^9/g) were isolated from the eclairs and SED was detected in the eclairs.

Milk is seldom involved, primarily because of the care used in handling the fluid milk supply, which generally involves pasteurization and refrigeration. One outbreak occurred in Kentucky where over 850 schoolchildren became ill after drinking 2% chocolate milk. The milk inadvertently was held for several hours at a warm temperature before pasteurization. No staphylococci were isolated, but SEA was detected in the pasteurized milk. Staphylococci, which grew and produced SEA during the warm holding period, were eliminated by pasteurization, but the heat-resistant SEA remained in the milk.

The foods involved in other countries vary with the diet as well as with the local conditions. In Japan, rice balls are commonly consumed and once were the major item involved in staphylococcal food poisoning. Rice is an excellent growth medium for staphylococci. Rice balls used to be prepared manually and were not usually refrigerated, but now they are made with machines to avoid human handling. In some European countries, such as Poland and the Czech Republic, ice cream made by small producers is a common cause of staphylococcal food poisoning. In Brazil the two foods most frequently involved are cream-filled cakes and a white cheese generally produced on the farm or in small establishments.

B. Source of Contamination

Staphylococci are ubiquitous in nature, and can be found in the air, dust, water, and on humans and animals. The main reservoirs of these organisms are the nasal cavity and skin of humans. Food handlers are the primary source of contamination of foods implicated in food poisoning. Staphylococci can be introduced into the food if they cough or sneeze while preparing foods or do not wash their hands properly prior to food handling. Individuals with some kind of staphylococcal infection, such as an infected cut on their hands, will introduce a larger inoculum into the food. The larger the inoculum, the shorter time the food needs to be held at a warm temperature before sufficient growth and enterotoxin production occur to cause problems. Most foods involved in staphylococcal food-poisoning outbreaks are cooked, and are recontaminated in the final preparation for serving.

Animals can be a source of contamination. They carry staphylococci in their noses and on the skin. Some animal diseases, such as mastitis, are caused by staphylococci. However, animal carriers are of less importance in staphylococcal food poisoning than human carriers, even though raw meats and milk are usually contaminated with staphylococci. Cooking and pasteurization will destroy the organism. In addition, staphylococci are poor competitors and do not grow well in the presence of other organisms naturally present in raw foods. An exception is when staphylococci are present in much higher numbers than the other organisms, as in milk from a mastitic animal. Two children died after drinking milk from a goat with mastitis.

Occasionally equipment has been implicated as a source of contamination. An outbreak resulted from contamination of baked ham with a meat slicer. The same strain of enterotoxigenic *S. aureus* was isolated from the ham and the slicer.

V. Isolation and Identification

Staphylococcal food poisoning is caused by the ingestion of enterotoxins produced by the staphylococci in the food. Therefore methods are needed for detection of both the organisms and the enterotoxins. Staphylococci have to grow in the food to produce enterotoxin; however, not all staphylococci are enterotoxigenic, therefore their presence alone is not sufficient evidence to incriminate a particular food as the vehicle for food poisoning. The presence of a large number of staphylococci in the food may indicate poor handling or sanitation. On the other hand, absence of staphylococci in the food does not necessarily indicate that enterotoxin is absent because the organisms are easily destroyed by heat, whereas the enterotoxins are not. The staphylococci may represent remnants of a population that has produced sufficient enterotoxin in the food to cause illness.

A. Detection of *S. aureus*

The isolation of coagulase-positive staphylococci from foods is of greatest concern because *S. aureus* is the species most often involved in food poisoning. In addition to *S. aureus*, *S. intermedius*, *S. hyicus*, *S. delphini*, and *S. schleiferi* subspecies *coagulans* can produce coagulase. Some strains from each of these species can produce enterotoxin.

No specific method may be useful in every case to isolate the staphylococci from the wide variety of foods in which they are found. As a result, a combination of selective and enrichment media are used that will support the growth of the staphylococci and at the same time suppress the growth of other microflora present. When low numbers of staphylococci are expected in a food sample, a most probable number procedure is generally used. A three-tube most probable number (MPN) method using trypticase soy broth with 10% NaCl and 1% sodium pyruvate has been accepted as the official method for recovery of coagulase-positive staphylococci from a wide variety of foods. However, thermally stressed cells of *S. aureus* cannot grow in this medium. Food samples likely to contain a small population injured cells can be incubated in double-strength trypticase soy broth before spread plating on Baird-Parker agar.

For detecting small numbers of *S. aureus* in raw food ingredients and nonprocessed foods expected to contain large numbers of competing organisms, incubation in trypticase soy broth containing 10% NaCl and 1% sodium pyruvate is carried out before plating on Baird-Parker agar plates. For detecting relatively large numbers of staphylococci, the food extract is plated directly on Baird-Parker agar.

Typical colonies of *S. aureus* on Baird-Parker agar are circular, smooth, convex, moist, ≥1.5 mm in size on uncrowded plates, gray-black to jet black in color with entire margins and off-white edges. An opaque zone forms around the colony owing to lipase activity of the organisms. In addition, proteases produced by the organisms frequently create a clear halo that extends beyond the opaque zone. Representative colonies are tested for coagulase and TNase production. An additional positive anaerobic mannitol fermentation test would confirm *S. aureus*. The number of colonies on the triplicate plates represented by the *S. aureus*-positive colonies is multiplied by the dilution factor, and the result reported as the number of *S. aureus* per gram of food.

Agglutination kits employing the clumping factor, protein A, and specific antigens of *S. aureus* are available for identification of *S. aureus* strains. However, these kits are designed primarily for use in the clinical field where large numbers of staphylococci are being examined and where large numbers of coagulase-negative species are encountered also. The clumping factor test would not be satisfactory because *S. intermedius* and some coagulase-negative species can be positive for this factor.

An alternative method has been proposed in which P agar (contains peptone, sodium chloride, yeast extract, and glucose) supplemented with acriflavin and the β-galactosidase test is used. Of the coagulase-positive species, only *S. aureus* will grow on the supplemented P agar and is negative with the β-galactosidase test. *S. intermedius* does not grow on the modified P agar and is positive with the β-galactosidase test, whereas *S. hyicus* is negative by both tests.

Another method that has been proposed to identify *S. aureus* from nonclinical sources employs an immunoenzymatic assay using a monoclonal antibody prepared against *S. aureus* endo-β-acetyl-glucosaminidase, an enzyme produced by all isolates of this species. Comparison of this method with six kits available for identification of *S. aureus* showed it to be specific for *S. aureus*, whereas the kits were positive for *S. aureus*, *S. intermedius*, *S. schleiferi*, and *S. lugdunensis*.

B. Detection of the Enterotoxins

The first test developed for the detection of staphylococcal enterotoxins was the monkey-feeding assay, which is still being used as a bioassay today. It is the only reliable method for the detection of unidentified staphylococcal enterotoxins. Of the biologically active substances produced by the staphylococci, only enterotoxins cause emesis when administered orally to monkeys. It is accepted that any staphylococcal product that produces an emetic reaction in monkeys is an enterotoxin. Assays are performed by administering solutions of the enterotoxins (up to 50 ml) into the stomachs of monkeys (2–3 kg) by catheter. The animals are observed for 5 hours. If vomiting occurs within that time, the sample is considered to contain enterotoxin.

The intravenous injection of cats and kittens also proved useful for the detection of the enterotoxins. When materials other than the purified enterotoxins are injected intravenously, it is necessary to inactivate any interfering substances by treatment with trypsin or pancreatin. Cats are not as reliable as monkeys because they are subject to nonspecific reactions.

After each enterotoxin has been identified and purified, specific antibodies to each can be produced. Then specific and sensitive immunoassays can be developed for their detection. All of the current methods for the detection of enterotoxin are based on the use of specific antibodies to the enterotoxins. The detection of enterotoxin in foods requires methods that are sensitive to less than 1 ng/g of food. The quantity of enterotoxin present in foods involved in food-poisoning outbreaks may vary from less than 1 ng/g to greater than 50 ng/g. In addition, not all the enterotoxin can be recovered from the food matrix during the extraction process, especially if the food contains low enterotoxin concentrations. The more sensitive the detection method, the simpler the method and the less time needed for extraction and concentration of the enterotoxin from the food.

1. Gel diffusion methods

The first serological tests developed were immunodiffusion assays based on the reaction of the enterotoxin with the specific antibodies in gels to form a line of precipitation. Antibody and enterotoxin samples placed in separate wells cut in the agar diffuse toward each other, and a precipitin line forms where the concentrations of the antibody and enterotoxin are optimal. The most common assays are some form of either the Ouchterlony gel plate or the microslide. The optimum sensitivity plate (OSP) method, which is a modification of the Ouchterlony plate test, permits detection of 0.5 μg/ml, and is adequate in sensitivity to detect most enterotoxigenic staphylococci. The sensitivity can be increased to 0.1 μg/ml by a fivefold concentration of the culture supernatant fluids. This method is not sensitive enough for detection of enterotoxin in foods.

The microslide method is the most sensitive of the gel diffusion (0.05–0.1 μg/ml) methods, but care is needed in preparing the slides. Frequently, the results are difficult to interpret, and experience is very important in using this method successfully. The original methods for detection of enterotoxin in foods used the microslide. However, it is necessary to extract from 100 g of food and concentrate the extract to 0.1–0.2 ml to achieve the concentration required for detection by the microslide. The method is therefore very cumbersome and time consuming, although it is still being used by the US Food and Drug Administration as the standard to evaluate newer and more sensitive enterotoxin detection methods.

2. Radioimmunoassay

The radioimmunoassay (RIA) was the first sensitive method developed for the detection of enterotoxins at levels of less than 1 ng/ml. The assay is based on the competition for specific antibody molecules between radiolabeled enterotoxin standards and unlabeled enterotoxin in the sample. The sensitivity is 0.3 ng/ml or approximately 0.5 ng/g of food. However, because of the health risks and costs involved in handling radioisotopes, the RIA has not been used widely.

3. Enzyme-linked immunosorbent assay

The enzyme-linked immunosorbent assay (ELISA) is the most commonly used method currently for detection of staphylococcal enterotoxins. The method involves the use of an enzyme coupled either to the enterotoxin or to the specific antibody and depends on the development of a color by the reaction of the enzyme on a suitable substrate. The most common type of ELISA is the double-antibody sandwich method in which the antibody immobilized on to a surface is reacted with the sample containing enterotoxin. The antibody-enterotoxin complex is then treated with the enzyme-antibody conjugate. This type of procedure is preferred because the amount of enzyme, and thus the color developed from the enzyme-substrate reaction, is directly proportional to the amount of enterotoxin in the sample. This eliminates the need for highly purified enterotoxins as crude or only partially purified enterotoxin is needed for the preparation of a standard curve. In a competitive ELISA format, the enzyme is conjugated to the enterotoxin, which competes with enterotoxin present in the sample for binding to the antibody. The amount of color developed is inversely proportional to the amount of enterotoxin present in the sample. The most commonly used enzymes are alkaline phosphatase and horseradish peroxidase.

Instead of coupling the enzyme directly to the antibodies, one can use biotinylated antibodies and avidin–alkaline phosphatase. Avidin binds specifically to biotin and a sensitivity of 0.1 ng/ml for SEA has been reported. Another modification of the ELISA, the enzyme-linked fluorescence assays (ELFA) uses fluorogenic substrates. The majority of users of ELISA methods employ microtiter plates or strips to which the antibodies are adsorbed. Other formats include polystyrene balls and nitrocellulose paper attached to wells in plastic sticks. ELISAs can detect less than 1 ng/ml. A detection limit of as low as 0.1 ng/ml has been reported. The sensitivity of the assay varies depending on the specific ELISA used, the enterotoxin type, and the food from which the toxin is extracted.

4. Reversed passive latex agglutination

In the reversed passive latex agglutination (RPLA) method, specific antibodies are attached to latex particles. When these particles are added to a sample containing enterotoxin, the latex particles agglutinate and the reaction is easily visible. The sensitivity is less than 1 ng/ml.

5. Nucleic acid-based detection methods

The nucleotide sequences of genes encoding all of the identified staphylococcal enterotoxins have been determined, hence DNA probes specific to those genes can be synthesized and used for detection. Detection of enterotoxigenic staphylococci in cultures and foods by the polymerase chain reaction (PCR) technique has been reported for SEA, SEB, SEC, SED, and SEE. However, this method cannot distinguish viable from nonviable cells, since DNA from dead cells also would be amplified. In addition, inhibitors of the polymerase have been reported to be present in many different types of food matrices and may generate false-negative results. While the PCR technique can be used for the detection of enterotoxigenic staphylococci, positive results only indicate the presence of the organism. It does not necessarily indicate the presence of enterotoxins in the food sample.

6. Diagnostic kits

Several diagnostic kits utilizing ELISA, ELFA, or RPLA formats have been developed for the detection of the enterotoxins. The sensitivity of all the kits is less than 1 ng/ml. The enterotoxins that can be detected depend on the kit, and include SEA, SEB, SEC, SED, and SEE. Some kits can be used to detect and identify individual enterotoxins, while others can only be used to determine if enterotoxin is present but not the type of enterotoxin. The latter are useful for screening purposes. Currently, there are no commercial kits that can detect the more recently identified SEG, SEH, SEI, and SEJ.

VI. Treatment and Prevention

There is no treatment for staphylococcal food poisoning, primarily because the illness develops so rapidly and is of such a short duration. The symptoms are relatively mild and the person usually recovers in a few hours. However, in severe cases where vomiting and diarrhea are excessive, intravenous administration of fluids may be necessary to restore the salt balance.

The staphylococci are ubiquitous and are impossible to eliminate from our environment. At least 30–50% of individuals carry these organisms in their nasal passages, throats, or on their hands. Any time a food is exposed to human handling, there is a possibility the food will be contaminated with staphylococci. About 30–50% of these may be enterotoxin producers. Food handlers should practice good personal hygiene and should be educated in proper food handling practices. Individuals with any type of infection should not be permitted to handle food. Heating the food normally would ensure against food poisoning unless the food was held unrefrigerated for several hours before heating. If enterotoxin is produced in the food, heating may not be sufficient to inactivate it. In many cases, foods are not processed further after handling, and unless proper care is taken, the organisms may grow and produce enterotoxin. Staphylococci do not grow at temperatures below 6.7°C and above 47.8°C. It is important to keep susceptible foods refrigerated at all times except when being prepared and while being served. Refrigeration should be done in such a manner to facilitate quick cooling of the entire food mass.

VII. Summary

Staphylococcal food poisoning is one of the most common types of foodborne disease worldwide. It is an intoxication, resulting from the ingestion of food containing one or more preformed staphylococcal enterotoxins. Many cases are not reported because the symptoms are generally mild and of short duration with no sequelae. The species primarily involved is *S. aureus*, although other staphylococci also may produce enterotoxins. Nine serological types of staphylococcal enterotoxins (SEA, B, C, D, E, G, H, I, and J) have been identified, and are differentiated based on their reactions with specific antibodies. The enterotoxins are a group of related, simple proteins of low molecular weight with varying degrees of sequence homology. They exhibit many biological activities and are classified as superantigens because of their ability to stimulate a high percentage of T cells. Current methods for detection of the identified enterotoxins are based on the reaction between the enterotoxins and specific antibodies. Unidentified enterotoxins exist, and the monkey-feeding assay is the only reliable method for their detection.

Staphylococci are widespread in nature, with humans and warm-blooded animals as their primary reservoirs. Humans are the major reservoir for *S. aureus*, and food handlers constitute the primary contamination source. Staphylococcal food poisoning usually occurs when a food is contaminated after processing, and then held at a warm temperature for a prolonged period. Good personal hygiene, proper food-handling practices, and appropriate holding temperatures for foods will reduce the risk of staphylococcal food poisoning.

Bibliography

Bergdoll, M. S. (1989). *Staphylococcus aureus*. In 'Foodborne Bacterial Pathogens' (M. P. Doyle, ed.), pp. 463–523. Marcel Dekker, New York.

Bergdoll, M. S. (1999). Detection of staphylococcal enterotoxins: overview. In 'Encyclopedia of Food Microbiology' (R. Robinson, C. Batt and P. Patel, eds), pp. 2076–2083. Academic Press, London.

Jablonski, L. M. and Bohach, G. A. (2001). *Staphylococcus aureus*. In 'Food Microbiology. Fundamentals and Frontiers', 2nd ed. (M. P. Doyle, L. R. Beuchat and T. J. Montville, eds), pp. 411–434. American Society for Microbiology, Washington, DC.

Lancette, G. A. and Bennett, R. W. (2001). *Staphylococcus aureus* and staphylococcal enterotoxins. In 'Compendium of Methods for the Microbiological Examination of Foods', 4th ed. (F. P. Downes and K. Ito, eds), pp. 387–403. American Public Health Association, Washington, DC.

Su, Y.-C. and Wong, A. C. L. (1997). Current perspectives on detection of staphylococcal enterotoxins. *J. Food Prot.* **60**, 195–202.

Botulism

17

Nina Gritzai Parkinson and Keith Ito

I. Introduction

The sporeforming bacterium *Clostridium botulinum* produces a neurotoxin, which is one of the most potent toxins known to man. Its distribution around the world, the heat resistance of the spore, the toxin and the illnesses attributed to the toxin, and the modes of transmission of spores and toxin are well documented. Recent studies have implicated the toxin in cases of infant deaths and it is now considered a potential cause of some cases of sudden infant death syndrome (SIDS). The toxin is also being used for medicinal purposes for conditions such as strabismus and blepharospasm. Thermal processes for commercially prepared low-acid foods have been successful in minimizing outbreaks of *C. botulinum*, but demand by consumers for 'fresh' packaged and minimally processed foods has introduced new concerns.

II. Characteristics of the Organism

C. botulinum is a Gram-positive, anaerobic, rod-shaped, motile by peritrichous flagella, sporeforming bacterium. The spores are usually oval and subterminal and swell the cell, producing a typical tennis-racket shape. The species produces a potent neurotoxin.

Foodborne Diseases 2nd Edn
ISBN: 0-12-176559-8

There are seven toxin types, which are recognized by their antigenic specificity. The toxins are typed as A, B, C, D, E, F, and G. The type C has two antigenic subtypes C_α and C_β. Toxin production by type C and D organisms is phage mediated; the specific type of toxin produced is determined by the specific phage that infects the culture. Although each of the toxins is similar in its reaction in a given host, there are some differences among toxins produced by strains of the same toxin type. Cross-reactions can occur between toxin types, for example, between types E and F, and types C and D.

C. botulinum can be divided into four distinct groups based upon their DNA homology and the reactions they have to specific substrates. Group I includes the proteolytic strains. These include all the type A and the proteolytic strains of types B and F. The organisms in this group digest gelatin, milk, and meat. They produce ammonia and hydrogen sulfide. Glucose is fermented, but mannose and sucrose are not. Fermentation byproducts in peptone yeast extract glucose broth include acetic and butyric acid, along with small amounts of other acids and alcohols. Hydrocinnamic acid and hydrogen gas are produced in trypticase soy broth. The strains are sensitive to chloramphenicol, tetracycline, and penicillin G. The optimum temperature for growth is 30–40°C. The minimum temperature for growth is 10–12°C, and the maximum growth temperature is 45–50°C. Growth is inhibited by 10% salt and a water activity of 0.94 (NaCl solution). Although plasmids have been found in this group, they have not been associated with the production of toxin.

Group II includes the nonproteolytic strains of types B and F and type E. The organisms in this group do not digest milk or meat, but gelatin is digested. Glucose is fermented, as are mannose and sucrose. Butyric and acetic acids and hydrogen gas are produced as fermentation byproducts when the organisms are grown in peptone yeast extract glucose broth. The strains are sensitive to chloramphenicol, penicillin G, and tetracycline. The optimum temperature for growth is 27–37°C. The minimum growth temperature is 3.3°C, and the maximum temperature is 40–45°C. Growth is inhibited by 5% salt and a water activity of 0.97 (NaCl solution). Although plasmids have been found in this group, they have not been associated with the production of toxin. Plasmids have been found which might relate to the production of boticin, an inhibitor of the organism's growth.

Group III includes types C and D. Organisms in this group digest gelatin. Milk and meat are digested by most, but not all, of the strains. Glucose is fermented, while sucrose usually is not. Mannose fermentation is variable. Growth in peptone yeast extract glucose broth produces butyric, propionic and acetic acids. Hydrogen gas is produced. The strains are susceptible to chloramphenicol, penicillin G, and tetracycline. The optimum temperature for growth is 30–37°C. The minimum growth temperature is 15°C, and the maximum temperature of growth is 50°C. Growth is inhibited by 3% salt. Toxin production is mediated by phage in this group, and the specific toxin type is dependent on the specific phage that is infecting the culture.

Group IV contains type G. Gelatin and milk are rapidly digested, and meat is digested slowly. Ammonia and hydrogen sulfide are produced. In peptone yeast extract broth, acetic, butyric, and isovaleric acids are among the acids produced. Hydrogen gas is formed. Glucose, sucrose, and mannose are not fermented. The optimum growth temperature is 30–37°C. The minimum growth temperature is 12°C. The strain is susceptible to chloramphenicol, penicillin G, and tetracycline. Plasmids have been found, and they are associated with toxin formation.

III. Characteristics of the Disease

The botulinum neurotoxin is produced when *Clostridium botulinum* is allowed to germinate and multiply, which occurs under anaerobic and other favorable conditions. It can affect man and animals, usually by ingestion of the toxin. The neurotoxin is a protein with a molecular mass of about 150 kDa. The molecule consists of a heavy chain of about 100 kDa and a light chain of about 50 kDa. The chains are joined by a disulfide bridge. The toxin molecules are formed intracellularly and have a low toxicity. The toxin becomes extracellular due to cell lysis and/or secretion. This exposes the molecule to proteolytic enzymes that cleave the molecule to form the highly toxic two-chained molecule. The neurotoxin inhibits the release of the neurotransmitter acetylcholine from the motor neurons to the muscles and ultimately affects the peripheral nervous system. Clinical symptoms of botulism can begin 12–48 hours after toxin ingestion, with weakness, dizziness, and dryness of the mouth. Nausea and vomiting may also occur. Neurologic features soon develop, including blurred vision, inability to swallow, difficulty in speech, descending weakness of skeletal muscles and ultimately respiratory paralysis.

In humans, there are four clinical forms of botulinum poisoning: foodborne, infant, wound and a fourth classification currently called 'adult infectious botulism'. In cases of foodborne botulism, the toxin is produced in a food where the organism has been allowed to grow, typically due to improper storage conditions or to poor thermal processing procedures, which result in inadequate destruction of the spores. When the food containing the toxin is ingested, it causes an intoxication (transmission via food will be discussed in more detail later in the chapter). In cases of infant botulinum and in adult infectious botulism, the spore is ingested or inhaled, it colonizes in gastrointestinal flora that is unable to inhibit its growth, and toxin is produced (infant botulism will be discussed in more detail later in this chapter). In cases of wound botulism, the vegetative cells or spores are introduced through a wound, where the organism finds an anaerobic environment and produces toxin. At this time there is an increasing number of wound botulism cases attributed to intravenous drug use.

IV. Transmission via Food

The types of foods involved in botulism vary according to food preservation and eating habits in different regions, as well as with food preparation techniques. *C. botulinum* spores are found all over the world; however, there appears to be a distinct distribution of the different types of *C. botulinum* and the types of illnesses seen in different parts of the world. Types A and B are found in the soil in temperate climates, while type E is found in colder aquatic sediments.

Proteolytic type A spores are found predominantly in soils of western US, South America, and southern European countries and China. They favor neutral to alkaline soils with low organic content, hence their virtual absence from the eastern US and from Europe, where the soil is heavily farmed. They are typically associated with illnesses caused by vegetables. In the US, proteolytic Type A is most common west of the Mississippi River and proteolytic Type B is most common east of the Mississippi River. Distribution of Type B spores seems more uniform, but those found in the soils of the Eastern US are proteolytic, while those found in the Central Europe are nonproteolytic. Meat frequently serves as the vehicle of the organism in foodborne illnesses in Central Europe.

Type E spores are found mainly in aquatic sediments of colder regions of the northern hemisphere, such as Alaska, Canada, Scandinavia, the countries of the former Soviet Union and also Japan. Type E is mostly associated with fish and marine mammals. Type E levels are positively correlated with low bottom oxygen content, depth and absence of bioturbation activity.

The organism is unable to grow in foods at pH below 4.6, although it has been shown to grow in laboratory media below this pH. For this reason, in the US, foods with pH values above 4.6, termed 'low-acid' foods, must be processed according to specific regulations.

Owing to strict federal regulations for commercially canned foods, the incidence of botulism cases has decreased in the past 50 years. In the US, the majority of outbreaks are caused by home-prepared foods. An increasing number of cases are attributed to restaurants and delicatessens where poor handling procedures (temperature abuse, cross-contamination) can affect a large number of people. Table 17.1 summarizes recent outbreaks, most of which have resulted from heating food enough to kill vegetative cells (including spoilage organisms) but not spores, and then holding the food at a temperature that allowed germination of the spores, outgrowth of the organism, and production of toxin.

New technologies in commercially processed foods are presenting new challenges in the research on the outgrowth of *C. botulinum* and its ability to germinate under certain conditions. These processes include hot and cold smoking, vacuum packaging,

Table 17.1 Examples of recent *C. botulinum* outbreaks

Year	Product	Country	People affected	Probable cause
1996	Acidified cream cheese	Italy	8	Temperature abuse
1994	Baked potato	US	30	Temperature abuse
1994	Beef stew	US	1	Temperature abuse
1994	Bean dip	US	1	Temperature abuse
1994	Clam chowder soup	US	2	Temperature abuse
1989	Hazelnut flavored yogurt	England	27	Underprocessed ingredients
1985	Garlic in oil	Canada and US	36	Temperature abuse Inadequate growth controls
1983	Sautéed onion	US	28	Temperature abuse

slow cold-fermentations, and refrigerated processed foods of extended durability (REPFEDs).

V. Isolation and Identification of the Organism and its Toxin

The presence of the organism and the toxin is usually determined from food samples. Clinical samples consisting of feces, blood, or vomitus may also be collected for examination. The need for examination occurs when a presumed diagnosis of botulism is made. The best confirmation of the diagnosis is by laboratory analysis of the sample obtained.

A. Sample Processing

Care must be taken in handling the sample because of the high toxicity associated with botulinum toxin. In general, food samples should be refrigerated until tested. If an intact swollen container is the sample, and it is not in danger of bursting, there is no need to refrigerate the sample. A clinical sample of blood or feces should be collected as soon as botulism is suspected. The sample must be collected before antiserum is given to the patient.

The sample should be identified and labeled appropriately. In the case of a food material, the type of food, label information, container code, container type and size, manufacturer (or home-preserved versus commercially manufactured), and the container condition are examples of the information that should be recorded. The clinical sample should indicate, for example, the type of sample, the time the sample was taken, and the name of the patient.

Because botulinum toxin is heat labile, samples and cultures should be refrigerated; they should also be handled carefully to avoid cross-contamination. Precautions should be taken to avoid the creation of aerosols. Mouth pipetting is never appropriate; a mechanical pipetting device should be used. All glassware, utensils, and materials that come in contact with the sample should be treated as hazardous. The sample and associated materials should be autoclaved or otherwise decontaminated before disposal.

B. Culture Methods

Specific procedures for the laboratory analysis are available in the *Compendium of Methods for the Microbiological Examination of Foods* and the US Food and Drug Administration (FDA) *Bacteriological Analytical Manual*.

Liquid samples may be placed directly into culture medium. Solid samples should be placed in a mortar or a stomacher and mixed thoroughly with an equal amount of gelatin phosphate buffer. A sample should be reserved for toxin testing, another for culturing, and the rest for later analysis.

The sample is examined by wet mount, using a phase-contrast microscope to observe the morphology of any organisms present to determine the existence of rod-shaped bacteria and the presence of typical tennis racket-shaped cells with spores. If the microscopic examination supports the presumptive presence of clostridial cells, the sample should be cultured and toxin tested.

A sample of 1–2 g is placed in tubes of cooked meat medium (CMM) for the detection of proteolytic types or in trypticase peptone glucose yeast extract (TPGY) medium for the nonproteolytic types. The medium is heated in flowing steam for at least 15 minutes in order to remove entrapped oxygen before inoculation with the sample. The cultures are incubated anaerobically at 35°C for the CMM and at 26°C for the TPGY. Anaerobiosis is obtained by placing the inoculated culture tube in an anaerobic chamber or by topping the medium with vaspar or agar. Incubate for 5 days for maximum toxin production. If no growth is observed after 5 days of incubation, the incubation period is extended to at least 10 days before discarding the cultures. Culture tubes should be observed for the presence of turbidity and gas; tubes with gas and turbidity are examined microscopically for the presence of clostridial-type cells and spores. Note the presence of a foul or fecal odor as an indication of proteolysis. If clostridial-type cells are present with gas formation, the sample needs to be tested for toxin.

To isolate the organism in pure culture from the sample or a subculture medium, an equal volume of the material should be mixed with absolute ethanol in a sterile tube. The mixture is held at room temperature for an hour, then cultured on *Clostridium botulinum* isolation (CBI) agar or on liver veal egg yolk agar. As an alternative, for the proteolytic types, the material can be heated for 10 minutes at 80°C, before culturing on the isolation agar. Incubate the plates anaerobically for 48 hours and select colonies that show a pearly layer (surface iridescence) for transfer. Types C, D, and E colonies may also have a zone of yellow precipitate around them. To grow the isolates, the colonies are cultured as described above.

C. Toxin Detection

The sample reserved for toxin testing should be centrifuged to separate the solids and the microorganisms from the fluid containing the toxin. It may be necessary to filter the fluid before testing for toxin. If nonproteolytic types are suspected, the sample should be trypsinized before it is tested for toxin. To a sample adjusted to pH 6.2, sufficient trypsin solution is added to obtain a tenfold dilution of the trypsin. The trypsin is prepared by adding 1 g of trypsin (Difco 1 : 250) to 20 ml of sterile distilled water. The trypsinized sample is incubated at 37°C for 1 hour with gentle agitation. One portion of the trypsinized sample is heated at 100°C for 10 minutes to destroy the toxin that might be present.

Toxin is detected by intraperitoneal injection into mice. Mice are injected with 0.5 ml of the sample. Mice are protected with antiserum for the specific botulinal toxin type, by intraperitoneal injection, 30–60 minutes before injection with the sample. Mice are also injected with sample that has been heated at 100°C for 10 minutes. This heating destroys the toxin, so these mice should survive, as should the mice protected with the specific antitoxin to the toxin in the material tested. All mice are observed for symptoms of botulism for at least 48 hours. Typical symptoms include labored breathing, weakness of the limbs, and ruffled fur. The observed symptoms are important in the evaluation of the presence of botulinum toxin. Death without typical symptoms is not evidence for the presence of botulinum toxin.

D. Alternative Methods

A number of rapid methods for the detection of the organism have been investigated. Several investigators have proposed use of the polymerase chain reaction (PCR) to determine the presence of types A, B, E, and F. Ribotyping has been evaluated for proteolytic and nonproteolytic types. Pulsed-field gel electrophoresis (PFGE) has been tested on nonproteolytic types. Chemical analyses, such as chromatographic analysis for fatty acids, have also been used. Several rapid methods have been proposed for detection of toxin. A PCR procedure for the detection of type A–F neurotoxin genes has been used for the rapid detection of botulism. A number of enzyme-linked immunosorbent assays (ELISAs) have been developed for the detection of botulinum toxins. These methods are nearly as sensitive and reproducible as the mouse bioassay.

VI. Treatment and Prevention

A. Illness and Treatment

Early diagnosis of the disease can lead to prompt treatment, and reduced morbidity and mortality. Early diagnosis also prevents other cases from occurring. Treatment is dependent upon the speed of diagnosis and how far the patient has progressed in the course of the disease. Early administration of antitoxin is essential to slow the onset of the toxin's effect. It is most effective within the first 24 hours of attack. Depending upon the severity of the attack, a tracheotomy and the use of mechanical respiratory aid may be necessary. A saline enema is sometimes necessary to clear the colon of the toxin. Vomiting may be induced to clear the contaminated food from the stomach. In wound botulism, the wound should be treated medically. This may require surgery to remove the source of the infection. Antimicrobials are usually not prescribed unless there are signs of infections by other organisms. In recent years, effective treatment regimens, such as the use of antitoxin, mechanical breathing apparatus, and intensive medical care, have markedly reduced the mortality from botulism, from about 60% 40 years ago to about 10% today.

B. Prevention

Botulinum toxin is easily destroyed by heat. Thus, foods brought to boiling for several minutes before ingestion will not have toxin present to cause illness. Because of its ubiquity in the environment, *C. botulinum* spores are presumed to be present on most foods. In many foods the organism is destroyed during processing. The spores may be present in other foods, but be inhibited from growing in the food and producing toxin.

In some instances – infant botulism or in adults with underlying gastrointestinal disease – the ingestion of the spores may result in the colonization of the gut. Thus, these individuals must take care with respect to foods that might contain the spores.

Botulism intoxication is prevented by not allowing *C. botulinum* to grow and produce the toxin. Growth is prevented by destroying or inhibiting the organism. Heat is the most effective means available for destroying the organism. The vegetative cells of all the types of *C. botulinum* are rapidly killed by pasteurization. Their resistance is similar to that of other vegetative cells.

C. Spore Destruction

The spores of the proteolytic strains are more resistant than those of the nonproteolytic strains. This high heat resistance has led to the 'botulinum process' or '12-D cook' used by the low-acid canned food industry as a minimum thermal process to insure the destruction of *C. botulinum* in the product. Resistance is affected by a number of factors, including the number of spores heated, how the spores were grown, the product in which the spores are heated and the material in which the heated spores are allowed to grow. The resistance varies widely, but at 115.6°C the *D*-value, or decimal reduction time (i.e. the time required for 90% kill), ranges from about 0.14 to 0.83 minutes for type As and from about 0.11 to 1.1 minutes for proteolytic type Bs (Table 17.2). Inasmuch as the Group I strains are the most heat-resistant of their species, these values are the basis for specifying a '12-D process'. If a temperature other than 115.6°C is to be used, the concept of *z*-value (i.e. the change of temperature required for a 10-fold change of the *D*-value) is applied. *D*-values for the nonproteolytic types are given at 82.2°C because these are the least heat-resistant members of the species; nevertheless, they are substantially more heat-stable than almost all non-sporeforming bacterial foodborne pathogens.

The *D*- and *z*-values compiled in Table 17.2 are only examples to indicate the ranges of resistance of the various types. The actual resistance in a specific product may vary within or outside these ranges, and thus should be determined in a specific product to establish the specific resistance in that product.

Table 17.2 Heat effects on various types of *Clostridium botulinum*

Group	Proteolysis	Serotype	D-value		z-value (range, °C)
			Temperature (°C)	Range (minutes)	
I	Yes	A	115.6	0.14–0.83	7.6–11.6
		B	115.6	0.11–1.1	7.9–11.3
		F	110	1.3–1.8	10–12
II	No	B	82.2	1.4–74	6.5–11
		E	82.2	0.3–1.2	7.3–11
		F	82.2	0.25–40	9.5–12
III	Yes	C, D	104	0.02–0.9	9.5–11.5
IV	Yes	G	110	~0.5	~20

C. botulinum spores are relatively resistant to ionizing radiation. The spores of the various types have similar resistance. The *D*-values range from about 0.8 to 4 kGy when tested in water or phosphate buffer. The resistance in foods is variable, depending upon the food and the temperature of radiation. The resistance is greater at temperatures below about $-80°C$ than at refrigeration temperatures. The toxin is not inactivated by radiation at doses that might be used for preserving foods.

C. botulinum spores are more resistant than vegetative cell contaminants to the effects of sanitizers. However, the spores are not unusually resistant to the effects of sanitizers as compared to other spore-forming bacteria. A 99.9% reduction can be obtained with 4.5 ppm free chlorine in 3–12 minutes when tested in phosphate buffer at pH 6.5, at room temperature. The proteolytic types are more resistant than the non-proteolytic types. A similar reduction is obtained when spores are exposed to about 6 ppm ozone for about 2 minutes, about 130 ppm free chlorine from chlorine dioxide for about 14 minutes and about 1500 ppm iodophors for about 30 minutes. Chlorine solutions are more effective when the temperature is higher and the solution at pH <7. Care must be taken to ensure that the chlorine stays in solution rather than turning into the gaseous form at the lower pH or higher temperature.

D. Preventing Growth

Since the toxin is formed when *C. botulinum* is allowed to multiply, if growth can be prevented, toxin is not formed. Refrigeration temperatures inhibit the organism's growth, particularly the proteolytic types. However, because of their ability to grow down to 3.3°C, the nonproteolytic types are capable of growth at temperatures usually achieved in the normal distribution of refrigerated foods. Even though the nonproteolytic strains are capable of growth at these temperatures, growth occurs slowly. Thus, shelf-lives of products can be extended with proper temperature controls.

Low water activity and low pH can also inhibit the outgrowth of *C. botulinum* spores. Low water activity is achieved with the addition of salt or sugars. The water activity needed to inhibit the proteolytic strains is about 0.93, whereas a water activity below about 0.97 inhibits the nonproteolytic strains. A pH of 4.6 or less is inhibitory to the outgrowth of *C. botulinum* spores. This pH can be achieved naturally as in acidic fruits or artificially by the addition of acids, usually organic acids. The organic acids tend to be better inhibitors of spore outgrowth than the inorganic acids. If acidification is used to inhibit the outgrowth, all components of the food must be properly acidified to pH 4.6 or less. In addition, care must be taken to prevent the growth of spoilage organisms, such as some yeasts, molds and bacteria, which can raise the pH to levels where *C. botulinum* might grow and produce toxin. There are reports that, in laboratory media with high concentrations of protein precipitates, *C. botulinum* outgrowth and toxin production have been observed at pH <4.6.

Preservatives serve an important role in inhibiting the growth of *C. botulinum* spores. The curing of meat has long relied upon the presence of sodium nitrite and salt to control *C. botulinum*. The nitrite imparts the unique cured meat flavor and keeps the typical pink color of the meat. The nitrite also prevents the growth of *C. botulinum*. In recent years, because of the concern for the formation of nitrosamines, other additives have been explored as replacements to nitrite. Although levels have been reduced, nitrite has not been replaced, as no acceptable replacement has been found. As with many foods, the inhibition of *C. botulinum* in cured meats is dependent upon the interaction of a number of inhibitory factors, such as nitrite, salt, and pH, all working together. Many other preservatives, such as sorbates, nisin, parabens and sodium lactate, either alone or in combination, have been evaluated as inhibitors of *C. botulinum* outgrowth and found to be potentially useful.

The design of foods to prevent or impede the outgrowth of *C. botulinum* spores usually involves a strategy of destruction and/or inhibition. If inhibition is chosen, usually a combination of factors is used so that the hurdles, which must be overcome for growth to occur, are sufficient to assure that growth will not occur before the food is eaten.

VII. Infant Botulism

This type of botulism was first recognized in 1976. It is the most common type of botulism in the US, accounting for about 79% of botulinum cases in 1983. Typically, it affects infants under 12 months of age, and more typically between 5 and 29 weeks of age. It is believed that the spores enter the infant's system through the mouth or nose; and because of an immature gut physiology or inadequate development of gut flora, the germination of the spores and subsequent growth are not suppressed, as they are in healthy adults. Sources of spores include honey, food, carpet soil/dust and dust associated with new construction. Of all potential sources, honey is the one dietary reservoir of *C. botulinum* spores thus far definitively linked to infant botulism by both laboratory and epidemiological studies. Honey accounts for up to 15% of the cases of infant botulism.

Constipation is usually the earliest clinical sign, followed by poor feeding, lethargy, a weak cry, decreased sucking, and generalized lack of muscle tone, noticeably characterized by a floppy head. The clinical spectrum of infant botulism includes mild, outpatient cases, and sudden unexpected death indistinguishable from typical sudden infant death syndrome.

The diagnosis of botulism in many patients is established by identification of *C. botulinum* toxin and organisms in the infants' feces. Evidence was also obtained that ingested spores of *C. botulinum* had produced the toxin in the infants' intestinal tract. Almost all cases of infant botulism in the US have been caused by Group I proteolytic type A or type B strains. Unusual strains of *C. baratii* and *C. butyricum* that make botulinum-like toxins E and F have also caused infant botulism.

VIII. Therapeutic Applications

Purified botulism toxin is the first bacterial toxin to be used as a medicine. The FDA licensed botulinum toxin as Oculinum in December 1989 for treating two eye conditions – blepharospasm and strabismus – characterized by excessive muscle contractions. It is now marketed under the trade name Botox.

Small doses of the toxin are injected into the affected muscles. As happens with botulism, the toxin binds to the nerve endings, blocking the release of the chemical acetylcholine, which would otherwise signal the muscle to contract. The toxin thus paralyzes or weakens the injected muscle, but leaves the other muscles unaffected.

More recent research shows promise in using the toxin in other neurological diseases that involve abnormal muscle posture and tension, including torticollis (contractions of the neck and shoulder muscles), oromandibular dystonia (clenching of the jaw muscles), and spasmodic dysphonia (results in speech that is difficult to understand) and migraine headaches. Other therapeutic applications are being investigated.

IX. Summary

Because the toxin produced by *C. botulinum* is so lethal, understanding the organism's growth and toxin production capabilities is critical to safe production of foods all over the world. In addition, understanding how to test for the organism and its toxin, and how to recognize and treat patients afflicted with the illness is imperative. Considerable progress has been made in these areas in the past century. As new food-processing technologies are developed, food scientists will have to continue to do research on how these technologies affect this organism.

Bibliography

Arnon, S. S. (1998). Infant botulism. *In* 'Textbook of Pediatric Infectious Disease', 4th ed. (R. D. Feigin and J. D. Cherry, eds), pp. 1570–1577. W.B. Saunders Company, Philadelphia.

Bell, C. and Kyriakides, A. (2000). '*Clostridium botulinum*. A Practical Approach to the Organism and its Control in Foods'. Blackwell Science, Oxford.

Cato, E. P., George, W. L. and Finegold, S. M. (1986). Section 13, Genus *Clostridium*. *In* 'Bergey's Manual of Systematic Bacteriology' (P. H. Sneath, N. S. Mair, M. E. Sharpe and J. G. Holt, eds), pp. 1157–1160. Williams and Wilkins, Baltimore, MD.

Hauschild, A. H. W. and Dodds, K. L. (1993). '*Clostridium botulinum*. Ecology and Control in Foods'. Marcel Dekker, New York.

Shapiro, R. L., Hatheway, C. and Swerdlow, D. L. (1998). Botulism in the United States: a clinical and epidemiologic review. *Ann. Intern. Med.* **129**, 221–228.

Smith, L. D. S. and Sugiyama, H. (1988). 'Botulism: the Organism, its Toxin, the Disease'. Charles C. Thomas, Springfield, IL.

Solomon, H. M., Johnson, E. A., Bernard, D. T., Arnon, S. S. and Ferreira, J. S. (2001). Chapter 33. *Clostridium botulinum* and its toxins. *In* 'Compendium for the Microbiological Examination of Foods', 4th ed. (F. P. Downs and K. Ito, eds), pp. 317–324. American Public Health Association, Washington, DC.

US Food and Drug Administration (1998). 'Bacteriological Analytical Manual'. 8th ed. AOAC International, Gaithersburg, MD.

Bacillus cereus Food Poisoning

18

Mansel W. Griffiths and Heidi Schraft

I. Introduction

Bacillus cereus belongs to a genus of Gram-positive, endospore-forming facultatively anaerobic bacteria. The resistance of its spores to a number of adverse conditions has resulted in widespread distribution of the organism. It has been isolated from air, soil and water, as well as animal and plant material. Indeed, it was originally isolated from air in a cowshed by Frankland and Frankland in 1887. The ubiquitous nature of this organism makes contamination of food materials a common occurrence.

The genus *Bacillus* is extremely heterogeneous, but may be divided into six sub-groups. Based on spore morphology, *B. cereus* is classified in the *B. subtilis* group. Four species within this group, *B. cereus*, *B. mycoides*, *B. thuringiensis*, and *B. anthracis*, are closely related both phenotypically and genotypically – so much so that the three latter species should be considered as subspecies of *B. cereus*. Of the 34 species in the genus, only two, *B. anthracis* and *B. cereus*, are recognized as common pathogens, but only *B. cereus* has been implicated in foodborne diseases. As early as 1906, an outbreak of foodborne illness was caused by a spore-forming bacillus that Lubenau called *Bacillus peptonificans*. The characteristics of the organism closely resembled those of *B. cereus*, but it was not until 1950 that sufficient evidence was provided by Steinar Hauge in Norway to prove conclusively that *B. cereus* was a human pathogen.

Foodborne Diseases 2nd Edn
ISBN: 0-12-176559-8

However, it is increasingly acknowledged that other *Bacillus* spp. can occasionally cause food poisoning in humans. These species include *B. licheniformis*, *B. subtilis*, *B. pumilus*, *B. brevis*, and *B. thuringiensis*.

As well as causing gastrointestinal disease, *B. cereus* is a virulent ocular pathogen and is responsible for conjunctivitis, panophthalmitis, keratitis, iridocyclitis, and orbital abscess. It can cause a number of other opportunistic infections, including infections of the respiratory tract and wound infections.

II. Characteristics of *Bacillus cereus* Food Poisoning

Two distinct types of illness have been attributed to the ingestion of foods contaminated with *B. cereus*: a diarrheal illness and an emetic illness (Table 18.1). In some *B. cereus* outbreaks, there appears to be a clear overlap of the diarrheal and emetic syndromes.

A. Incidence of Illness

In general, the symptoms of *B. cereus* food poisoning are transient and mild. This undoubtedly means that it is vastly under-reported. Owing to the ubiquitous nature of the organism, it may also be that much of the population is partly protected through immunity acquired by continuous exposure. Another factor that may be responsible for the relatively low incidence of *B. cereus* food poisoning is the large cell numbers required to cause illness. To reach these levels of contamination, the food may require storage for extended periods likely to result in obvious spoilage. The incidence and nature of foodborne illness caused by *B. cereus* vary from country to country. In North America and Europe the diarrheic syndrome is more common, but in Japan the emetic type of illness is reported ten times more frequently than diarrheal illness. It has been estimated that there are about 84 000 cases of foodborne illness caused by *B. cereus* annually in the US at a per-case cost of $430 and a total cost of $36 million. However, the annual average numbers of *B. cereus* foodborne disease outbreaks

Table18.1 Foodborne illnesses associated with *Bacillus cereus*

Property	Diarrheal syndrome	Emetic syndrome
Incubation period	8–16 hours	0.5–5 hours
Duration of illness	12–24 hours (sometimes >24 hours)	6–24 hours
Infective dose	10^5–10^7 cells ingested	10^5–10^8 cells/g of food
Symptoms	Abdominal pain, watery diarrhea, occasional nausea	Nausea, vomiting, occasional diarrhea
Foods commonly implicated	Meat products, soups, milk and milk products, vegetables, puddings and sauces	Rice, pasta, noodles, pastries

reported to the Centers for Disease Control and Prevention (CDC) in the US between 1988 and 1992 was 4.2 (0.9% of all foodborne illness outbreaks) resulting in an annual average of 86.6 cases (0.6% of all cases). The corresponding figures for Canada between 1975 and 1986 were 10.5 annual average number of outbreaks (1.3% of all outbreaks) and 63.6 annual cases (1% of all cases). The percentages of reported foodborne outbreaks and cases attributable to *B. cereus* in North America, Europe, and Japan vary between 1–22% for outbreaks and 0.7–33% for cases. The Netherlands and Norway purport to have the greatest problem due to *B. cereus*.

B. Outbreaks Caused by *Bacillus* spp. Other than *B. cereus*

B. licheniformis and *B. subtilis* have been repeatedly isolated from foods linked to illness. *B. licheniformis* infection predominantly results in diarrhea with occasional vomiting, whereas *B. subtilis* causes vomiting frequently accompanied by diarrhea. The implicated foods include mainly cooked meat dishes for *B. licheniformis* and cooked meat products, stuffed poultry, pizza and wholemeal bread for *B. subtilis*. *B. pumilus* infection results in diarrhea, vomiting, and dizziness and has been associated with meat products, cheese sandwiches, and canned tomato juice. *B. thuringiensis* was isolated from the stools of four patients suffering from gastroenteritis, and no other enteric pathogen was isolated from three of these individuals.

III. Characteristics of the Organism

Arguably the most important characteristic of *Bacillus* spp. is their ability to form refractile endospores. These spores are more resistant than vegetative cells to heat, drying, food preservatives, and other environmental challenges. Strains of *Bacillus* produce catalase, which, in addition to the aerobic production of spores, distinguishes *Bacillus* from *Clostridium*. The bacteria of the genus *Bacillus* are usually free living, that is, not host adapted, and their spores are widely distributed throughout nature.

The spores of *B. cereus* are ellipsoidal, central to subterminal, and do not distend the sporangia. *B. cereus* can be distinguished from other closely related species by: (1) a positive egg-yolk lecithinase reaction; (2) inability to produce acid from mannitol; (3) ability to grow anaerobically; (4) a positive Voges-Proskauer reaction; and (5) resistance to 0.001% lysozyme. The cells of *B. cereus* are typically large, reaching 3–5 μm in length and 1.0–1.2 μm in width; they often occur in chains. *B. cereus* cells are motile by peritrichous flagella. The mol % G + C of the DNA has been reported to be approximately 32–38. Two other *Bacillus* species, *B. thuringiensis* and *B. mycoides*, are closely related to *B. cereus*, with a similar range of mol % G + C. Only two phenotypic characteristics distinguish the three species: mycoidal growth of *B. mycoides*, and crystalline parasporal inclusion bodies present in *B. thuringiensis*. However, both characteristics are plasmid encoded and may be lost during subculturing. Thus, several researchers have suggested that the three species belong in one group.

A. The Spore

Spores from strains commonly associated with food poisoning have a $D_{95°C\ (203°F)}$ of approximately 24 minutes, with other strains showing a wider range of heat resistance ($D_{95°C\ (203°F)}$ of 1.5–36 minutes). The z-value varies between about 6°C (10.8°F) and 9°C (16.2°F). It has been suggested that strains involved in food poisoning may have higher heat resistance and therefore be more apt to survive cooking. The spore is also more resistant to irradiation; the dose for a 90% reduction in count lies between 1.25 and 4 kGy. The corresponding dose for vegetative cells is 0.17–0.65 kGy.

Spore germination can occur over the temperature range of 5°C (41°F) to 50°C (122°F) in cooked rice and between −1°C (30.2°F) and 59°C (138.2°F) in laboratory media. Glycine or a neutral L-amino acid and purine ribosides induce germination. L-Alanine is the most effective amino acid stimulating germination.

Their hydrophobic nature, coupled with the presence of appendages on their surface, enables spores to adhere to several types of surfaces, including epithelial cells. This adhesiveness makes them difficult to remove during cleaning and sanitation of food contact surfaces. When spores are located in the space between seals and seal surfaces where water may be excluded, their heat resistance increases significantly.

B. Vegetative Growth

Early in the growth cycle, vegetative cells are Gram-positive, but cells may become Gram-variable when in late log or stationary phase. Colonies on agar media have a dull or frosted appearance.

The emergence of psychrotrophic strains of the organism now means that the temperature range for growth is between approximately 5°C (41°F) and 50°C (122°F). Generation times at 7°C (44.6°F) lie between 9.4 and 75 hours, but in boiled rice at 30°C (86°F) the generation time can range between 26 and 57 minutes. Several workers have shown that *B. cereus* is capable of producing enterotoxin during growth at refrigeration temperatures, but the practical significance of this is uncertain. Several strains can grow slowly at sodium chloride concentrations of 10%. The minimum water activity for growth is between 0.91 and 0.93. *B. cereus* has an absolute requirement for amino acids as growth factors, but vitamins are not required. The organism grows over a pH range of approximately 4.4–9.3. These limits for growth are not absolute and are dependent on several factors, including strain.

IV. Pathogenesis

A. Diarrheal Syndrome

B. cereus was proven to be responsible for the diarrheal syndrome after a volunteer experiment of Hauge in 1955, who ingested vanilla sauce with high numbers of *B. cereus* and subsequently developed diarrhea and abdominal pain. However, causative agents for the disease still have not been fully identified. To test for diarrhea-causing activity of purified *B. cereus* proteins, animal models such as the rabbit ileal loop test (RIL) and the vascular permeability test (VP) are employed. Cell culture assays, relying on Vero cells, are also commonly used to detect enterotoxic activities. It seems likely, on the basis of such tests, and results of gene cloning and sequencing studies, that at least three enterotoxigenic proteins are produced by *B. cereus*. Results of experiments conducted in the 1980s indicated that *B. cereus* enterotoxicity was caused by a multicomponent protein. These findings are supported by results of Wong and coworkers at the University of Wisconsin, who described a three-component enterotoxin, hemolysin BL, consisting of three proteins termed B, L_1 and L_2. All three components are apparently required to produce maximal fluid accumulation in the RIL assay. In contrast, Granum's research group in Norway claims that a 40-kDa protein fraction they purified from *B. cereus* is toxic to Vero cells on its own. The Japanese research group of Shinagawa has also purified a protein from *B. cereus*, which is enterotoxigenic on its own. This protein, Wong's hemolysin B and Granum's 40-kDa protein are most likely identical. The biological activity ascribed to a single protein by Granum and Shinagawa is probably due to contamination of the preparation with small amounts of hemolysin L_1 and L_2. Agata and coworkers described yet another protein, enterotoxin T, which showed positive reactions in VP and RIL tests. The protein and the encoding DNA sequence are not related to the previously described three proteins. Additional research will have to demonstrate whether this protein plays any role in *B. cereus* diarrheal disease.

Only limited work has been carried out on the mode of action of the enterotoxin. It reverses absorption of fluid, Na^+ and Cl^- by epithelial cells, causes malabsorption of glucose and amino acids as well as necrosis and mucosal damage. It has been suggested that the effects on fluid absorption were due to stimulation of adenylate cyclase.

Prevailing expert opinion has been that *B. cereus* enterotoxin is preformed in foods, causing a classical foodborne intoxication. However, two outbreaks have been described where people were suffering from more severe symptoms than those normally associated with *B. cereus* diarrheal syndrome. In the first outbreak, 17 out of 24 people were affected after eating stews, and three of the patients were hospitalized, one for 3 weeks. The infective dose observed in this outbreak was 10^4–10^5 cells, somewhat lower than that usually associated with *B. cereus* (Table 18.1). The time to onset of symptoms for these patients was more than 24 hours, also longer than commonly observed for *B. cereus*. The second outbreak affected competitors at a skiing event in Norway where 152 out of 252 people became ill after drinking milk. The illness was restricted to the skiers between the ages of 16 and 19, and the older coaches and officials did not exhibit symptoms. Again, the incubation period was greater than 24 hours in some cases and the patients were ill for periods between from 2 to several days. Perhaps this more severe form of *B. cereus* gastroenteritis results from adhesion of spores to the epithelial cells, followed by enterotoxin production within the intestinal tract. The longer incubation time seen in these cases may be due to the time required for germination of the spores.

The view that the diarrheal syndrome is caused by ingestion of cells rather than preformed toxin is supported by the fact that the toxin is rapidly degraded at pH 3 and

Table 18.2 Properties of *Bacillus cereus* toxins

Property	Diarrheal	Emetic
Structure	Three-component protein complex? More than one enterotoxin?	Cyclic peptide
Heat stability	Destroyed at 55°C (131°F) after 20 minutes	Stable for 90 minutes at 126°C (259°F)
pH stability	Unstable <4 and >11	Stable between 2 and 11
Effect of proteolytic enzymes	Inactivated by pronase and trypsin	No effect
Common assay procedures	Immunoassays, rabbit ileal loop, cytotoxicity assay	Monkey challenge, vacuolation of HEp-2 cells
Mode of action	Produced in gut? Reverses fluid and ion flux in intestines	Preformed in food; acts on 5-HT$_3$ receptors

hydrolyzed by the proteolytic enzymes of the digestive tract. There is evidence, however, that the toxin may be protected from such denaturation when it is present in a food matrix.

B. The Emetic Syndrome

For many years the structure of the emetic toxin was elusive, partly because there was no convenient assay. It was thought to be a lipid and was known to be nonantigenic and extremely resistant to heat, alkaline, and acid pH and proteolysis (Table 18.2). The emetic toxin has now been identified as a ring-structured peptide composed of three repeat sequences of four amino acids and/or oxy-acids. The dodecadepsipeptide, termed cereulide, has a molecular weight of 1.2 kDa and the following composition:

$$[\text{D-}O\text{-leu-D-ala-L-}O\text{-val-L-val}]_3$$

It closely resembles the potassium ionophore, valinomycin. It is not yet known if cereulide is a modified gene product or, as is more likely, produced enzymatically. On the cellular level, cereulide seems to cause swelling of mitochondria. Emesis is probably produced by binding of the preformed toxin to 5-HT$_3$ receptors stimulating the vagus afferent nerve. Experiments with monkeys have found that about 70 μg of cereulide are necessary to cause vomiting.

A severe incident of *B. cereus* emetic disease involved a 17-year-old boy and his father, who had acute gastroenteritis after eating spaghetti and pesto that had been prepared 4 days earlier. Within 2 days, the boy developed liver failure and died. The father had hyperbilirubinemia and rhabdomyolysis, but he recovered. High levels of *B. cereus* emetic toxin were found in the pan used to reheat the food and in the boy's liver and bile. *B. cereus* was also cultured from both intestinal contents and pan residues. It was concluded that liver failure was due to the emetic toxin causing inhibition of hepatic mitochondrial fatty-acid oxidation.

V. Transmission via Food

Bacillus cereus has been isolated from a wide variety of foods, including milk and dairy products, meat and meat products, pasteurized liquid egg, rice, ready-to-eat vegetables and spices. Based on the ubiquitous distribution of the organism, it is virtually impossible to obtain raw products that are free from *B. cereus* spores.

A large percentage of *B. cereus* isolated from milk have been shown to be potentially pathogenic. In a Canadian study, 85% of psychrotrophic *Bacillus* spp. isolated from milk were enterotoxigenic. Other studies have confirmed that a high percentage of dairy *B. cereus* are cytotoxigenic (50–90% of strains). In a Norwegian study, 59% of *B. cereus* isolates from dairy products were enterotoxigenic, 15% were psychrotrophic and 7% were both psychrotrophic and enterotoxigenic. Although some outbreaks of foodborne illness associated with pasteurized milk have been reported, a study carried out in the Netherlands concluded that the probability of healthy adults becoming ill from *B. cereus* in pasteurized milk is small and there was no evidence of infection when *B. cereus* numbers in the milk were below 10^5 colony forming units (CFU)/ml. In the Dutch study, the milk was stored at 7.5°C (45.5°F) for up to 14 days; however, if pasteurized milks were subjected to temperature abuse at 10°C (50°F) during their expected shelf-life, all the milks examined were found to contain *B. cereus* and in about 40% of cartons there was evidence of enterotoxin production.

A. Diarrheal Syndrome

Several foodstuffs have been identified as vehicles for transmission of the diarrheal syndrome caused by *B. cereus*. The first well-documented outbreak described by Hauge in 1955 involved a vanilla sauce. Subsequently, outbreaks linked with several different foods, including meat and meat dishes, vegetables, cream, spices, poultry and eggs, have been described. Up to 50% of such foods are commonly contaminated with *B. cereus*, and a high level of asymptomatic fecal carriage of *B. cereus* has been found in humans. Therefore, to prove that a particular food is responsible for an outbreak, food isolates and *B. cereus* detected in patients must display common epidemiological features, e.g. identical serotype, phage type, biotype, or genotype. Also, the food isolate should be tested for production of enterotoxin.

B. Emetic Syndrome

The emetic syndrome associated with *B. cereus* was first identified in the UK following five incidents in 1971 resulting from the consumption of contaminated fried rice from Chinese restaurants. In the following 13 years, there were 192 further outbreaks involving more than 1000 cases. Incidents of emetic illness are almost invariably associated with farinaceous (starchy) foods. In fact, starch promotes the growth of *B. cereus* and production of emetic toxin. The vehicle in about 95% of all emetic-type incidents has been Cantonese-style cooked rice served by Chinese restaurants. The practice of saving portions of boiled rice and allowing them to dry at room temperature for several hours before subsequent frying with beaten egg is thought to be the main reason for the problem. Spores of *B. cereus* survive the cooking process, germinate in the cooked rice at room temperature and the vegetative cells then proliferate rapidly, especially when beef, chicken, or egg is added. A small proportion of outbreaks have been linked to other rice dishes, pastries (e.g. vanilla-flavored cake slices), pasteurized cream, milk pudding, chicken supreme, cooked pasta and, perhaps most alarmingly, reconstituted infant formula. A survey of *B. cereus* isolated during outbreaks of food poisoning revealed that 72% of emetic strains did not hydrolyze starch and 99% of *B. cereus* in serovar H-1 produced cereulide. In contrast, none of the *B. cereus* strains capable of starch hydrolysis produced emetic toxin.

VI. Isolation and Identification

A. Cultural Methods

Selective media for the isolation and identification of *B. cereus* rely on the organism's ability to produce a lecithinase (phospholipase C) and its inability to ferment mannitol. Two common media used are mannitol-egg yolk-polymyxin (MYP) and polymyxin-pyruvate-egg yolk-mannitol-bromothymol blue agar (PEMBA). Typical colonies growing on PEMBA are a distinctive turquoise to peacock blue and are surrounded by an egg-yolk precipitate of the same color. *B. cereus* colonies on MYP have a violet–red background and are surrounded by a zone of egg-yolk precipitate. A medium (VRM) not requiring polymyxin and supposedly capable of differentiating between *B. cereus* and *B. thuringiensis* has been reported as well. Because of the high numbers usually being sought, enrichment is not necessary, but when needed, either Robertson's heated cooked meat medium or trypticase-soy-polymyxin broth is recommended. Following its isolation from food, various confirmatory tests are carried out. The US Food and Drug Administration tests for anaerobic glucose fermentation, nitrate reduction, Voges-Proskauer reaction, citrate utilization, tyrosine decomposition and resistance to lysozyme. All these tests are positive for *B. cereus*. The Central Public Health Laboratory in the UK tests for motility, hemolysis, rhizoid growth, susceptibility to γ-phage and fermentation of ammonium salt-based glucose but not mannitol, arabinose, or xylose. A rapid *Bacillus* identification system is available from bioMérieux called the API 50CHB test and determines the ability to assimilate 49 carbohydrates. It is claimed that the system can identify potential emetic strains.

B. Typing Methods

Several methods have been investigated for epidemiological typing of *B. cereus*. These include: serotyping of spore, flagellar and somatic antigens; phage typing; plasmid analysis; and biotyping. The analysis of the fatty-acid composition of *B. cereus* using gas chromatography has proved useful in epidemiological investigations. A number of workers has also used random amplified polymorphic DNA (RAPD) to investigate the sources of contamination of milk by *B. cereus*. RAPD is based on a randomly primed polymerase chain reaction for amplification of arbitrary fragments of the genome to produce unique genetic fingerprints of strains. There is evidence that this technique can be used to differentiate emetic strains from other *B. cereus*. An automated ribotyping instrument, the RiboPrinter, developed by Qualicon, Inc., has also been used to genotype *B. cereus*.

C. Noncultural Detection Methods

An immunoassay to detect spores of *B. cereus* has been documented, but there is little information on the practical application of this technique. Schraft and Griffiths in 1995 described a very sensitive PCR assay for species belonging to the *B. cereus* group. The assay amplifies a portion of the cereolysin gene.

D. Detection of Toxins

Traditionally diarrhea-causing enterotoxin has been detected by fluid accumulation in ligated rabbit ileal loop assays or by the vascular permeability reaction. To circumvent the use of animals, Jackson developed a fluorescent immunodot assay using antiserum raised in rabbits against either purified enterotoxin or culture supernatant. The antibody raised against the pure enterotoxin had a sensitivity of 50 ng and was specific to *B. cereus* culture filtrates. Commercial immunoassays for *B. cereus* enterotoxin are now available. The BCET-RPLA of Oxoid uses reversed passive latex agglutination. The presence of enterotoxin is detected by agglutination of latex particles sensitized with the antitoxin. The Oxoid kit detects the L_2 component of the enterotoxin, and there has been some concern about the specificity of the assay. A second commercial immunoassay, TECRA BDE Visual Immunoassay, is marketed by Bioenterprises Pty Ltd and has a sandwich ELISA format. The antibody used in the kit is active against two proteins, which are produced simultaneously to the enterotoxin. Neither of these proteins is toxic.

The emetic toxin has traditionally been detected using primate feeding trials, but Hughes and coworkers in 1988 determined that the ability to cause vacuolation of HEp-2 (larynx carcinoma) cells in tissue culture was specific to culture filtrates of emetic strains of *B. cereus*. The identification of the emetic toxin will probably lead to the development of improved diagnostic tests.

VII. Treatment and Prevention

The symptoms of *B. cereus* infection are usually mild and self-limiting, and would not normally require treatment. Exceptions have been described above.

Because *B. cereus* can almost invariably be isolated from foods and can survive extended storage in dried food products, it is not practical to eliminate low numbers of spores from foods. Control against food poisoning should be directed at preventing germination of spores and minimizing growth of vegetative cells. To accomplish this, foods should be maintained above 60°C (140°F) or rapidly and efficiently cooled to less than 7°C (44.6°F), and be thoroughly reheated before serving.

VIII. Summary

B. cereus is commonly isolated from many foods including meat, vegetables, and dairy products. It causes two types of illness: an emetic syndrome, and a diarrheal syndrome. Desserts, meat dishes, and dairy products are most frequently the vehicles for transmission of the diarrheal form of the illness, whereas rice is the main vehicle of the emetic illness. The emergence of psychrotrophic strains of the organism may mean that *B. cereus* will be of increasing concern to the food industry in the future.

Bibliography

Andersson, A., Ronner, U. and Granum, P. E. (1995). What problems does the food industry have with the spore-forming pathogens *Bacillus cereus* and *Clostridium perfringens*? *Int. J. Food Microbiol.* **28**, 145–155.

Fermanian, C., Fremy, J.-M. and Lahellec, C. (1993). *Bacillus cereus* pathogenicity: a review. *J. Rapid Meth. Autom. Microbiol.* **2**, 83–134.

Granum, P. E. (2001). *Bacillus cereus*. *In* 'Food Microbiology: Fundamentals and Frontiers', 2nd ed. (M. P. Doyle, L. R. Beuchat and T. J. Montville, eds), pp. 373–381. American Society for Microbiology Press, Washington, DC.

Johnson, E. A. (1990) *Bacillus cereus* food poisoning. *In* 'Foodborne Diseases' (D. O. Cliver, ed.), pp. 127–135. Academic Press, San Diego, CA.

Kramer, J. M. and Gilbert, R. J. (1989). *Bacillus cereus* and other *Bacillus* species. *In* 'Foodborne Bacterial Pathogens' (M. P. Doyle, ed.), pp. 21–70. Marcel Dekker, New York.

Mycotoxins

Fun S. Chu

I. Introduction

Mycotoxin is a convenient generic term describing the toxic substances formed during the growth of fungi. 'Myco' means 'fungal (mold)', and 'toxin' represents 'poison'. In contrast to the bacterial toxins, which are mainly proteins with antigenic properties, the mycotoxins encompass a considerable variety of low-molecular-weight compounds with diverse chemical structures and biological activities. Like most microbial secondary metabolites, the functions of mycotoxins for the fungi themselves are still not clearly defined. In considering the effect of mycotoxins on the animal's body, it is important to distinguish between *mycotoxicosis* and *mycosis*. Mycotoxicosis is used to describe the action of mycotoxin(s) and is frequently mediated through a number of organs, notably the liver, kidney, lungs, and the nervous, endocrine, and immune systems. On the other hand, mycosis refers to a generalized invasion of living tissue(s) by growing fungi.

Mycotoxins and mycotoxicoses are an especially significant problem for human and animal health because, under certain conditions, foodstuffs can provide a favorable medium for fungal growth and toxin production. Because of the relative stability of mycotoxins to heat and other treatments, they may remain in foods and feeds for a long period. Mycotoxins have caused great economic losses because not only is there an outright loss of crops and animals when a severe outbreak occurs; but even lower mycotoxin levels in feed affect animal health by causing feed refusal and an increased

Foodborne Diseases 2nd Edn
ISBN: 0-12-176559-8

susceptibility to infection. This may result in declines in reproductive success, in the production of milk and eggs, and in the quality of animal products.

The mycotoxin problem is actually an old one. Ergotism and mushroom poisoning, for example, have been known for centuries. Outbreaks of other toxicoses associated with the ingestion of moldy foods and feeds by humans and animals have also been recorded in this century. A well-documented example is the disease called alimentary toxic aleukia (ATA) that resulted in more than 5000 deaths in humans in the Orenberg district of the USSR during World War II. The cause of this outbreak was later determined to be trichothecene mycotoxins produced by fungi growing on grain allowed to stand in the field during winter. Since the discovery in the early 1960s of aflatoxins, highly potent carcinogens produced by *Aspergillus flavus* and *A. parasiticus*, research has focused new attention on mycotoxins. Developments in the last three decades have disclosed many new fungal poisons that are attracting attention because of their association with foods and animal feeds, and because of their diverse toxic effects. It is beyond the scope of this book to review all literature on this subject. Instead, the discussion will focus on selected mycotoxins that most frequently contaminate our foods, their modes of action, and potential hazard to humans as well as on possible preventive measures.

II. Production of Mycotoxin by Toxicogenic Fungi

Invasion by fungi and production of mycotoxins in commodities can occur: under favorable conditions in the field (preharvest); at harvest; and during processing, transportation and storage. Fungi that are frequently found in the field include *Aspergillus flavus*, *Alternaria longipes*, *Alternaria alternata*, *Claviceps purpurea*, *Fusarium moniliforme*, *F. graminearum*, and a number of other *Fusarium* spp. Species most likely introduced at harvest include *F. sporotrichioides*, *Stachybotrys atra*, *Cladosporium* spp., *Myrothecium verrucaria*, *Trichothecium roseum*, as well as *A. alternata*. Most penicillia are storage fungi. These include *Penicillium citrinum*, *P. cyclopium*, *P. citreoviride*, *P. islandicum*, *P. rubrum*, *P. viridicatum*, *P. urticae*, *P. verruculosum*, *P. palitans*, *P. puberulum*, *P. expansum*, and *P. roqueforti*, all of which are capable of producing mycotoxins in grains and foods. Other toxicogenic storage fungi are *A. parasiticus*, *A. flavus*, *A. versicolor*, *A. ochraceus*, *A. clavatus*, *A. fumigatus*, *A. rubrum*, *A. chevallieri*, *F. moniliforme*, *F. tricinctum*, *F. nivale*, and several other *Fusarium* spp. Thus, most of the mycotoxin-producing fungi belong to three genera: *Aspergillus*, *Fusarium*, and *Penicillium*. However, not all species in these genera are toxicogenic.

Genetics and environmental and nutritional factors greatly affect the formation of mycotoxins. In the field, weather conditions, plant stress, invertebrate vectors, species and spore load of infective fungi, variations within plant and fungal species, and microbial competition all significantly affect mycotoxin production. During storage and transportation, water activity (a_w), temperature, crop damage, time, blending with moldy components, and a number of chemical factors, such as aeration (O_2, CO_2 levels), types of grains, pH, and presence or absence of specific nutrients and inhibitors are important. In general, mold growth in grains or foods is necessary before production of toxin occurs, and optimal conditions for toxin formation generally have a narrower window than those for mold growth. The production of fumonisin by *F. moniliforme* on maize is a good example. When a_w was lowered from 1.0 to 0.95, there was no change in fungal growth but fumonisin levels only decreased threefold. However, a 300-fold reduction in fumonisin production was found when a_w was lowered from 1.0 to 0.90, with only a 20% decrease in fungal growth.

III. Natural Occurrence and Toxic Effects of Selected Mycotoxins

Some of the most frequent naturally occurring mycotoxins, the toxin-producing fungi, and their major toxic effects in human and animals are shown in Table 19.1. It is apparent that, whereas some mycotoxins (e.g. aflatoxin; AF) are produced only by a few species of fungi within a genus, others (e.g. ochratoxin A; OA) are produced by fungi across several genera. A number of commodities can be contaminated with different types of mycotoxins. In general, most mycotoxins affect specific organs, but due to diverse structures, trichothecenes (TCTCs) affect many organs. Among the mycotoxins, AF, OA, fumonisin (Fm), deoxynivalenol (DON), and several other TCTCs are the most frequent contaminants in foods and feeds. Only these mycotoxins will be discussed in some detail.

A. Aflatoxins

1. General considerations

Aflatoxins have chemical structures containing dihydrofuranofuran and tetrahydrofuran fused with a substituted coumarin. At least 16 different structurally related toxins have been found; the structures of four major AFs, e.g. B1, G1, B2, and G2 are shown in Fig. 19.1. The toxins are primarily produced by *A. flavus* and *A. parasiticus* in a number of important agricultural commodities in the field and during storage. Recent studies have shown that some *A. nominus* and *A. tamarii* strains are also AF producers. The optimal temperatures and a_w for the growth of *A. flavus* and *A. parasiticus* are around 35–37°C (range from 6° to 54°C) and 0.95 (range from 0.78 to 1.0), respectively; for aflatoxin production, they are 28–33°C and 0.90–0.95 (0.83–0.97), respectively. Because the four major toxins were originally isolated from fungal cultures of *A. flavus*, the first few letters of the name of the fungus ('A' and 'fla') were used to name the toxin. These toxins fluoresce either blue or green under ultraviolet light, and this distinguishes the B or G types of toxins. AFB1 is most toxic in this group and is one of the most potent naturally occurring carcinogens; other AFs are less toxic (B1 > G1 > B2 > G2). Aflatoxin B1 is synthesized by the fungus via the polyketide pathway. Starting with polyketide precursors such as acetate, at least 20 steps involving 17 postulated enzymes, have been identified. The genes for almost all the key

Table 19.1 Natural occurrence of selected common mycotoxins

Mycotoxins	Major producing fungi	Typical substrate in nature	Biological effect
Alternaria[a] mycotoxins	*Alternaria alternata*	Cereal grains, tomato, animal feeds	A, Hr, M
Aflatoxin B_1 and other aflatoxins	*Aspergillus flavus* *A. parasiticus*	Peanuts, corn, cottonseed, cereals, figs, most tree nuts, milk, sorghum, walnuts	A, C, H, I, M, T
Citrinin	*Penicillium citrinum*	Barley, corn, rice, walnuts	Nh, C(?), M
Cyclochlorotine	*P. islandicum*	Rice	A, H, C
Cyclopiazonic acid	*A. flavus* *P. cyclopium*	Peanuts, corn, cheese	Cv, I, M, Nr
Deoxynivalenol	*Fusarium graminearum*	Wheat, corn	I, Nr
Fumonisins	*F. moniliforme* *F. verticillioides*	Corn, sorghum, rice	A, C, Cv, H, Nr, R
Luteoskyrin	*P. islandicum* *P. rugulosum*	Rice, sorghum	A, C, H, M
Moniliformin	*F. moniliforme*	Corn	Cv, Nr
Ochratoxin A	*A. ochraceus* *P. viridicatum* *P. verrucosum*, etc.	Barley, beans, cereals, coffee, feeds, maize, oats, rice, rye, wheat	A, I, Nh, T
Patulin	*P. patulum* *P. urticae* *A. clavatus*, etc.	Apples, apple juice, beans, wheat	C(?), D, Nr, T
Penicillic acid	*P. puberulum* *A. ochraceus*, etc.	Barley, corn	M, Nr
Penitrem A	*P. palitans*	Feedstuffs, corn	Nr
Roquefortine	*P. roqueforti*	Cheese	Nr
Rubratoxin B	*P. rubrum* *P. purpurogenum*	Corn, soybeans	A, H, T
Sterigmatocystin	*A. versicolor* *A. nidulans*, etc.	Corn, grains, cheese	C, H, M
T-2 toxin	*F. sporotrichioides*	Corn, feeds, hay	A, ATA, Cv, D, I, T
12–13-Epoxytrichothecenes other than T-2 and deoxynivanenol	*F. nivale*, etc.	Corn, feeds, hay, peanuts, rice	A, D, I, Nr
Zearalenone	*F. graminearum*, etc.	Cereals, corn, feeds, rice	G, M

A, apoptosis; ATA, alimentary toxic leukemia; C, carcinogenic; C(?), carcinogenic effect is still questionable; Cv, cardiovascular lesion; D, dermatoxin; G, genitotoxin and estrogenic effects; H, hepatotoxic; Hr, hemorrhagic; I, immuno-modulation effects; M, mutagenic; Nh, nephrotoxin; Nr, neurotoxins; R, respiratory; T, teratogenic.
[a] The optimal temperatures for the production of mycotoxin are generally between 24° and 28°C, except for T-2 toxin, which is generally produced maximally at 15°C.

enzymes and a regulatory gene (*aflR*), which encodes a 47 kD protein, AFLR, that has been shown to be required for transcriptional expression of structural genes, have been cloned and characterized. These genes are clustered within an approximately 60-kb region of the fungal genome.

Because AFB1 frequently contaminates several major commodities, extensive studies have been done on its toxicity and biological and biochemical effects. The main target organ of AF is the liver. Typical symptoms of aflatoxicoses in animals include proliferation of the bile duct, centrilobular necrosis and fatty infiltration of the liver, generalized hepatic lesions and hepatomas. The susceptibility of animals to AFB1 varies considerably with species, with a decrease in sensitivity (LD_{50} in mg/kg) in the following order: rabbits (0.3), ducklings (0.34), mink (0.5–0.6), cats (0.55), pigs

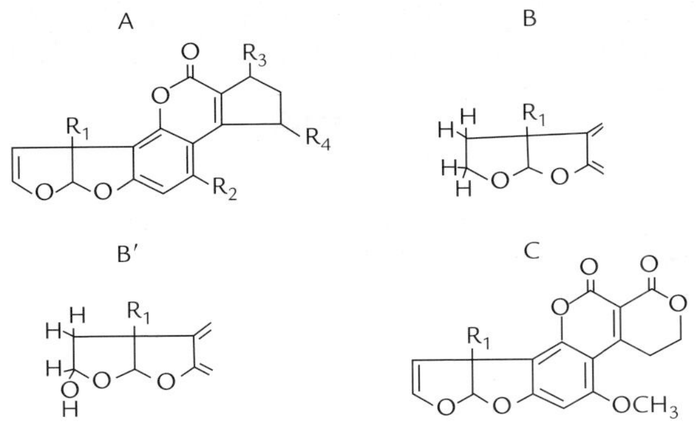

Aflatoxin	S	R_1	R_2	R_3	R_4
B_1(F)	A	H	OCH_3	=O	H
M_1(F,M)	A	OH	OCH_3	=O	H
P_1(M)	A	H	OH	=O	H
Q_1(M)	A	H	OCH_3	=O	OH
R_0(M)	A	H	OCH_3	OH	H
R_0H_1(M)	A	H	OCH_3	OH	OH
B_2(F)	AB	H	OCH_3	=O	H
B_{2a}(F,M)	AB'	H	OCH_3	=O	H
M_2(F,M)	AB	OH	OCH_3	=O	H
G_1(F)	C	H	—	—	—
G_2(F)	BC	H	—	—	—
G_{2a}(F,M)	B'C	H	—	—	—
GM(F,M)	BC	OH	—	—	—

Figure 19.1

(6–7 kg size, 0.6), trout (0.8), dogs (1.0), guinea pigs (1.4–2.0), sheep (2.0), monkeys (2.2), chickens (6.3), rats (5.5 for male, 17.9 female), mice (9.0), and hamsters (10.2). In addition to acute hepatotoxic effects, carcinogenic effects of AFB1 are also of concern. For example, a 10% incidence of hepatocarcinoma was observed in male Fisher rats fed a diet containing only 1 µg AFB1/kg over a period of 2 years. Rats, rainbow trout, monkeys, and ducks are most susceptible, and mice are relatively resistant to carcinogenic effects of AFB1. Consumption of AFB1-contaminated feed by dairy cows results in the excretion of AFM1 in milk. AFM1, a hydroxylated metabolite of AFB1, is about ten times less toxic than AFB1; but its presence in milk is of concern for human health. Because of their presence in foods and strong evidence of their association with human carcinogenesis, aflatoxins are still a serious threat to human health even after more than 40 years of research.

2. Aflatoxins in human foods

Aflatoxins have been found in corn, peanuts and peanut products, cotton seeds, rice, pistachios, tree nuts (Brazil nuts, almonds, pecans), pumpkin seeds, sunflower seeds and other oil seeds, copra, spices, and dried fruits (figs, raisins). Among these products, frequent contamination with high levels of AF in peanuts, corn, and cottonseed, mostly due to infestation with mold in the field, are of the most concern. Soybeans, beans, pulses, cassava, sorghum, millet, wheat, oats, barley, and rice are resistant or only moderately susceptible to AF contamination in the field. It should be reiterated that resistance to AF contamination in the field does not guarantee that the commodities are free of AF contamination during storage. Inadequate storage conditions, such as high moisture and warm temperatures (25–30°C), can create conditions favorable for the growth of fungus and production of AF. High levels of AFB1 have been reported in some lots of rice, cassava, figs, spices, pecans, and other nuts.

The potential hazard of AFs to human health has led to worldwide monitoring programs for the toxin in various commodities as well as regulatory actions by nearly all countries. Levels varying from zero tolerance to 50 ppb have been set for total AFs. Most countries, including US, have a regulatory level around 20 ppb in foods. However, considerably lower limits for AF in foods (2.0 ppb for AFB and 4.0 ppb for total AFs) have been established by the European Economic Community in 1999. For AFM1 in dairy products, a level between zero tolerance to 0.5 ppb has been used. To avoid contamination of milk and other dairy products with AFM, rigorous programs regulating AFB1 in feed have also been established. Most governments set a lower tolerance level for AFs in the feed for dairy cows. In the US, the level of AFs in feed for dairy cows is 20 ppb; but for other animals, it is 100 ppb.

3. Aflatoxin and human carcinogenesis

Whereas AFB1 has been found to be a potent carcinogen in many animal species, the role of AF in carcinogenesis in humans is complicated by hepatitis B virus (HBV) infections in humans. Epidemiological studies have shown a strong positive correlation between AF levels in the diet and primary hepatocellular carcinoma (PHC) incidence in some parts of the world, including certain regions of the People's Republic of China, Kenya, Mozambique, Philippines, Swaziland, Thailand, and Transkei of South Africa. Adducts of AF (i.e. AFB1-DNA and AF-albumin) as well as several AF metabolites, mainly AFM1, have been detected in serum, milk, and urine of humans in these regions. However, the prevalence of HBV infection is also correlated to liver cancer incidence in these regions. Recent studies indicated that AF exposure is the more important factor. Since multiple factors are important in carcinogenesis, environmental contaminants such as AFs and other mycotoxins may, either in combination with HBV or independently, be important etiological factors. Recent data on the enhancement of mutation of *p53* gene suggest the synergistic effect of these two risk factors for PHC in humans.

B. Ochratoxins

1. General considerations

Ochratoxins, a group of dihydroisocoumarin-containing mycotoxins (Fig. 19.2), are produced by a number of fungi in the genera *Aspergillus* and *Penicillium*, including *A. sulphureus*, *A. sclerotiorum*, *A. melleus*, *A. ochraceus*, *A. awamori*, *P. viridicatum*,

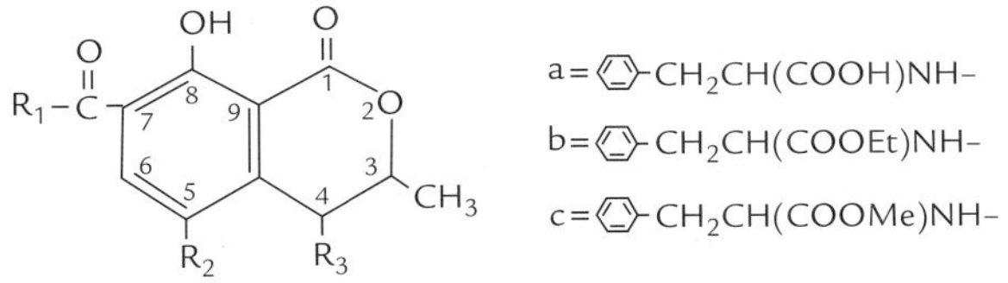

Ochratoxin	R_1	R_2	R_3
A	a	Cl	H
B	a	H	H
C	b	Cl	H
A Methyl ester	c	Cl	H
B Methyl ester	c	H	H
B Ethyl ester	b	H	H
4-OH-OA	a	Cl	OH
α	OH	Cl	H
β	OH	H	H

Figure 19.2

P. palitans, *P. commune*, *P. variabile*, *P. purpurescens*, *P. cyclopium*, and *P. chrysogenum*. *Aspergillus ochraceus* and *P. viridicatum*, two species that were first reported as OA producers, occur frequently in nature. Because of its distinct chemotype, *P. viridicatum* has been reclassified as *P. verrucosum*. Other fungi, such as *Petromyces alliceus* (isolated from onion), *A. citricus*, and *A. fonsecaeus* (both in *A. niger* group), have also been found to produce OA. Because most OA producers are storage fungi, preharvest fungal infection and OA production is not a serious problem; the toxins are generally produced in grains during storage in temperate climate regions. Although most OA producers can grow in a range from 4°C to 37°C and at a_w as low as 0.78, optimal conditions for toxin production are narrower with temperature at 24–25°C and a_w values >0.97 (minimum a_w for OA production is about 0.85). Worldwide, OA occurs primarily in cereal grains (barley, oats, rye, corn, wheat) and mixed feeds, and levels higher than 1 ppm have been reported. OA has been found in other commodities, including beans, coffee, nuts, olives, cheese, fish, pork, milk powder, wine, beer, and bread. The presence of OA residues in animal products is of concern because it binds tightly to serum albumin and has a long half-life in animal tissues and body fluids. Thus, OA can be carried through the food chain; and the natural occurrence of OA in kidneys, blood serum, blood sausage and other sausage made from pork has been reported.

2. Toxic effects

Ochratoxin A, the most toxic member (LD_{50} about 20–25 mg/kg in rats) and also the most commonly found toxin in this group, has been found to be a potent nephrotoxin, causing kidney damage, including degeneration of the proximal tubule, in many animal species. Liver necrosis and enteritis were also observed in these animals. Other than acute toxic effects, OA also acts as an immunosuppressor and a teratogen in test animals. Although OA has never been shown to be mutagenic, a weak genotoxic effect has been demonstrated in several systems. OA is considered to be a weak nephrocarcinogen because a high level of toxin and an extended period of exposure are necessary to induce the tumors. A dose-related induction of renal tubular cell tumors was found in Fisher rats (F344/N) with a significant increase in renal tubular cell tumors at levels of 70 and 210 µg OA/kg body weight/day; but cancer incidence was not significantly different from the control at lower levels (21 µg OA/kg body weight/day). Recent studies showed that OA also causes liver cancer in rats.

3. Ochratoxin A and human health

The role of OA in human pathogenesis at present is still speculative. OA has long been considered to be associated with the nephropathy of people residing in certain Scandinavian and Balkan regions, and more recently in Tunisia, where exposure may be 'endemic'. The pathological lesions in humans are similar to those observed in pigs with endemic porcine nephropathy, which is due to the consumption of feeds contaminated with OA. Since a high proportion of the patients suffering from endemic nephropathy in the Balkan region develop tumors of the renal pelvis and ureter, the involvement of nephropathy in tumor development was suggested. OA has been found in human serum in Tunisia and a number of European countries including Bulgaria, Poland, Yugoslavia, and Germany. In certain areas of the Balkans and Tunisia, OA levels in the food and in human serum in endemic regions are higher than in the nonendemic areas. OA also has been found in human milk and kidneys in some endemic regions.

Human exposure to OA could occur through consumption of OA-containing cereals or animal foods. In a recent workshop on the impact of OA on human health [see a series of papers published in *Food Additives and Contaminants* Vol. 14 (suppl. 1–3), 1996], the mean daily intake of OA by humans in European Union member countries was 1.8 ng of OA/kg body weight (range 0.7–4.7), as calculated from data on OA levels in foods and 0.9 ng/kg (range 0.2–2.4) when calculated from the levels of OA in human blood. The main dietary sources (55%) were cereals and cereal products (0.2–1.6 µg OA/kg). Coffee (mean level 0.8 µg/kg), beer, pork, blood products, and pulses also contribute to the OA intake in humans. These findings re-emphasize the possible involvement of OA in human carcinogenesis and the health hazard of exposure to OA in humans. Among 77 countries, which have regulations for different mycotoxins, eight have specific regulations for OA, with limits ranging from 1 to 20 µg/kg in different foods.

C. Fumonisins

1. General considerations

Fumonisins (Fms) are a group of toxic metabolites produced primarily by *Fusarium verticillioides* (previously *F. moniliforme*), one of the most common fungi colonizing corn throughout the world. More than 11 structurally related Fms (B1, B2, B3, B4, C1, C4, A1, A2, etc.), have been found since the discovery of FmB1 (diester of propane-1,2,3-tricarboxylic acid of 2 amino-12,16-dimethyl-3,5,10,14,15-pentahydroxyicosane) in 1987 (Fig. 19.3). Several hydrolyzed derivatives of Fms, resulting from removal of the tricarballylic acid and other ester groups, have also been found in nature. *F. proliferatum*, another common naturally occurring species, also produces Fms. Although

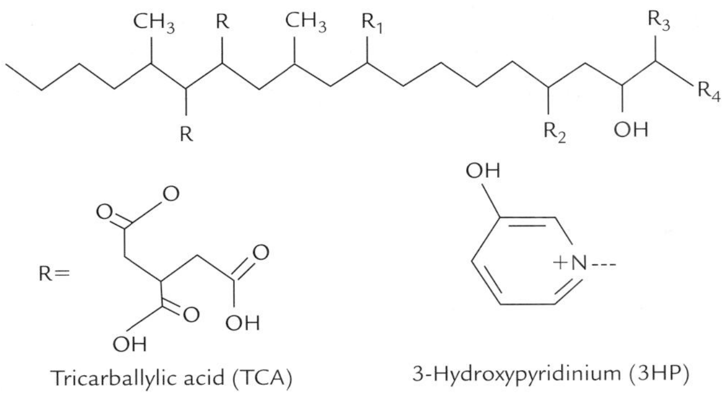

Figure 19.3

Tricarballylic acid (TCA)   3-Hydroxypyridinium (3HP)

Fumonisin	R_1	R_2	R_3	R_4
FmB1	OH	OH	NH_2	CH_3
FmB2	H	OH	NH_2	CH_3
FmB3	OH	H	NH_2	CH_3
FmB4	H	H	NH_2	CH_3
FmA1	OH	OH	$NHCOCH_3$	CH_3
FmA2	H	OH	$NHCOCH_3$	CH_3
FmC1	OH	OH	H	H
FmC4	H	H	H	H
FmP1	OH	OH	3HP	CH_3
FmP2	H	OH	3HP	CH_3
FmP3	OH	H	3HP	CH_3

F. anthophilum, *F. napiforme*, and *F. nygamai* are capable of producing Fms, they are not commonly isolated from food and feed. In addition to fusaria, *Alternaria alternata* f. sp. *lycopersici* produces a group of host-specific toxins, named AAL toxins, which have a chemical structure similar to the Fms and induce similar toxic effects. Production of FmB has been found in some *A. alternata* f. sp. *lycopersici* cultures. Likewise, some Fm-producing isolates also produce small amounts of AAL toxins.

Although the biosynthetic pathway for the fumonisins has not been elucidated, labeling studies indicate that the polyketide synthase (PKS), alanine or other amino acid, and methionine are involved. Detailed genetic analyses of the fungi have uncovered 15 genes that are associated with fumonisin production, and these genes are also clustered together. A PKS gene (*Fum*5) was shown to be involved in the formation of polyketide. Genes *Fum*2 and *Fum*3 are associated with the interconversion of different fumonisins.

Fumonisins are most frequently found in corn, corn-based foods, and other grains (such as sorghum and rice). The level of contamination varies considerably with different regions and year (range from 0 to >100 ppm, generally below 1 ppm). FmB1 is the most common fumonisin in naturally contaminated samples; FmB2 generally accounts for one-third or less of the total. Although production of the toxin generally occurs in the field, continued production of toxin during postharvest storage also contributes to the overall levels. In the laboratory, the highest yield was achieved in corn culture and less in rice culture; peanuts and soybeans are poor substrates for toxin production. The toxin is very stable to heat as well as resistant to other treatments.

Acid hydrolysis causes the loss of tricarballylic acid, but the hydrolyzed products may still have toxic effects. Transmission of FmB1 to the edible portion of meat or egg is unlikely to occur because of the rapid excretion of the toxin in animals. No significant amount of FmB1 is transmitted to milk.

2. Toxicologic effects of FmB1

Fumonisin B1 is primarily a hepatotoxin and carcinogen in rats (Class 2B carcinogen). Earlier studies showed that feeding culture material from *F. moniliforme* to rats resulted in cirrhosis and hepatic nodules, adenofibrosis, hepatocellular carcinoma–ductular carcinoma, and cholangiocarcinoma. Later studies indicated that the kidney is also a target organ, and tubular nephrosis was found both in rats and in horses in a field case associated with equine leukoencephalomalacia (ELEM). Results from these earlier observations were confirmed in studies using purified Fms. Although FmB1 was originally found to be a potent cancer promoter, subsequent studies showed that all of the three major Fms, i.e. FmB1, FmB2 and FmB3, are carcinogens and all have cancer initiation and promoting activities in rats. The effective dose of FmB1 for cancer initiation in rat liver depends both on the levels and the duration of exposure. In cell culture systems, FmB1 acts as a mitogen, cytotoxic to the cells and able to alter the expression of genes associated with the cell cycle. It displays *in vivo* genotoxicity, but has no mutagenic effect in the *Salmonella* system.

Fumonisin B1 was identified as an etiological agent responsible for ELEM in horses and other Equidae (donkeys and ponies) and for porcine pulmonary edema (PPE). ELEM was originally found to be a seasonal disease occurring in late fall and early spring. The disease is characterized primarily by neurotoxic effects, including uncoordinated movements and apparent blindness showing as violent blundering into stalls and wall. Liver damage has also been reported. Death sometimes occurs with no nervous symptoms. Clinically, the disease is characterized by edema in the cerebrum of the brain. The levels of FmB1 and FmB2 in feeds associated with confirmed cases of ELEM ranged from 1.3 to 27 ppm. In pigs, PPE occurs only at high FmB levels (175 ppm), while liver damage occurs at much lower concentrations, with a NOAEL of <12 ppm. Cardiovascular function is altered by FmB1. The FmB1-induced PPE is caused by left-side heart failure and not by altered endothelial permeability. In cattle, renal injury, hepatic lesions and alteration of sphingolipid in various organs is observed. High levels of Fm are also needed to cause disease in poultry. Similar to AAL toxin, Fms are also toxic to some plants: jimsonweed, black nightshade, duckweed, and tomatoes with the *asc/asc* genotype are most susceptible.

3. Impact on human and animal health

While Fms are commonly detected in corn-based foods and feeds, the impact of low levels of Fms in human foods is not clear. Current data suggest that they may have more impact on the health of farm animals than on humans. To protect animal and human health, guidelines for Fm levels in feeds to be used for various animals have been established by the US Food and Drug Administration (FDA) in 2000. The FmB1 levels be limited to 5, 20, 60 100, 30, and 10 ppm, in corn and corn byproducts to be used for horses and rabbits, catfish and swine, ruminants and mink, poultry, breeding stock (ruminants, poultry and mink) and others (dogs and cats), respectively. The FmB1 levels are to be limited to 1, 10, 30, 50, 15, and 5 ppm in the total ration for the above animal species, respectively (e.g. 1 ppm for horse and rabbit).

Several reports indicated that significantly higher levels of Fms, sometimes together with AFs and TCTC mycotoxins, were present in corn samples collected from areas with high rates of human esophageal cancer. The ability of *F. moniliforme* to produce nitrosamines (carcinogens) suggests that a combination of several etiological agents in the moldy corn, including FmB1 and possibly other mycotoxins such as AFB1, may play an important role in carcinogenesis in humans. Currently, only Switzerland regulates the Fm level (1.0 ppm) in foods.

D. Selected Important Trichothecene Mycotoxins

1. General considerations

Trichothecenes are a group of naturally occurring toxic tetracyclic sesquiterpenoids produced by many species of fungi in the genera *Fusarium* (most frequently), *Myrothecium*, *Trichoderma*, *Trichothecium*, *Cephalosporium*, *Verticimonosporium*, and *Stachybotrys*. The term TCTC is derived from trichothecin, the first compound to be isolated in this group. All the mycotoxins in this group contain a common 12–13-epoxytrichothecene (six-membered oxygen-containing ring) skeleton and an olefinic bond with different side-chain substitutions (Fig. 19.4). Based on the presence of a macrocyclic ester or ester–ether bridge between C-4 and C-15, TCTCs are generally classified as macrocyclic (type C) or nonmacrocyclic. The nonmacrocyclic TCTCs are further divided into two types: type A TCTCs, including T-2 toxin, HT-2 toxin, neosolaniol (NESO), diacetoxyscirpenol (DAS), and T-2 tetraol (T-4ol), which contain a hydrogen- or ester-type side chain at the C-8 position, and type B TCTCs, including DON, nivalenol (NIV), and fusarenon-x (FS-x, or 4-acetyl-niv), which contain a ketone. Type C group TCTCs, including roridins and verrucarins, contain a macrocyclic ring. In addition to fungi, extracts from a Brazilian shrub, *Baccharis megapotamica*, also contain macrocyclic TCTCs.

Similar to other sesquiterpenes, TCTCs are biosynthesized via the mevalonate pathway: the TCTC skeleton is formed by cyclization of farnesyl pyrophosphate via the intermediate trichodiene by an enzyme trichodiene synthase. The isovaleroxy side chain in T-2 toxin is derived from leucine. Several key enzymes in the biosynthetic pathway

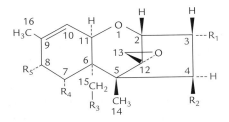

Figure 19.4

have been identified. At least six genes (*Tri1* to *Tri6*) involved in the biosynthesis of TCTC have been cloned, and these genes are clustered together on a chromosome. Genes *Tri3*, *Tri4*, and *Tri5*, which encode a transacetylase (15-*O*-acetyltransferase), a cytochrome P-450 monooxygenease, and trichodiene synthase, respectively, are contained within a 9-kb region, while *Tri5* and *Tri6* (a regulatory gene) are in a 5.7 kb region.

The structural diversity of TCTC mycotoxins results in a variety of toxic effects in animals and humans. Unlike AFB1 and OA, where primary effects and clinical manifestations are well defined in the liver and kidney, TCTC mycotoxicoses affect many organs, including the gastrointestinal tract, hematopoietic, nervous, immune, hepatobiliary, and cardiovascular systems. Ingestion of foods and feeds that contain TCTC mycotoxins causes many types of mycotoxicoses in humans and animals, including moldy corn toxicosis, scabby wheat toxicosis (or red-mold, or akakbi-byo disease, or scabby barley poisoning), feed refusal and emetic syndrome (swine), fusaritoxicoses, hemorrhagic syndrome, and alimentary toxic aleukia (ATA). More than 100 TCTC have been identified in the laboratory; only a few selected toxins that have been found in foods are described herein.

2. T-2 toxin and other type A TCTCs

T-2 toxin, a highly toxic type A TCTC, was isolated in the middle 1960s. It was originally found to be produced by a strain of *F. tricinctum* isolated from moldy corn and later found to be produced by *F. sporotrichioides* (major), *F. poae*, *F. sulphureum*, *F. acuminatum*, and *F. sambucinum*. Unlike most mycotoxins, which are usually synthesized near 25°C, the optimal temperature for T-2 toxin production is around 15°C. Higher temperatures (20–25°C) are needed for the production of related metabolites, such as H-T2 toxin. Although T-2 toxin occurs naturally in cereal grains, including barley, corn, corn stalk, oats, wheat, and mixed feeds, contamination with T-2 toxin is less frequent than with deoxynivalenol. However, T-2 toxin (LD_{50} in mice: 2–4 mg/kg mice) is much more toxic to animals, perhaps also to humans, than DON (LD_{50} in mice: 50–70 mg/kg).

Almost all the major TCTCs, including T-2 toxin, are cytotoxic and cause hemorrhage, edema, and necrosis of skin tissues. Inflammatory reactions near the nose and mouth of animals are similar to some lesions found in humans suffering from ATA disease. The severity of lesions is also related to chemical structure. Compared with other types of TCTCs, group A toxins (T-2 toxin) are less toxic than macrocyclic toxins such as verrucarin A but more toxic than type B toxins, such as NIV and DON. Neurologic dysfunctions, including emesis, tachycardia, diarrhea, refusal of feed/anorexia, and depression have also been observed. T-2 toxin and some TCTCs also induce pathological lesions in the gastrointestinal system. However, the major lesion of T-2 toxin is its devastating effect on the hematopoietic system in many mammals, including humans. Typically, there is a marked initial increase in circulating white blood cell counts, especially lymphocytes, followed by a rapid decrease to 10–75% of normal values. Platelet counts are also reduced. There is also extensive cellular damage in the bone marrow, intestines, spleen, and lymph nodes, and in severe cases, complete atrophy of bone marrow and marked alteration of plasma coagulation factors. T-2 toxin and related TCTCs are the most potent immunosuppressants of the known mycotoxins and cause significant lesions in lymph nodes, spleens, thymus, and the bursa of Fabricius. The heart and pancreas are other target organs for T-2 toxin intoxication. Urinary and hepatobiliary lesions induced by T-2 toxin and DAS are secondary.

3. Deoxynivalenol

Deoxynivalenol is a major type B TCTC mycotoxin produced primarily by *F. graminearum* and other related fungi such as *F. culmorum* and *F. crookwellense*. Because DON causes feed refusal and emesis in swine, the name 'vomitoxin' is also used. Although DON is considerably less toxic than most other TCTC mycotoxins, the level of contamination of this toxin in corn and wheat is generally high, usually above 1 ppm, sometimes greater than 20 ppm. Contamination of DON in other commodities, including barley, oats, sorghum, rye, safflower seeds, and mixed feeds has also been reported. DON has been found in cereal grains in many countries, including Australia, Canada, China, Finland, France, Germany, Hungary, India, Italy, Japan, South Africa, the UK, the US, and many others.

Although inadequate storage may lead to the production of some TCTC mycotoxins, infestation of fusaria in wheat and corn in the field is of most concern for the DON problem. With wet and cold weather during maturation, grains are especially susceptible to *F. graminearum* infection, which causes so-called 'scabby wheat' and simultaneously produces the toxin. The optimal temperature for DON production is about 24°C. Outbreaks of DON in winter wheat in the US, Finland, and Canada usually occur when continental chilly and humid weather favoring the fungal infection is followed by a humid summer favorable for toxin production. Depending on the weather conditions, the infestation of *F. graminearum* in wheat and corn and subsequent production of toxins in the field varies considerably regionally and from year to year. Thus, the levels of DON in these commodities are difficult to predict.

Toxicologically, DON induces anorexia and emesis both in humans and animals. Swine are most sensitive to feed contaminated with DON. Because of the frequent occurrence of high levels of DON in wheat and corn, its stability, and reported food poisoning outbreaks in humans, contamination of cereals with DON is a major concern of both the government, and food and feed industries. Contamination of DON in wheat and corn may be associated with other toxic effects because other fusarium toxins, including zearalenone and other TCTCs, may also be present. Other type B TCTCs, such as nivalenol and acetylated DON, which are more toxic than DON to test animals, occur naturally in some parts of the world.

4. The impact of TCTC on human and animal health

Because of their toxicity and their frequent presence in foods and feeds, TCTCs are potentially hazardous to human and animal health. However, among the many types of TCTC mycotoxicoses, only ATA and scabby wheat toxicosis have been demonstrated in human populations. The former (ATA) was attributed to the consumption of overwintered cereal grains colonized by *F. sporotrichioides* and *F. poae*; it caused the deaths of hundreds of people in the USSR between 1942 and 1947. Later studies indicated that T-2 toxin and related TCTCs were the primary cause. The signs and symptoms of ATA disease, which include skin inflammation, vomiting, damage to hematopoietic tissues, leukocytosis, and leukopenia, are similar in intoxicated humans and animals.

DON has been found to be primarily responsible for outbreaks of scabby wheat toxicosis, which are common in several countries, but these toxicoses rarely cause death. Between 1961 and 1985, 35 outbreaks involving 7818 cases were caused by consumption of foods made from either scabby wheat or moldy corn in China. The symptoms, which occurred between 15 minutes and 1 hour after eating foods made from moldy corn meal, included nausea (90%), emesis (61%), headache and drowsiness (78%), and 5–6% had abdominal pain, diarrhea, and a low fever. People generally recovered within 2–4 days. Analysis of the leftover moldy corn revealed that the samples had 0.34–93.8 ppm of DON; no T-2 toxin or NIV was found. Similar cases have been reported for people consuming scabby wheat flour.

Although DON is not as toxic as other TCTCs, the level of contamination in wheat and corn is high, and intoxication of humans by DON occurs more often. The tolerable daily intakes of DON for adults and infants were estimated to be 3 μg/kg and 1.5 μg/kg body weight, respectively. Consequently, a tolerance level of 1 ppm for DON in grains for human consumption has been set by a number of countries, including the US. In Canada, the guideline for DON in the uncleaned soft wheat used for nonstaple foods is 2 ppm, but 1 ppm for the infant foods.

TCTCs may also be involved in the so-called 'sick building syndrome' (SBS) in humans. *Stachybotrys atra* was isolated from a badly water-damaged Chicago suburban home where the occupants complained about headaches, sore throats, hair loss, flu symptoms, diarrhea, fatigue, dermatitis, and general malaise. Several TCTCs (verrucarins B and J, satratoxin H, trichoverrins A and B) were found in the *S. atra*-contaminated materials of this home. T-2 toxin, DAS, roridin A, and T-2 tetraol were isolated from the dust samples from the air ventilation system of other suspected cases of SBS in three urban Montreal office buildings. *S. chartarum,* an indoor mold, has been associated with pulmonary hemorrhage cases in the Cleveland, Ohio, area. A remediation program involving removal of all contaminated wallboard, paneling, and carpeting in the water-damaged areas of the home, as well as spraying sodium hypochlorite on all surfaces during remediation, appeared to be effective. Air samples taken from postremediation buildings showed no detectable levels of *S. chartarum* or related toxicity. Several studies have shown that other molds and mycotoxins are present in mold-damaged buildings. Thus, it is important to identify both the mold and mycotoxins to establish the cause of SBS.

E. Other Selected Mycotoxins

In addition to the mycotoxins discussed above, a number of other mycotoxins occur naturally. The impacts of some of these mycotoxins on human and animal health are discussed in the following sections.

1. Other mycotoxins produced by *Aspergillus*

Sterigmatocystin (ST) is a naturally occurring hepatotoxic and carcinogenic mycotoxin produced by fungi in the *Aspergillus*, *Bipolaris*, and *Chaetomium* genera and *Penicillium luteum*. Structurally related to AFB1, ST is known to be a precursor of AFB1. Rapid progress in understanding the biosynthesis of AFB1 has been made through the studies on the biochemical and genetics of biosynthesis of ST in recent years. Although the carcinogenicity of ST is less (10–100 times) than that of AFB1 in test animals, this mycotoxin has been found to be mutagenic and genotoxic. ST occurs naturally in cereal grains such as barley, rice, and corn; coffee beans; and foods such as cheese. It has also been found in pickled foodstuffs in Linxian, a high esophageal cancer-incidence region of the People's Republic of China and in foods from Mozambique, where high PHC incidence was reported. Toxigenic fungi have been isolated from patients with esophageal cancer, and these strains are capable of producing ST in many commodities. Evidence for the role of ST in human carcinogenesis is indirect and inconclusive.

Aspergillus terreus and several other fungi (e.g. *A. flavus* and *A. fumigatus*) have been found to produce tremorgenic toxins, territrems, aflatrem, and fumitremorgin. *Aspergillus terreus*, *A. fumigatus*, and *Trichoderma viride* also produce gliotoxin, an epipolythiopiperazines-3,6-diones-sulfur containing piperazine antibiotic, which may have immunosuppressive effects in animals. In addition, *A. flavus*, *A. wentii*, *A. oryzae*, and *P. atraovenetum* are capable of producing nitropropionic acid (NPA), a mycotoxin causing apnea, convulsions, congestion in the lungs and subcutaneous vessels, and liver damage in test animals. This toxin was also identified as an etiological agent for 'deteriorated sugarcane poisoning (SP)', a fatal food poisoning that occurred in China. However, the fungi contaminating the sugar cane and producing NPA were *Arthrinium sacchari*, *Arth. saccharicola*, and *Arth. phaeospermum*.

2. Other mycotoxins produced by *Penicillium*

Other than OA, penicillia produce many mycotoxins with diverse toxic effects. Cyclochlorotine, luteoskyrin (LS), and rugulosin (RS) have long been considered to be possibly involved in the yellow rice disease during World War II. They are hepatotoxins and also produce hepatomas in test animals. However, incidents of food contamination with these toxins have not been well documented. Several other mycotoxins, including patulin (PT), penicillic acid (PA), citrinin (CT), cyclopiazonic acid (CPA), citreoviridin, and xanthomegnin, which are produced primarily by several species of penicillia, have attracted some attention because of their frequent occurrence in foods.

PT and PA are produced by many species in the genera *Aspergillus* and *Penicillium*. A heat-resistant fungus, *Byssochlamys nivea*, frequently found in foods also produces PT. Both toxins are hepatotoxic and teratogenic. Patulin is frequently found in damaged apples, apple juice, apple cider, and sometimes in other fruit juices and feed. PA has been detected in 'blue eye corn' and meat. Owing to its highly reactive double bonds, which readily react with sulfhydryl groups in foods, patulin is not very stable in foods containing these groups. Nevertheless, PT is considered a health hazard to humans. At least ten countries have regulatory limits, most commonly at a level of 50 μg/kg, for PT in various foods and juices. Frequently associated with the natural occurrence of OA is citrinin, also a nephrotoxin, produced by *P. citrinum* and several other penicillia and aspergilli. Recent discovery of the capability of *Monocuus ruber* and *M. purpureus*, two molds frequently used in the preparation of certain Asian foods, to produce CT re-emphasized the potential hazard of CT to human health. One of the mycotoxins closely associated with the natural occurrence of AF in peanuts is CPA, which causes hyperesthesia and convulsions as well as liver, spleen, pancreas, kidney, salivary gland, and myocardial damage. The toxin was originally found to be produced by *P. cyclopium*; but a number of other penicillia (*P. crustosum*, *P. griseo-fulvin*, *P. puberulum*, *P. camemberti*) and aspergilli (*A. versicolor*, *A. flavus*, but not *A. parasiticus* and *A. tamarii*) also produce CPA. Natural occurrence of CPA in corn, peanuts, and cheese has been reported. *Penicillium rubrum* and *P. purpurogenum* produce two highly toxic hepatotoxins (LD_{50} 3.0 mg/kg mice, IP) called rubratoxins A (RA, minor) and B (RB, major), which are complex nonadrides fused with anhydrides and lactone rings. Rubratoxin B has synergistic effects with AFB1.

In addition, penicillia produce many mycotoxins with strong pharmacological effects on neurosystems. For example, *P. crustosum* and *P. cyclopium* produce tremorgenic indoloditerpenes called penitrems A–F. Penitrem A, the major toxin in this group, causes tremorgenic effects in mice. Roquefortines A–C (C is most toxic), which are produced by *P. roqueforti* and several other penicillia, have neurotoxic effects in animals and have been found in cheese. Tremorgens in the paspalitrem group (paspalicine, paspalinine, paspalitrem A and B, paspaline and paxilline) are produced by *Claviceps paspali* and some penicillia.

3. Other mycotoxins produced by *Fusarium*

Other than TCTCs and Fm, some fusaria produce other mycotoxins. Zearalenone [6-(10-hydroxy-6-oxo-*trans*-1-undecenyl)-β-resorcyclic acid lactone], is a mycotoxin produced by the scabby wheat fungus, *F. graminearum* (roseum), which also produces DON. Zearalenone (ZE), also called F-2, is a phytoestrogen causing hyperestrogenic effects and reproductive problems in animals, especially swine. Natural contamination with ZE primarily occurs in cereal grains such as corn and wheat. Contamination with this mycotoxin, sometimes together with DON, in feed may result a large economic loss in the swine industry. *Fusarium moniliforme* also produces several other mycotoxins, including fusarins A–F, moniliformin, fusarioic, fusaric acid, fusaproliferin, and beauvericin in addition to Fms. Although the impact of these mycotoxins on human health is still not known, fusarin C (FC) has been identified as a potent mutagen and is also produced by *F. subglutinans*, *F. graminearium*, and several other fusaria. Among many fungi, *F. moniliforme* is also most capable of reducing nitrates to form potent carcinogenic nitrosamines. These observations further suggest that the contamination of foods with this fungus could be one of the etiological factors involved in human carcinogenesis in certain regions of the world.

4. Mycotoxins produced by other species

Other mycotoxins are produced by *Alternaria*, a plant pathogen and another genus of fungi common in our environment. These toxins can be produced in both preharvest and postharvest commodities. *Alternaria alternata* and other *Alternaria* species produce many types of mycotoxins (AM), including dibenzo-[*a*]-pyrone types of mycotoxins: alternariol, alternariol monomethyl ether (AME), altenuene, isoaltenuene and altenuisol, tetramic acid metabolites tenuazonic acid (TzA) and related compounds, and perylene derivatives altertoxins (ATX) I, II, (also called stemphyltoxin II), III, and stemphyltoxin. Although most of those compounds are relatively nontoxic, AME has been shown to be positive in the Ames test at relatively high concentrations. TzA is a protein synthesis inhibitor, and is capable of chelating metal ions and forming nitrosamines. This mycotoxin is also produced by *Phoma sorghina* and *Pyricularia oryzae* and may be related to 'onyalai', a hematological disorder in humans living south of the Sahara in Africa. Although no extensive survey has been conducted to determine the occurrence of these mycotoxins in human foods, limited studies indicate that AME and TzA could be high in apple and tomato products. As mentioned earlier, the recent discovery of the structural and functional similarity between fumonisins and AAL toxin further shows the importance of mycotoxins produced by fungi in the *Alternaria* family.

Sporidesmines, a group of hepatotoxins discovered in the 1960s, are also worthy of mention. These mycotoxins, causing facial eczema in animals, are produced by *Pithomyces chartarum* and *Sporidesmium chartarum*, and are very important economically to the sheep industry. Slaframine, a significant mycotoxin produced by *Rhizoctonia leguminicola* (in infested legume forage crops), causes excessive salivation or slobbering in ruminants as a result of blocking acetylcholine receptor sites.

F. Classic Examples of Intoxication by Fungal Metabolites

Two classic examples related to mycotoxins are ergotism and mushroom poisons. Because these intoxications are a result of ingestion of a fungal body containing toxic metabolites, these poisons sometimes are excluded from the modern discussion of mycotoxins. Ergots and poisonous mushrooms still can be unintentionally introduced into the food chain in modern days. Some phytoalexins elaborated by plants in response to either fungal infection or other damage have also been found toxic to humans and animals. A brief discussion of ergotism and some of these phytoalexins is included in this chapter; for mushroom poisoning, see Chapter 14.

1. Ergotism

Ergotism is a human disease that results from consumption of the ergot body, the sclerotium of the fungus, in rye or other grains infected by members of the genus *Claviceps*, e.g. *C. purpurea* and *C. paspali*. Two types of ergotism have been documented. In the convulsive type, the affected persons have general convulsions, tingling sensation of muscles, and sometimes the entire body is racked by spasms. Epidemics occurred between 1581 and 1928 in European and other countries. Although it has been suggested that the consumption of ergots that cause convulsive ergotism played a role in the 'Salem witchcraft' incidents, this is still a controversial issue. In the gangrenous type of ergotism, the affected parts became swollen and inflamed with violent, burning pains, hence, the 'fire of St Anthony'. In general, the affected area became numb first, turned black, then shrank, and finally became mummified and dry. Outbreaks occurred in the Middle Ages and into the 19th century. In some areas of France, grain contained as much as 25% ergots, and patients died as a result of ingestion of ~100 g of ergots over a few days. Between 1770 and 1771, about 8000 people died in one district alone in France. Usually, 2% ergots in grain is sufficient to cause an epidemic. European and most other countries have a regulatory limit of 0.1–0.2% ergots in flour. Biochemically, ergotisms are due to intoxication by ergoline alkaloids present in the sclerotia of *C. purpurea*. The most active components are amides of D-lysergic acid, including both cyclic-type peptides and nonpeptide amides of ergot alkaloids. These alkaloids, causing smooth muscle contraction and blocking neurohormones, have both vasoconstriction and vasodilation effects, and also affect the central nervous system. Thus, they have some therapeutic uses.

2. Toxic metabolites in fungal-damaged foods of plant origin

Phytoalexins are a group of compounds produced in plants as a result of physical or biological damage, which alters the metabolism of the host plant. Some of these compounds are toxic to humans and animals. One example is the fungal-infected sweet potato. Before World War II, Japanese investigators found that sweet potatoes infected by black rot mold, *Ceratocystis fimbrita* (*Ceratostomella fimbriata*), were toxic to animals. A number of compounds, including ipomeamarone, ipomeanine, β-furoic acid and batatic acid, which contained a furan moiety with a side chain attached at the 3- or β-position, were isolated. Ipomeamarone (IPM-one) gives a bitter taste to the sweet potato and is most abundant. It is considered to be a hepatotoxin but may also be toxic to the lungs. Ipomeanol and IPM-one are produced by *F. solani javanicum* (in the US). Although these metabolites have diverse toxic effects in farm animals, their impact on humans is not known. Another group of compounds of possible important are some nitroso compounds produced by some commonly occurring fungi. A number of nitrosamines have been identified in foods, and it was postulated that the presence of these nitrosamines in foods in high esophageal cancer regions in China may play a role in human carcinogenesis.

IV. Modes of Action of Mycotoxins

Owing to the diversity of their chemical structures, mycotoxins exhibit a variety of biological effects, including both acute and chronic toxic effects, as well as carcinogenic, mutagenic/genotoxic, and teratogenic effects, in both prokaryotic and eukaryotic systems. As shown in Table 19.1, the toxic effects of most mycotoxins are very organ-specific. Once mycotoxins are ingested, absorbed, and transported to the target organs, the toxins may either react directly with the target organs to exert toxic effects or be metabolized to an active intermediate (activated) and then exert their toxic effects. The toxins or their metabolites may also be detoxified and then excreted from the body. The biological effects of mycotoxins are greatly affected by metabolic activities in different animals under various conditions. For example, AFB1 is metabolically activated before the formation of AFB–DNA adducts, which exert carcinogenic effects. However, the activated AFB1 can also conjugate with proteins and interact with glutathione, and then be excreted from the body. Glutathione S-transferase serves as a key enzyme in the detoxification process for AFB1. In addition, other types of mixed function oxygenases (cytochrome P-450) can also metabolize AFB1 to various hydroxylated metabolites, including AFM1 and AFQ1, and demethylase converts AFB1 to AFP1. The cytosolic steroid reductase can reversibly convert AFB1 to aflatoxicol, which serves as a reservoir for AFB1.

For other mycotoxins, metabolism is usually related to detoxification. For example, metabolic deacetylation and de-epoxidation of TCTCs forms the hydroxylated derivatives and de-epoxide-TCTCs that are less toxic. Hydrolysis of OA by proteolytic enzymes leads to the nontoxic metabolites. Hydroxylation of T-2 at the C-3′ and OA at the C-4 position by the microsomal enzymes has also been demonstrated. Most hydroxylated mycotoxins can be excreted directly or form conjugates with glucuronide or sulfate, and then be excreted from the body. Thus, various factors affect the kinetics of formation of adducts and detoxification, greatly affecting the toxicity of mycotoxins. Other factors, including sex and species of the animal (genetics), environmental factors, nutritional status, and mycotoxin synergism, etc., can directly or indirectly modulate toxic effects. Modulation of the immune system also plays an important role in the overall toxic effects; mycotoxins can be either immunosuppressive (most often) or immunostimulatory.

A. Interactions of Mycotoxins with Macromolecules

1. Noncovalent interaction of mycotoxins with macromolecules

Mycotoxins and their metabolites can either react directly with macromolecules noncovalently or form a covalent bond(s) with macromolecules to exert their toxic effects. Noncovalent interactions usually occur between mycotoxins and enzymes, hormone receptors, or other macromolecules. These interactions are reversible and generally involve competition with binding sites of bioactive macromolecules. For example, OA interacts strongly with serum albumin, which leads to a long half-life for the toxin in animal and human bodies. Both AFB1 and AFG1 also bind with serum albumin, but their affinities to albumin are not as high as OA. Zearalenone and aflatoxicol interact with estrogenic receptors and steroid receptors, respectively. Both AFB1 and AFG1 and their stable epoxides are capable of intercalating with double-stranded d(ATG-CAT)2 or d(GCATGC)2 and B-DNA. Gel-shifting analysis of the binding of both mycotoxins with plasmid pBR322 revealed that intercalation of AF between the base pairs was involved. Such intercalation plays a significant role in subsequent covalent attachment of the carcinogen at the N-7 position of guanine in DNA. Noncovalent binding of TCTC mycotoxins with ribosomes and, specifically with peptidyl transferase, is a key step in the inhibition of protein synthesis. The interaction of TCTC and OA with some immunomodulators and receptors in suppressor cells is considered to be one of the mechanisms by which these mycotoxins cause immunosuppression.

2. Formation of mycotoxin–DNA adducts

The mode of action of AFB1 is one of the best examples demonstrating the importance of activation of a mycotoxin for its interaction with macromolecules. AFB1 must first be activated by mixed-function oxidases, specifically cytochrome P-450 1A2 and 3A4, to a putative short-lived AFB-8,9-exo-epoxide, which is then bound to DNA, RNA, and protein to exert its toxic, carcinogenic, and mutagenic effects. Adduct formation of this epoxide with DNA occurs through nucleophilic attack, primarily at the N-7 guanine position of G:C-rich regions of DNA, which then produces G-C to T-A or to A-T nucleotide substitutions and causes mutations. The AFB1-N7 guanine adduct is very unstable with only the open-ring form stable enough to be considered biologically important. Analysis of this adduct in human urine permits quantitative evaluation of adduct formation after exposure to AFB1.

Aflatoxin-induced G:C mutations, both G to T and G to A, have been implicated in the inactivation of human *p53* tumor-suppressor gene, and a high frequency of mutations has been found at the third position of codon 249 of *p53*. Prevalence of codon 249 mutation, i.e. AGG to AGT (arginine to serine), was found to be around 50% in human primary hepatocellular carcinoma (PHC). In contrast, the prevalence of such mutations is low in non-PHC patients. The identification of mutations/inactivation of p53 at this site, has been used as a biomarker for AFB-induced liver cancers in humans. Sterigmatocystin acts by a mechanism similar to AFB in carcinogenesis, through the formation of DNA adducts and mutation of the P53-suppressor gene. Although OA–DNA adducts have been found in kidney, liver, and several other organs of mice and rats, neither the adduct structures nor their toxic effects have been defined.

3. Covalent interactions of mycotoxins with enzymes/proteins

Several mycotoxins have a reactive group(s) and thus readily react with proteins and enzymes, either specifically or nonspecifically. PT and PA, for example, have been shown to react covalently with $-SH$ and also possibly $-NH_2$ groups of proteins through addition or substitution reactions with their α, β-unsaturated double bonds conjugated to the lactone ring. In the model systems, PT forms the same type of adducts with glutathione and N-acetyl-L-cysteine (NAC). However, free cysteine forms markedly different adducts, including mixed thiol/amino types of adducts involving the α-NH_2 group. Although the inhibitory effect of PT and PA on certain $-SH$ enzymes, such as lactate and alcohol dehydrogenase, aldolase, and several bacterial enzymes, was thought to be due to the blocking of $-SH$ and $-NH_2$ groups at the active sites of the enzymes, inhibition of lactate dehydrogenase by both PT and PA is reversible. Gliotoxin has also been found to bind covalently with $-SH$ enzymes. Binding of membrane proteins with TCTC mycotoxins may be due to their interactions with the $-SH$ groups in the proteins. Both T-2 toxin and fusarenon X interact with a number of '$-SH$ enzymes', thus causing a reduction in enzyme activity.

Although it has been shown that an activated AFB-epoxide reacts with the nucleophiles, its role in covalent interactions with specific enzymes is less clear. Several aflatoxin metabolites have been shown to interact with enzymes and proteins. Both aflatoxins B2a and G2a, which react readily with the $-NH_2$ groups of proteins to form Schiff bases, have been shown to be very effective inhibitors of DNase *in vitro*.

The activated AFB1 and AFG1 also react with albumin to form albumin adducts through the same Schiff-base mechanism, followed by Amadori rearrangement and subsequent condensation with another aldehyde group. The presence of this adduct in human serum has been used as one of the indexes for human exposure to AFs.

B. Inhibition of Specific Enzymes and Protein Synthesis

1. Inhibition of specific enzymes

Mycotoxins inhibit many enzyme systems both *in vivo* and *in vitro*, but most of these effects are secondary. For example, deoxyribonuclease and RNA polymerase activity are inhibited by aflatoxins, ST, PT, LS, TCTC, and α-amanitin in both *in vitro* and *in vivo* experiments. Ochratoxin A inhibits carboxypeptidase A, renal phosphoenolpyruvate carboxykinase (PEPCK), phenylalanine–tRNA synthetase and phenylalanine hydroxylase activity. Both PT and PA inhibit dehydrogenase through their interactions with the SH group at the enzyme's active center. Moniliformin binds strongly with pyruvate dehydrogenase and subsequently affects the enzymes in the TCA cycle. Several mycotoxins, including territrem B and slaframine, are acetylcholine esterase inhibitors. However, the inhibition of ceramide synthase (sphinganine/sphingosine *N*-acyltransferase) by Fm and AAL toxins is considered to be a primary effect because complex sphingolipids and ceramides are heavily involved in cellular regulation, including cell differentiation, mitogenesis, and apoptosis. Alteration of sphingolipid metabolism results in a dramatic increase in the free sphingolipid bases, i.e. sphinganine (Sa) and sphingosine (So) in tissues and a decrease in complex sphingolipid levels. Significant increases in the Sa/So ratio were found in serum of a pig fed a diet containing as little as 5 ppm total FmB. Thus, testing Sa/So ratios in serum and urine has been suggested as an early biomarker of FmB exposure in animals and humans. FmB1 was also found to inhibit cytochrome P-450 enzymes selectively with the most significant effect on CYP2C11. Such inhibition was considered to be due to a suppressed activity of protein kinase activity resulting from the inhibition of sphingolipid biosynthesis.

2. Inhibition of protein synthesis

A number of mycotoxins, including OA, CT, PR-toxin and TzA, inhibit protein synthesis, but most of these effects are considered secondary. Ochratoxin A inhibits protein synthesis by acting as a competitive inhibitor for phenylalanine–tRNA synthetase and its inhibitory effect on PEPCK is due to inhibition of the synthesis of this enzyme. Even aflatoxins are protein synthesis inhibitors, but this effect is secondary because aflatoxins primarily act at the transcriptional level.

The most potent protein inhibitory mycotoxins are TCTCs, which act at different steps in the translation process. Inhibition of protein synthesis is one of the earlier events in manifestation of TCTC toxic effects. Inhibitory effects vary considerably with the chemical structure of the side chain. In general, T-2, HT-2, NIV, FS-x, DAS, verrucarin A, and roridin A affect at the initiation step, whereas verrucarol, trichothecin, and crotocin affect elongation. Inhibition of the elongation step by DAS and FS-x has also been demonstrated. Some of the less well-known TCTCs, such as trichodermol and trichodermone, affect the termination step. In contrast, ZE stimulates protein synthesis by mimicking hormonal action.

C. Other Modes of Action

1. Modulation of immune systems

Although many mycotoxins affect the immune systems in some animals or *in vitro* systems, only AFB, OA, and TCTC mycotoxins modulate immune processes that may affect human and animal health. Depending on the nutritional status, toxin dose, and other factors, these mycotoxins can be either immunosuppressive (most often) or immunostimulatory. Animals receiving OA showed general immunotoxic signs, including lymphocytopenia and depletion of lymphoid cells, especially in the thymus, bursa of Fabricius and Peyer's patches. In general, TCTCs and OA affect both humoral and cell-mediated immunity (CMI), while AFB primarily affects CMI. Because TCTC mycotoxins such as T-2 toxin and DAS exert major toxic effects on the bone marrow, lymph nodes, thymus, and spleen in mammals, modulation of the immune systems by this group of mycotoxins has been most extensively studied.

Administration of T-2 toxin to animals results in reduction of their resistance to infection and decreased antibody formation. T-2 toxin or DAS may selectively affect subpopulations of T-suppressor cells or their precursors; the suppression of antibody synthesis might be due either to impairment of antibody-forming or T-helper cell activities. Immunosuppressive effects have also been observed for other TCTCs, including the macrocyclic types. However, DON has both immunostimulatory and immunosuppressive effects in experimental animals. Mice fed a diet containing more than 10 ppm of DON experienced thymic atrophy, decrease in antibody formation, and suppression of B- and T-cell proliferation. Serum and saliva IgA was significantly increased in mice fed high levels (25 ppm) of DON, whereas serum IgM and IgG decreased. The increased IgA production is related to IgA-mediated nephropathy in mice; thus, it was postulated that DON might be one of the etiologic agents in IgA nephropathy, which is the most common glomerulonephritis in humans worldwide. Depressed humoral immune responses, including lower serum IgM and IgG levels, and suppressed antibody responses to sheep red blood cells and other antigens in mice, have been observed in animals fed or injected with OA. The immunosuppressive effects of OA are prevented by phenylalanine both *in vitro* and *in vivo*.

Regarding CMI, both T-2 toxin and DAS inhibit proliferation of mitogen-stimulated human lymphocytes, especially T-cell populations. This effect is concentration-dependent, and in fact, low concentrations of T-2 toxin were stimulatory. Structure–activity studies showed that T-2, HT-2, and 3'-OH T-2 toxins were most effective in inhibition while 3'-OH HT-2, T-2 triol, and T-2 tetraol toxins were 50–100 times less effective. The macrocyclic TCTC roridin and verrucarin A were 75–100 times more potent than T-2 toxin. The differences in the immunotoxic effects of various TCTCs were attributed to differences in both the uptake and metabolism of toxins by the cells. Interleukin (IL)-1 and -2 production in spleen cell cultures was stimulated by trace amounts of TCTC. Hyperinduction of cytokines in T-helper cells by DON (most), cyclopiazonic acid, OA, and α-zearalenol, has been found; but PT and T-2 toxin were inhibitory. Induction of expression of mRNA of IL-2, 4, 5 and 6 in a T-cell model EL4.IL-2 by DON was found at levels required for partial or maximal protein synthesis inhibition. A single oral gavage with DON was sufficient to induce these mRNA levels in Peyer's patches and spleen. The effect in IL-6-deficient mice were refractory to VT-induced dysregulation of IgA production and development of IgA nephropathy. TCTC-induced cytokine superinduction could lead to the terminal differentiation of immunoglobulin-secreting cells via T-cell-mediated polyclonal differentiation of B-cells or its precursors. The Peyer's patch might be particularly sensitive to such dysregulation.

2. Induction of apoptosis

As in other research with naturally occurring toxicants, recent studies have shown that many mycotoxins cause apoptosis (APT), a regular programmed cell death characterized by cell shrinkage, membrane blebbing, and chromatin condensation together with fragmentation of DNA. Induction of APT by AFB1, OA, several TCTCs (T-2 toxin, roridin A, NIV, DON), CT, LS, RB, FmB1, and AAL toxin-TA, gliotoxin (GT) and wortmannin (WM), has been observed in a number of cell lines. The concentrations causing such induction vary with different toxins tested. Some mycotoxins may induce APT in one system but not in others. For example, no induction of APT was observed for cytochalasin A, ST, RS, cyclocholortine, fusaric acid, kojic acid, RB, FmB, and WM in HL-60 human promyeloytic leukemia cells, but induction of APT has been observed for RB, FmB, and WM in other cell lines. Depending on lymphocyte set, tissue source and glucocorticoid induction, DON can either inhibit or enhance apoptosis in murine T, B, and IgA cells.

Induction of APT by AFB1 and LS was attributed to their inhibitory effect on RNA synthesis, as a result of their binding with DNA and inhibitory effect on DNA-dependent RNA polymerase. Although suppression of protein synthesis was considered to be the mechanism for the induction of APT by TCTC, OA, and several other mycotoxins, this effect is by no means specific. The T-2 toxin-induced APT in epidermal cells of dorsal skin of hypotrichotic WBN/ILA-Ht rats was found to be associated with the induction of c-fos and perhaps also c-jun mRNAs. OA interacts specific cell types with distinct members of the mitogen-activated protein kinase family at concentrations where no acute toxic effect can be observed. Induction of APT via the c-jun amino-terminal-kinase pathway can explain some of the OA-induced changes in renal function as well as part of its teratogenic action. Exposure of human proximal tubule-derived cells and renal epithelial cell lines to low OA concentrations can lead to direct or indirect caspase 3 activation and subsequently to APT. Inhibition of phosphatidylinositol 3-kinase activity by WM was considered to cause apoptotic cell death for this mycotoxin, and APT by GT in spleen cells appears to be cyclic AMP (cAMP) mediated. Regulation of cellular Ca^{2+} ions plays an important role in the induction APT by several mycotoxins, including T-2, RB, FmB, and TA (one of the major AAL toxins). However, p53 was found not to be involved in the RB-induced APT signal transduction, exposure of cells to ST resulted in failure of G1 arrest. Thus, the carcinogenic effect of ST was considered to be mediated by failure of p53-mediated G1 checkpoint. In addition to the Ca^{2+} regulated effect, proteases were considered to be involved in the induction of APT by RB. ZE stimulated cytokine production and the growth of estrogen receptor-positive human breast carcinoma cell line MCF-7. It functioned as an antiapoptotic agent by increasing the survival of MCF-7 cell cultures undergoing apoptosis caused by serum withdrawal. The mitogen-activated protein kinase signaling cascade is required for ZE's effects on cell-cycle progression in MCF-7 cells.

While the effect of FmB1 on the immune system is less clear, the effect of FmB on APT in many cell lines has been extensively studied. Immune cells are one of the types of target cells. For example, an increase in nitric oxide and prostaglandin (PG) E-2 production was found when a murine macrophage cell line was grown in the presence of FmB. The kidney, liver, and tomato cells, and other animal or plant cells treated with FmB or TA showed typical apoptotic characters with the degree of injury being related to dose and time of exposure. The fundamental elements of APT, as characterized in animals, are conserved in plants. Inhibition of the ceramide synthase by FmB and related mycotoxins is considered to play a key role in the APT of the cells of the target organs/tissues. FmB diminished the cytotoxic effects of lipids on esophageal tumor cells. In the presence of FmB, the marked APT induced by PGA2 and the PG2- and AA-induced p53 levels were lowered. It repressed specific isoforms of protein kinase C and cyclin-dependent kinase (CDK) activity and induced expression of CDK inhibitors in monkey kidney cells (CV-1). Such effects lead to cell-cycle arrest and APT. Eight genes induced by FmB1 have been identified. The ability of FmB1 to alter gene expression and signal transduction pathways is considered to be necessary for its carcinogenic and toxic effects. Since FmB is neither genotoxic in bacterial mutagenesis screens nor affects unscheduled DNA synthesis, the apoptotic necrosis, atrophy, and consequent regeneration may play an essential role in its carcinogenic effects. FmB1 may be the first example of an apparently nongenotoxic (non-DNA reactive) agent producing tumors through such a mechanism.

3. Other mechanisms

Several other mechanisms have been postulated as modes of action for mycotoxins. Alteration of membrane structures by TCTCs and several other mycotoxins may directly affect cellular components. Modulation of translocation of protein kinase (PK) by mycotoxins has been suggested as another mechanism. For example, FmB may directly be involved in the activation of PKC. CPA alters Ca^{2+} hemostasis and induces charge alteration in plasma membranes and mitochondria; the reversible inhibition of reticulum Ca^{2+}-dependent ATPase has been considered as the primary cause. Paxilline, a reversible, noncompetitive inhibitor of the cerebellar inosital 1,4,5-triphosphate (InsP-3) receptor, inhibits the InsP-3-induced Ca^{2+} release. OA also may impair cellular Ca^{2+} cAMP homeostasis. In the renal epithelial cells, OA interferes with hormonal Ca^{2+} signaling, leading to altered cell proliferation. Secalonic acid D, a cleft palate-inducing mycotoxin produced by *P. oxalicum* and other fungi, reduces palatal cAMP levels. It inhibits the binding of the cAMP response elements to the binding protein, alters phosphorylation and leads to altered expression of genes involved in cell proliferation, an event critical for normal palate development.

Formation of free radicals by mycotoxins, including AF, anthraquinone type of mycotoxins, and OA has been considered as one of the mechanisms for their carcinogenic effect because the free radicals can interact with macromolecules to cause cellular damages. For example, modification of deoxyguanine and deoxyribose by LS and related anthraquinones was found to be due to the induction of free radicals. Enhancement of lipid peroxidation is considered as one of the manifestations of cellular damage in OA toxicity. Superoxide dismutase and catalase have been shown to have some protective effect for OA-induced nephrotoxicity in rats. OA-induced lipid peroxidation in the Vero cell was also found to be decreased in the presence of these two enzymes, but aspartame and piroxcam were less effective. The induction of free radicals in a bacterial model system was found to be regulated by Ca^{2+} ions.

V. Preventive Measures

A. Controlling Mycotoxin Problems Through Prevention of Toxin Formation

Owing to the widespread nature of fungi in the environment and high stability of the toxins, mycotoxin problems are difficult to control. The most effective approach is to prevent the formation of the toxins in the field as well as during storage. Some measures, including use of resistant varieties (most effective, but not all are successful), crop rotation, use of earlier harvest varieties, avoidance of overwintering in the field, bird and insect damage and mechanical damage, cleaning and drying grain quickly to less than 10–13% moisture, increased sanitation in the field, and use of mold inhibitors have been suggested. After harvest, crops should be stored in a clean area to minimize insect and rodent infestation, and moisture should be reduced by regular aeration to prevent mold contamination and growth. Rigorous quality control programs, including analysis of mycotoxins, examination of broken kernels, and removal of suspected contaminated feed should be established in the milling facility. Some of these practices have been implemented at different levels in farm practice and agricultural industries. Current research is aimed at development of more resistant crops and new biocontrol agents. One approach, which involves application of nontoxicogenic *Aspergillus* strains to the peanut, corn, or cotton field, has shown promising results in reducing AFB formation in the field.

In postharvest storage, some antifungal agents, including organic acids (sorbic, propionic, benzoic), antibiotics (neatamycin or pimaricin, nisin), phenolic antioxidants (butylated hydroxyanisole and butylated hydroxytoluene) and fumigants (methylbromide, chlorine) have been found to be partially effective in inhibiting the growth and toxin production of some fungi. Some spices and essential oils also appear to have some antifungal effects. Current research is aimed at identifying and characterizing naturally occurring antifungal agents and also attempting to transfer the genes that produce such agents to crops to enhance their natural resistance as well as to prevent toxin formation. Genetically modified crops resistant to insect infestation also have shown some ability to minimize fungal propagation and mycotoxin production.

B. Avoiding Human Exposure Through Rigorous Monitoring Programs

While it is impossible to remove mycotoxins completely from foods and feeds, effective measures to decrease the risk of exposure depend on a rigorous program of monitoring mycotoxins in foods and feeds. Consequently, governments in many countries have set limits for permissible levels or tolerance levels for a number of mycotoxins in foods and feeds. Official methods of analysis for many mycotoxins have been established. However, because of the diverse chemical structures of mycotoxins, the presence of trace amounts of toxins in a very complicated matrices that interfere with analysis, and the uneven distribution of the toxins in the sample, analysis of mycotoxins is a difficult task. In general, the samples are first ground to fine particles, extracted with appropriate solvent systems, followed by a clean up before they are subjected to separation and quantification and confirmation protocols. Thus, it is not uncommon that the analytical error can amount to 20–30%. To obtain reliable analytical data, an adequate sampling program and an accurate analytical method are both important.

To minimize the errors that might be introduced at each of the above steps, investigations on the development of new sensitive, specific, and simple methods for mycotoxin analysis have been conducted since aflatoxin was first discovered in the early 1960s. Such studies have led to many improved and innovative analytical methods for mycotoxin analysis in the last few years. Simplified and efficient clean-up procedures, such as solid-phase extraction cartridges, have been established. Better analytical quality control has been established by using 'certified reference materials'. New, more sensitive thin-layer chromatography (TLC), high-performance liquid chromatography (HPLC), and gas chromatography (GC) techniques are now available. Sensitive and versatile high-resolution mass spectroscopy (MS) and GC/tandem MS/MS are coming to the market. The MS methods have also been incorporated into HPLC systems. New chemical methods, including capillary electrophoresis and biosensors are emerging and have gained application for mycotoxin analysis. After a number of years of research, immunoassays have gained wide acceptance as analytical tools for mycotoxins within the last few years. Antibodies against almost all the mycotoxins are now available. Whereas quantitative immunoassays are still primarily used in the control laboratories, immunoscreening methods have been widely accepted as a simple approach to screening for mycotoxins in several commodities. Many immunoscreening kits, which require less than 15 minutes per test, are commercially available. These assays eliminate many of the above steps; a sample after extraction usually proceeds directly to analysis. The immunoaffinity method has also become popular as a clean-up method in conjunction with other chemical methods. Several immunoassay protocols have been adopted as the first action by the Association of Official Analytical Chemists (AOAC). Automation of mycotoxin analysis has became a reality by combining these newer techniques. Detailed protocols for mycotoxin analysis can be seen in several of the most recent reviews and books and the most recent edition of AOAC. The progress of research dealing with mycotoxin analysis can be seen from the numerous papers cited by the 'General Referee on Mycotoxins' in the 'Annual Report on Mycotoxins', which generally appears in the January/February issue for each year of the *International Journal of the Association of Official Analytical Chemists* (*J AOAC, International*).

C. Removal of Mycotoxins from Commodities Through Detoxification and Other Physical and Chemical Means

Detoxification may be desirable in controlling mycotoxins, but it is very difficult and in some cases not economically feasible. The toxicity of the detoxified commodities should be tested extensively. Although several detoxification methods have been established for aflatoxins, only the ammoniation process is an effective and practical method. Physical screening and subsequent removal of damaged kernels by air blowing, washing with water or use of specific gravity methods have been effective for some mycotoxins, including DON and FmB and AFB1. Other chemicals such as ozone, chlorine, and bisulfite have been tested and have shown some effect on some mycotoxins. Solvent extractions have been shown to be effective but are not economically feasible. Fumonisins can form Schiff's bases with reducing sugar such as fructose under certain conditions. However, cooking generally does not destroy mycotoxins.

D. Dietary modifications

Dietary modification greatly affects the absorption, distribution, and metabolism of mycotoxin; and this can subsequently affect its toxicity. For example, the carcinogenic effect of AFB1 is affected by nutritional factors, dietary additives, and anticarcinogenic substances. Diets containing chemoprotective agents and antioxidants such as ascorbic acid, butylated hydroxyanisole (BHA), butylated hydroxytoluene (BHT), ethoxyquin, oltipraz, penta-acetyl geniposide, Kolaviron biflavonoids, and even green tea, have also been found to inhibit carcinogenesis caused by AFB1 in test animals. Dietary administration of the naturally occurring chemopreventive agents, ellagic acid, coumarin or α-angelicalactone caused an increase in glutamate–cysteine ligase activity, a key enzyme for the synthesis of glutathione. The protective effects are due to shifting of metabolism to a detoxification route by formation of AFB1-glutathione conjugate, rather than formation of AFB1-DNA adducts. Ebselen possesses a potent protective effect against AFB1-induced cytotoxicity; such protective effect may be due to its strong capability in inhibiting intracellular reactive oxygen species formation and preventing oxidative damage. The toxic effect of OA to test animals was minimized when antioxidants such as vitamin C are added to the diet. Most mycotoxins have a high affinity for hydrated sodium calcium aluminasilicate (HSCAS or NovaSil) and other related products. Diets containing NovaSil and related absorbers have been found effective in preventing absorption of AFB1 and several other mycotoxins in test animals, thus decreasing their toxicity. Several other adsorbents such as zeolite, bentonite, and superactive charcoal have been found to be effective in decreasing the toxicity of other mycotoxins such as T-2 toxin. Aspartame, which is partially effective in decreasing the nephrotoxic and genotoxic effects of OA, competes with OA for binding to serum albumin. L-Phenylalanine was found to have some protective effect against the toxic effects of OA because it diminished OA's inhibitory effect on some of the enzymes discussed earlier. Vitamin E was found efficient in preventing the cytotoxicity induced by FmB.

VI. Concluding Remarks

Mycotoxins are a group of naturally occurring, low-molecular-weight, fungal secondary metabolites frequently contaminating agricultural commodities, foods, and feeds. Numerous molds can produce mycotoxins both preharvest and postharvest, but not all are toxigenic. The major mycotoxins that we are currently most concerned about are aflatoxins, ochratoxins, fumonisins, and some trichothecene mycotoxins such as deoxynivalenol. Whereas natural occurrence of toxic fungi and mycotoxins in foods and feeds has been reported and outbreaks of mycotoxicoses have been documented, there are still some mycotoxicoses that have not been well characterized. Production of mycotoxins is controlled by the genetics of the mold and by such environmental conditions as substrate, water activity (a_w) or moisture and relative humidity, temperature, time, atmosphere, and microbial interactions. Although some mycotoxins are known to cause certain mycotoxicoses, others are not. Many mycotoxins are highly toxic to animals and probably to humans.

Mycotoxins can cause both acute and chronic effects in prokaryotic and eukaryotic systems, including humans, and their toxicities vary considerably with toxin and animal species. Most of their effects are organ-specific, but some mycotoxins may affect many organs. Induction of cancers through initiation and promotion processing, by some mycotoxins in animals and possibly in humans, is one of major concerns for their chronic effects. Modulation of immune systems by some mycotoxins is another concern. Interaction with macromolecules through either noncovalent or covalent bindings or both is the basis of their mode of action. For some mycotoxins, such as aflatoxins, metabolic activation is necessary before their binding with specific macromolecules, but for others, metabolic activation is not necessary. Nevertheless, metabolism plays a key role in modulating toxicity because metabolism can lead to either activated metabolites or nontoxic metabolites for subsequent conjugation and excretion. Thus, factors affecting the metabolism of mycotoxins greatly affect the toxicity. Recent studies have shown that many mycotoxins induce or enhance apoptosis, and investigation of the mechanisms of mycotoxin-induced apoptosis has aided in our understanding of the mode of action of mycotoxins.

Almost all of the mycotoxins are very stable; only limited detoxification methods are currently available, and some of these are not economically feasible. Because the toxins are formed, both preharvest and postharvest, it is very difficult to control toxin formation. Nevertheless, good farm management could minimize the problems, and current research is directed to finding new biocontrol agents that could prevent toxin formation in the field. Dietary modification can also decrease the risk. Because most of these control measures are not very effective, rigorous programs for preventing human and animal exposure to contaminated foods and feed have been established, and effective methods for monitoring toxin levels in foods have been developed. Mycotoxin problems may be with us for a long time to come, and there are many mycotoxicoses that require further study. It is unrealistic to propose complete elimination of the problem; only through multiple approaches can we minimize the problem and enhance human and animal health.

Acknowledgments

The author thanks Professor Dean O. Cliver for providing excellent suggestions and Ms Ellin Doyle and Barbara Cochrane for their help in the preparation of this manuscript.

Abbreviations

AAL	*A. alternata* f. sp. *lycopersici*
AF	aflatoxin
AFB1	aflatoxin B1
AFG1	aflatoxin G1
AFM1	aflatoxin M1
AFP1	aflatoxin P1
AFQ1	aflatoxin Q1
AM	*Alternaria* mycotoxins
AME	alternariol monomethyl ether
AOAC	Association of Official Analytical Chemists
APT	apoptosis
ATA	alimentary toxic aleukia
ATX	altertoxin
BHA	butylated hydroxyanisole
BHT	butylated hydroxytoluene
CDK	cyclin-dependent kinase
CMI	cell-mediated immunity
CPA	cyclopiazonic acid
CT	citrinin
DAS	diacetoxyscirpenol
DON	deoxynivalenol

ELEM	equine leukoencephalomalacia
FC	fusarin C
FDA	US Food and Drug Administration
Fm	fumonisin
FmB1	fumonisin B1
FmB2	fumonisin B2
FmB3	fumonisin B3
FS-x	fusarenon-x
GC	gas chromatography
GT	gliotoxin
HBV	hepatitis B virus
HSCAS	hydrated sodium calcium aluminasilicate
HPLC	high-performance liquid chromatography
IgG	immunoglobulin G
IgM	immunoglobulin M
IL	interleukin
IP	intraperitoneal
IPM-one	ipomeamarone
InsP -3	inosital-1,4,5-triphosphate
LD_{50}	median lethal dose
LS	luteoskyrin
MS	mass spectroscopy
NAC	N-acetyl-L-cysteine
NESO	neosolaniol
NIV	nivalenol
NPA	nitropropionic acid
OA	ochratoxin A
PA	penicillic acid
PEPCK	renal phosphoenolpyruvate carboxykinase
PG	prostaglandin
PHC	primary hepatocellular carcinoma
PK	protein kinase
PKS	polyketide synthase
PPE	porcine pulmonary edema
PT	patulin
RA	rubratoxin A
RB	rubratoxin B
RS	rugulosin
Sa	sphinganine
SBS	'sick building syndrome'
So	sphingosine
SP	'deteriorated sugarcane poisoning'
ST	sterigmatocystin
TCTCs	trichothecenes
TLC	thin-layer chromatography

TzA   tenuazonic acid
T-4ol   T-2 tetraol
WM   wortmannin
ZE   zearalenone

Bibliography

Beasley, V. R. (1989). 'Trichothecene Mycotoxicosis, Pathophysiologic Effects', Vols I and II. CRC Press, Inc., Boca Raton, FL.

Bhatnagar, D., Lillehoj, E. B. and Arora, D. K. (1992). 'Handbook of Applied Mycology, Vol. V, Mycotoxins in Ecological Systems'. Marcel Dekker, New York.

Bhatnagar, D., Cotty, P. J. and Cleveland, T. E. (1993). Preharvest aflatoxin contamination. Molecular strategies for its control. *In* 'Food Flavor and Safety. Molecular Analysis and Design' (A. M. Spanier *et al.*, eds), American Chemical Society Symposium Series 528, 272–292.

Bondy, G. S. and Pestka, J. J. (2000). Immunomodulation by fungal toxins. *J. Toxicol. Environ. Health B Crit. Rev.* **3**, 109–143.

Busby, W. F. Jr and Wogan, G. N. (1979). Food-borne mycotoxins and alimentary mycotoxicoses. *In* 'Foodborne Infections and Intoxications' (H. Riemann and F. L. Bryan, eds), pp. 519–610. Academic Press, New York.

CAST (1989). 'Mycotoxins, Economic and Health Risks'. Council for Agricultural Science and Technology (CAST), Task Force Report no. 116. 92p. CAST, Ames, IA.

Chu, F. S. (1991). Mycotoxins: food contamination, mechanism, carcinogenic potential and preventive measures. *Mutat. Res.* **259**, 291–306.

Chu, F. S. (1995). Mycotoxin analysis. *In* 'Analyzing Food for Nutrition Labeling and Hazardous Contaminants' (I. J. Jeon and W. G. Ikins, eds), pp. 283–332. Marcel Dekker, New York.

Chu, F. S. (1997). Trichothecene mycotoxicosis. *In* 'Encyclopedia of Human Biology', 2nd ed., Vol. 8 (R. Dulbecco, ed.), pp. 511–522. Academic Press, New York.

Chu, F. S. (1998). Mycotoxins – occurrence and toxic effects. *In* 'Encyclopedia of Food and Nutrition – Food contaminants', 2nd ed. pp. 858–869. Academic Press, New York.

Cole, R. J. and Cox, R. H. (1981). 'Handbook of Toxic Fungal Metabolites'. Academic Press, New York.

Dorner, J. W., Cole, R. J. and Blankenship, P. D. (1992). Use of a biocompetitive agent to control preharvest aflatoxin in drought stressed peanuts. *J. Food Prot.* **55**, 888–892.

Dragan, Y. P., Bidlack, W. R., Cohen, S. M, Goldsworthy, T. L., Hard, G. C., *et al.* (2001). Implications of apoptosis for toxicity, carcinogenicity, and risk assessment: fumonisin B-1 as an example. *Toxicol. Sci.* **61**, 6–17.

Eaton, D. L. and Groopman, J. D. (1994). 'The Toxicology of Aflatoxins – Human Health, Veterinary and Agricultural Significance'. Academic Press, New York.

FAO/WHO (1997). Worldwide regulations for mycotoxins – 1995. A compendium. FAO Food and Nutrition paper. # 64. Food and Agriculture Organization of the United Nations, Rome.

Galvano, F., Piva, A., Ritieni, A. and Galvano, G. (2001). Dietary strategies to counteract the effects of mycotoxins: a review. *J. Food Prot.* **64**, 120–131.

Groopman, J. D., Kensler, T. W. and Links, J. M. (1995). Molecular epidemiology and human risk monitoring. *Toxicol. Lett.* **82**, 763–769.

Harris, C. C. (1996). The 1995 Walter Hubert lecture – Molecular epidemiology of human cancer: insights from the mutational analysis of the p53 tumour-suppressor gene. *Br. J. Cancer* **73**, 261–269.

International Agency for Research on Cancer (IARC) (1993). 'Some Naturally Occurring Substances: Food Items and Constituents, Heterocyclic Amines and Mycotoxins'. IARC Monographs on the Evaluation of Carcinogenic Risk to Humans, Vol. 56. IARC, Lyon.

Jackson, L., DeVries, J. W. and Bullerman, L. B. (eds) (1996). 'Fumonisins in Food'. Plenum, New York.

Kuiper-Goodman, T. (1995). Mycotoxins: risk assessment and legislation. *Toxicol. Lett.* **82**, 853–859.

Marasas, W. F. O., Nelson, P. E. and Toussoun, T. A. (1986). 'Toxigenic *Fusarium* Species: Identity and Mycotoxicology'. Pennsylvania State University Press, University Park, PA.

Marasas, W. F. O. (1995). Fumonisins: their implications for human and animal health. *Nat. Toxins* **3**, 193–198.

Merrill, A. H., Liotta, D. C. and Riley, R. T. (1996). Fumonisins: fungal toxins that shed light on sphingolipid function. *Trends Cell Biol.* **6**, 218–223.

Miller, J. D. (1995). Fungi and mycotoxins in grain: implications for stored product research. *J. Stored Product Res.* **31**, 1–16.

Miller, J. D. and Treholm, H. L. (1994). 'Mycotoxins in Grain: Compounds Other than Aflatoxin'. Eagan Press, St Paul, MN.

Park, D. L. (1993). Controlling aflatoxin in food and feed. *Food Technol.* **47**, 92–96.

Payne, G. A. and Brown, M. P. (1998). Genetics and physiology of aflatoxin biosynthesis. *Ann. Rev. Phytopathol. 1998* **36**, 329–362.

Pestka, J. J. and Bondy, G. S. (1990). Alteration of immune function following dietary mycotoxin exposure. *Can. J. Physiol. Pharmacol.* **68**, 1009–1016.

Riley, R. T. and Richard, J. L. (eds) (1992). Fumonisins: a current perspective and view to the future. *Mycopathologia* **117**, 1–124.

Riley, R. T., Voss, K. A., Yoo, H. S., Gelderblom, W. C. A. and Merrill, A. H., Jr (1994). Mechanism of fumonisin toxicity and carcinogenesis. *J. Food Prot.* **57**, 528–535.

Shank, R. C. (1981). 'Mycotoxins and Nitroso Compounds: Environmental Risks', Vols I and II. CRC Press, Inc., Boca Raton, FL.

Sharma, R. P. and Salunkhe, D. K. (1991). 'Mycotoxin and Phytoalexins in Human and Animal Health'. CRC Press, Inc., Boca Raton, FL.

Shier, W. T. (2000). The fumonisin paradox: a review of research on oral bioavailability of fumonisin B-1, a mycotoxin produced by *Fusarium moniliforme*. *J. Toxicol. Toxin Rev.* **19**, 161–187.

Sinha, K. K. and Bhatnagar, D. (1998). 'Mycotoxin in Agriculture and Food Safety'. Marcel Dekker, New York.

Smela, M. E., Currier, S. S., Bailey, E. A. and Essigmann, J. M. (2001). The chemistry and biology of aflatoxin B-1 from mutational spectrometry to carcinogenesis. *Carcinogenesis* **22**, 535–545.

Trenholm, H. L., Prelusky, D. B., Young, J. C. and Miller, J. D. (1988). Reducing Mycotoxins in Animal Feeds, Agriculture Canada Publication 1827E. Agriculture Canada, Ottawa, Canada.

Trucksess, M. W. and Pohland, A. E. (2000). 'Mycotoxin Protocols'. Humana, Totowa, NJ.

Ueno, Y. (1986). Toxicology of microbial toxins. *Pure Appl. Chem.* **58**, 339–350.

Van Egmond, H. P. (1989). 'Mycotoxins in Dairy Products'. Elsevier, London.

Wild, C. P. and Hall, A. J. (2000). Primary prevention of hepatocellular carcinoma in developing countries. *Mutation Res.* (*Rev. Mutation Res.*) **462**, 381–393.

Wogan, G. N. (1992). Aflatoxins as risk factors for hepatocellular carcinoma in humans. *Cancer Res.* (Suppl.) **52**, 2114s–2118s.

Chemical Intoxications

20

Steve L. Taylor

I. Introduction

In this chapter, the potential hazards associated with man-made chemicals in foods will be discussed. Naturally occurring toxicants are addressed in Chapter 14.

A. Incidence of Foodborne Disease Outbreaks of Chemical Etiology

Chemical intoxications account for about 5–7% of all acute foodborne disease outbreaks according to statistics compiled by the US Centers for Disease Control and Prevention (CDC). However, many of the causes of these chemical intoxications are naturally occurring chemicals. In Chapter 15, the importance of the seafood toxins such as ciguatera toxins, paralytic shellfish toxins, and histamine was emphasized. Naturally occurring mushroom toxins are another common cause of acute chemical intoxications associated with foods (see Chapter 14). Man-made chemicals actually account for a very small percentage of the foodborne disease outbreaks reported to the CDC. Most of these incidents have involved inadvertent or accidental contamination of foods with heavy metals or detergents.

    Of course, not all foodborne disease outbreaks of chemical etiology will be reported to the CDC. Reporting is not mandated. Some of the illnesses are either mild and

Foodborne Diseases 2nd Edn
ISBN: 0-12-176559-8

short-lived or difficult to ascribe to particular man-made chemicals in foods. Thus, incomplete reporting of acute intoxications associated with foodborne chemicals of man-made origin is a certainty. Also, the CDC do not attempt to define cause-and-effect relationships between man-made chemicals in foods and chronic diseases, such as cancer.

B. Types of Chemicals Involved in Chemical Intoxications Associated with Foods

The major categories of man-made chemicals that can occur in foods are agricultural chemicals (insecticides, herbicides, fungicides, fertilizers, feed additives, and veterinary drugs), food additives, chemicals migrating from packaging materials, and inadvertent or accidental contaminants including industrial and environmental pollutants. Chemicals produced by reactions occurring during the processing, preparation, storage, and handling of foods could also be considered man-made, because these processes occur through the intervention of humans. Table 20.1 lists some examples of chemicals in each of these categories that will be considered in this chapter.

II. Agricultural Chemical Residues

A. Insecticides

Insecticides are applied to foods and feeds to control insect infestation. In areas of the world where insecticides are not widely used, a substantial portion of the food is lost to insects. The major types of insecticides applied to food crops in the US are organochlorine compounds (e.g. heptachlor and chlordane), organophosphate compounds (e.g. parathion and malathion), carbamate compounds (e.g. carbaryl and aldicarb), botanical compounds (e.g. nicotine and pyrethrum), and inorganic compounds (e.g. arsenicals). The US Environmental Protection Agency (EPA) regulates the manner in which insecticides can be used. Insecticides are approved for use on only certain crops. The EPA also establishes tolerances for the amounts of insecticide residues allowed on raw food crops. The US Food and Drug Administration (FDA) has tolerances for the amounts of insecticide residues that can be present on processed foods.

Several characteristics of insecticides merit consideration when one is trying to assess their safety. Some insecticides will accumulate in the environment. Organochlorine compounds, such as DDT, are good examples of insecticides that accumulate in the environment. Concerns were voiced about this environmental accumulation because DDT is a weak animal carcinogen and because it may have adverse effects on certain types of wildlife. For example, DDT residues may have interfered with the reproductive efficiency of certain birds. Therefore, DDT use was banned in the US. The alternative has been to use insecticides that do not accumulate in the environment. However, these insecticides tend to be more acutely toxic to humans and other animals. Certain of the organophosphate insecticides, such as parathion, pose a substantial hazard to farm workers because of their degree of toxicity in humans and because they can be absorbed through the skin. However, the organophosphates and carbamates are

Table 20.1 Examples of man-made chemicals found in foods

Agricultural chemicals
Insecticides
    Organochlorines (heptachlor)
    Organophosphates (malathion, parathion)
    Carbamates (aldicarb, carbaryl)
    Botanicals (pyrethrum, nicotine)
    Inorganics (arsenicals)
Herbicides
    Chlorophenoxy compounds (2,4-D)
    Dinitrophenols
    Bipyridyls (paraquat, diquat)
    Substituted ureas (monuron)
    Carbamates (propham)
    Triazines (simazine)
Fungicides
    Captan
    Folpet
    Dithiocarbamates
    Pentachlorophenol
    Mercurials
Fertilizers
    Nitrogen fertilizers
    Sewage sludge
Feed additives and veterinary drugs

Food additives
GRAS ingredients
Direct additives
Color additives

Chemicals migrating from packaging materials
Monomers, plasticizers, and stabilizers
Compounds from printing inks
Inorganic chemicals – lead and tin

Chemicals produced during processing, preparation, storage, and handling of foods
Nitrosamines
Mutagens from heat processing of meats
Polycyclic aromatic hydrocarbons
Lipid oxidation products

Inadvertent or accidental contaminants
Industrial/environmental pollutants
    PCBs and PBBs
    Mercury
Chemicals from utensils
Accidents and errors

widely used because they do decompose rather rapidly in the outdoor environment. Recently, in Louisiana, concerns have surfaced regarding the indoor use of parathion, which decomposes much more slowly when not exposed to sunlight. Indoor use of parathion could pose a hazard to dwellers. Insects can develop resistance to various insecticides. Thus, the arsenal of insecticides must be rather large to prevent the recurrent exposures of the insects that would result in the development of resistance.

While the exceedingly low residue levels of insecticides found in most foods are not particularly hazardous, large doses of insecticides can be toxic to humans and other animals. Many of the common insecticides are neurotoxins. The organophosphates and carbamates are cholinesterase inhibitors. Cholinesterase is a key enzyme involved in the transmission of nerve impulses at the synapse. The neurotoxic mechanism of the organochlorines is unknown. The arsenicals are a particularly hazardous class of insecticides, which are fairly toxic to humans and which also tend to accumulate in the environment.

No food poisoning incidents have ever been attributed to the proper use of insecticides on foods. Several incidents have occurred as the result of improper use of insecticides, however. Several reasons exist for the low degree of hazard posed by insecticide residues on foods: (1) the level of exposure is very low if the insecticides are used properly; (2) some insecticides are not very toxic to humans; (3) some insecticides decompose rapidly in the environment; and (4) many different insecticides are used, which limits our exposure to any one particular insecticide. Problems can arise from the inappropriate use of certain insecticides. A good example is the outbreak of aldicarb intoxication from watermelons that occurred on the US West Coast several years ago. It is illegal to use aldicarb on watermelons because excessive levels of the insecticide are concentrated in the edible portion of the melon. Several farmers used aldicarb on watermelons illegally, resulting in several cases of aldicarb intoxication, and the recall and destruction of thousands of watermelons.

Recently, certain food-source plants have been genetically modified to contain insecticidal protein. The insecticidal protein comes from the bacterium, *Bacillus thuringiensis* (*Bt*), which has been used as a biological control agent by farmers for decades. *Bt* produces a protein that selectively kills certain crop-damaging insects. This protein is not hazardous to humans, so the incorporation of the *Bt* gene into transgenic plants has been approved for corn, potatoes, and other food-source plants.

B. Herbicides

Herbicides are applied to agricultural lands and crops to control the growth of weeds. Many different types of herbicides are used. Among the more important classes of herbicides are chlorophenoxy compounds (e.g. 2,4-D and 2,4,5-T), dinitrophenols (e.g. dinitroorthocresol), bipyridyl compounds (e.g. paraquat and diquat), substituted ureas (e.g. monuron), carbamates (e.g. propham), and triazines (e.g. simazine). Again, the EPA controls which herbicides can be used on various crops and establishes limits on the allowable residue levels.

In assessing the potential risks from the use of herbicides on foods, the toxicity of the herbicides, their environmental half-lives, and the presence of toxic impurities must be considered. Most herbicides are selectively toxic toward plants. Thus, they present very little hazard to humans. Exceptions are the bipyridyl compounds, such as paraquat and diquat. These bipyridyl compounds are nonselective herbicides; they are usually used to kill all vegetative growth in an area. The bipyridyl compounds are also rather toxic to humans, and their use should be closely monitored. Most herbicides do not accumulate in the environment. Highly toxic impurities have not been identified in most herbicides. An exception is the chlorophenoxy compound 2,4,5-trichloro-phenoxy acetic acid or 2,4,5-T. During the Vietnam War, large quantities of this herbicide, known as Agent Orange, were sprayed as part of the defoliation program. Some batches of Agent Orange contained a highly toxic class of impurities, the dioxins, including tetrachlorodibenzodioxin (TCDD).

Generally, herbicide residues in foods do not represent any hazard to consumers. No food-poisoning incidents have ever resulted from the proper use of herbicides on food crops. The lack of hazard from herbicide residues is associated with the low level of exposure, their low degree of toxicity and selective toxicity toward plants, and the use of many different herbicides which limits exposure to any particular herbicide. Recently, herbicide-tolerant plants have been produced through genetic modification. The availability of these transgenic varieties may increase the use of certain herbicides, especially glyphosate. However, glyphosate is selectively toxic to plants and is rapidly degraded in the environment, so increased consumer exposure to this herbicide is not a safety issue.

C. Fungicides

Fungicides are used to prevent the growth of molds on food crops. Some of the more important fungicides are captan, folpet, dithiocarbamates, pentachlorophenol, and mercurials. The hazards associated with fungicides are minuscule because our exposure to these chemicals is very low, most of the fungicides do not accumulate in the environment, and most fungicides are not very toxic to humans. Pentachlorophenol and the mercurials do persist in the environment. The mercurials, in contrast to other fungicides, are quite hazardous to unwary consumers. The mercurials are often used to treat seed grains to prevent mold growth during storage. The seed grains treated with the mercurials are usually colored pink or some other noticeable color. These treated grains are clearly intended for planting and not for ingestion. On several occasions, consumers have eaten these seed grains and developed mercury poisoning. Care must be taken to avoid the consumption of seeds treated with these fungicides.

D. Fertilizers

The most commonly used fertilizers are combinations of nitrogen and phosphorus compounds. Nitrogen fertilizers stimulate the growth of plants, but must first be oxidized to nitrate and nitrite in the soil. Both nitrate and nitrite can be hazardous to humans if ingested in large amounts. The major problem is the contamination of ground water with nitrate. This ground water is often used as drinking water on farms and in farming communities. Infants can suffer from methemoglobinemia, a blood disorder associated with an inability of hemoglobin to carry oxygen, if they routinely consume water that has nitrate residues exceeding 10 ppm. Some plants can accumulate nitrate. An example is spinach. If spinach, or other nitrate-accumulating plants, are allowed to grow on soil overly fertilized with ammonia or other nitrogen fertilizers, the spinach can accumulate hazardous levels of nitrate. This situation can become even more serious if nitrate-reducing bacteria are allowed to proliferate on these foods. In that case, some of the nitrate is converted to the more hazardous nitrite. Only a few instances of food poisoning have occurred in this manner, and all of these incidents have involved infants eating spinach or other plants containing excessive nitrate and/or nitrite. Commercial processors are aware of this risk and monitor for it. Actually, these incidents are the result of the inappropriate use of excessive amounts of nitrogen-based fertilizers, which should not occur with proper farming methods.

An increasingly important fertilizer is sewage sludge. Sewage sludge has the potential to carry any toxic chemicals that might be dumped into the sewage system. Most of the concerns have centered around residues of heavy metals, such as cadmium, from industrial activities. The residues of heavy metals are concentrated by some plants. Thus, sewage sludge should be monitored periodically for certain types of toxic chemicals.

The degree of hazard associated with the use of fertilizers on food-producing land is quite low. The few incidents that have occurred have involved excessive use of the fertilizers.

E. Feed Additives

Some chemicals are added to feed to serve as growth stimulants. Diethylstilbestrol (DES) was once allowed to be used as a growth promoter with beef cattle. However, when DES was shown to be carcinogenic, its use as a feed additive was banned. Diethylstilbestrol is definitely carcinogenic to humans; its use as a drug to prevent miscarriages in pregnant women has been directly linked to certain types of cancer in their offspring. However, no evidence exists to demonstrate that the extremely low levels of DES found in edible beef after its use as a growth promoter pose any carcinogenic hazard to humans.

F. Veterinary Drugs and Antibiotics

Antibiotics are widely used with food-producing animals. Penicillin is often used to treat mastitis in cows. Concerns over the effects of penicillin residues in milk on consumers with penicillin allergy led the FDA to enact a zero-tolerance for penicillin in milk. That action has not prevented penicillin use, but has resulted in the dumping of milk from treated cows during the treatment period and for a short time thereafter. The hazards associated with other antibiotics are generally limited to concerns about their role in the development of antibiotic-resistant bacteria. That particular concern is beyond the scope of this chapter.

III. Food Additives

Numerous substances are knowingly added to foods to provide a wide variety of technical benefits. Several thousand food additives exist, but many of these chemicals are used in rather small amounts. Food additives can be classified on the basis of their regulatory standing: (1) generally recognized as safe (GRAS) substances; (2) flavors and extracts; (3) direct additives; and (4) color additives.

Generally recognized as safe substances are those food ingredients that were in common use before the latest version of the Food, Drug, and Cosmetic (FD&C) Act, which was enacted in 1958. The 1958 Act required FDA approval of any newly developed food additives, but recognized the long history of safe use of many additives. Over 600 chemicals are on the GRAS list, including such materials as sucrose, salt, butylated hydroxytoluene (BHT), and spices. Most of the common food additives are on the GRAS list because they were in common use before 1958. From a legal standpoint, GRAS chemicals are not actually additives, but that distinction is seldom made by the consumer. Reviews of the safety of the GRAS substances have been conducted since 1958 and deficiencies in our information on their toxicity have been identified and corrected in some cases. Substances, or certain uses of substances, can be removed from the GRAS list if the FDA acquires evidence of some hazard to consumers.

A large number of food additives would fall into the category of flavors and extracts. The Flavor and Extract Manufacturers Association (FEMA) keeps a list of accepted flavors and extracts. FEMA evaluates the safety of the chemicals on the list. In essence, the FEMA list is a GRAS list for flavors and extracts. Over 1000 chemicals are on the FEMA list, although some of these chemicals and extracts are no longer used.

Direct food additives are those new food additives that have been approved by the FDA since 1958. Actually, very few approvals for new food additives have been granted in recent years. Extensive safety data are required, so the process of obtaining FDA approval for a new food additive can be rather costly. Aspartame, a non-nutritive sweetener, is perhaps the most notable direct food additive in use in the US. Since approval of aspartame as a sweetener for certain types of uses a few years ago, the consumption of this new additive has become substantial. More recently, Olestra, a noncaloric fat replacer, was approved for certain uses such as the deep-fat frying of chips. Olestra use promises to be considerable. Olestra was approved despite some concerns over mild gastrointestinal disturbances following ingestion of large doses and its possible interference with the absorption of fat-soluble vitamins. The fortification of foods with additional fat-soluble vitamins may mitigate this concern.

Color additives are regulated in a separate part of the FD&C Act. New color additives must be approved by the FDA in much the same way that new food additives are approved. Some color additives in common use before 1958 have been banned because of more recent concerns about their chronic toxicity. A good example is FD&C Red #2.

The degree of hazard associated with the presence of additives in our foods is quite small. Several factors account for the low degree of hazard. First, our level of exposure to most food additives is rather low. This is especially true for many of the flavoring ingredients. Second, the oral toxicity of food additives tends to be extremely low, especially with regard to acute toxicity. Some concerns have arisen about the chronic toxicity of some food additives. High doses of some food additives have been shown to cause cancer in animals. One of the best examples is saccharin, which caused bladder cancer in animals fed extremely high doses of the sweetener. Questions exist regarding how properly to extrapolate these results to humans, who typically ingest much lower levels of the ingredient. Thus, the carcinogenicity of saccharin to humans at typical levels of intake is uncertain. A third reason for the low hazard associated with food additives is the established safety of many additives. Most food additives have been subjected to some safety evaluations in laboratory animals. Thus, the toxicity of food additives is often well known and exposure can be limited to levels far below any dose that would be hazardous. By contrast, the toxicity of naturally occurring chemicals in foods is often not known, and we cannot be certain that hazardous circumstances will not exist.

The safety of some food additives has been questioned. Most of the questions have revolved around evidence for weak carcinogenic activity in laboratory animals. Some additives, such as FD&C Red #2 and cyclamate, have been banned. Warning labels are required for saccharin. These are but a few examples. In addition to carcinogenicity, other concerns have arisen, such as the role of sugar in dental caries and abnormal behavioral reactions, the role of monosodium glutamate (MSG) in asthma, and the role of aspartame in headaches and other behavior and neurological reactions. While a detailed discussion of these issues is beyond the scope of this chapter, suffice it to say that many of these assertions have been questioned and remain to be validated.

A few food additives have caused acute illness under certain conditions of exposure. These intoxications are usually the result of either excessive consumption of the additive or ingestion by an individual who has an abnormal sensitivity to the additive. Misuse of food additives by consumers, or food processors, has also created hazardous situations on occasion. These types of situations will be discussed later in this chapter.

Dietetic food diarrhea is a good example of an intoxication resulting from the excessive consumption of a food additive. The hexitols and sorbitol are widely used sweeteners. They are especially common in candy and chewing gum because they do not promote tooth decay. These sugar alcohols are not as easily absorbed as sugar, but once absorbed, they are equally as caloric as sugar. Because of their slow absorption, these sweeteners can cause an osmotic-type diarrhea when excessive amounts are consumed. Several cases have been reported where consumers were ingesting more than 20 g of these sweeteners per day. The levels of hexitols and/or sorbitol used in foods will vary, but in one case the ingestion of 12 pieces of hard candy over a short period of time provided 36 g of sorbitol and resulted in diarrhea.

Sulfite-induced asthma is a good example of an extreme sensitivity to a food additive that afflicts only a small percentage of the population. Sulfites are widely used food additives with a number of desirable technical properties. The acceptable daily intake for sulfites is 72 mg/kg body weight. However, a small percentage of asthmatics, perhaps 1–2% of the 9 million asthmatics in the US, are exquisitely sensitive to sulfites. Ingestion of certain sulfited foods can induce asthma in these individuals; other consumers eat these same foods with no ill effects. A few deaths have been attributed to the ingestion of sulfites by such consumers. As a result, the FDA rescinded the GRAS status of sulfites for use on raw vegetables other than potatoes in 1986, because the use of sulfites on these products was often not labeled and sensitive individuals would not be aware of the hazard. Sulfite-sensitive individuals can avoid many of the hazards associated with sulfited foods by reading labels and avoiding foods that contain this additive.

IV. Chemicals Migrating from Packing Materials

Chemicals migrating from packaging materials do not represent a significant hazard. A variety of chemicals can migrate from packaging materials into foods, including plastics monomers, plasticizers, stabilizers, printing inks, and others. However, the level of exposure to these chemicals is extremely low and most of these chemicals are not particularly toxic. US law requires that packaging materials be tested to determine the degree of migration.

Concerns have arisen about residues of lead migrating into foods from soldered cans. Lead is a well-known toxicant that can affect the nervous system, the kidney, and the bone. Lead-soldered cans have been phased out as a result of this concern.

V. Chemicals Produced During Processing, Storage, Preparation, and Handling of Foods

Countless chemical reactions occur during processing, storage, preparation, and handling of foods. Literally millions of chemicals are formed as a result of these reactions. The toxicity of these chemicals has not been established in most cases. A thorough discussion of this subject is beyond the scope of this chapter. However, the formation of nitrosamines should serve as a good example of this type of situation.

Nitrosamines are formed by the reaction of nitrites with secondary amines. The amines are common, naturally occurring components of many foods. Nitrite is a food additive used in the curing of meats, but the majority of nitrite entering the gastrointestinal tract arises from other sources, such as water, plants (especially if fertilized with nitrogen-based fertilizers), nitrogen oxides in polluted air and the open flames used in some types of food processing, and saliva. Nitrosamine formation is favored by heating and by acidic conditions. Among the cured meats, nitrosamine formation in bacon is most critical because of the high temperatures used in frying. Most nitrosamines are carcinogenic. Nitrosamines can alkylate DNA and act as initiators of the carcinogenic process. The risk associated with nitrosamines in foods can be lessened by lowering the nitrite levels used in the curing of meats and by the addition of vitamin C, vitamin E, or antioxidants to cured meats. These antioxidants diminish nitrosamine formation. The removal of nitrites entirely from cured meats would not eliminate this possible hazard because many other sources of nitrite exist. Also, the removal of nitrite from cured meats would increase the risk of botulism.

VI. Inadvertent or Accidental Contaminants

A. Industrial and/or Environmental Pollutants

Industrial and/or environmental pollutants often migrate into foods in small amounts. Occasionally hazardous levels of such chemicals enter the food supply, causing food-borne illness. Obviously, such gross cases of pollution should be prevented, but occasional incidents are likely to occur even with diligent preventive measures.

The contamination of foods with polychlorinated or polybrominated biphenyls (PCBs or PBBs) has occurred on several occasions. The most infamous incident involved the accidental inclusion of PBBs in dairy feed in Michigan. The PBBs are lipid soluble so they were concentrated in the milk. As a result of this incident, many cows and their milk had to be destroyed. Consumers of this milk continue to have PBB residues in their fatty tissues many years later. The ultimate consequences of this incident remain unclear, but it was obviously unfortunate and unnecessary. A separate incident involving a leaking transformer in a facility that produced poultry feed resulted in the contamination of the feed with PCBs. This incident resulted in the destruction of chickens, eggs, and egg-containing products. Pollution of Lake Michigan by PCBs has reached such a level that fish in that lake are routinely contaminated with PCBs. Commercial fishing on Lake Michigan has ceased as a result. Recreational fishing continues, even though the consequences of frequent consumption of these fish are uncertain.

Perhaps the most famous incident involving industrial pollution occurred in Minamata Bay, Japan. An industrial firm was dumping mercury-containing wastes into the bay. Bacteria in the sediment converted the inorganic mercury to highly toxic methyl mercury. Fish in the bay became contaminated with the methyl mercury. The ultimate result was over 1200 cases of mercury poisoning among consumers of these fish. Symptoms included tremors and other neurotoxic effects and kidney failure.

B. Chemicals from Utensils

Potentially toxic heavy metals will leach from containers and utensils into beverages if the beverages are acidic. Acute heavy-metal intoxication is a relatively common cause of foodborne disease. It almost always results from the contact of acidic beverages with containers or utensils containing heavy metals, such as copper, zinc, cadmium, or tin. Lead contamination can also occur but does not usually result in acute intoxications.

Copper poisoning, characterized primarily by nausea and vomiting, can occur from faulty check valves in soft-drink vending machines. The check valves prevent contact between the acidic, carbonated beverage and the copper tubing that delivers the water.

Several outbreaks have resulted from faulty check valves, and each outbreak has the potential to involve hundreds of people. Zinc intoxication can result from the storage of acidic beverages in galvanized buckets. Refrigerator shelves containing cadmium residues have caused problems when the shelves were employed as barbecue grills. Tin intoxication has resulted from the placement of acidic juices in unlined cans. Juices are usually packaged in cans with a lining that prevents contact between the acidic beverage and the tin plate. Glazed pottery and painted glassware have been recalled from sale because of the presence of lead in the glazes and paints and the fear of lead poisoning.

C. Accidental Contaminants

Occasionally, accidents will result in foodborne intoxications. Some of these accidents are errors of ignorance, such as the use of galvanized containers to store acidic beverages. These accidents could be avoided if the individuals involved were aware of the risks. Other accidents are not so clearly preventable.

Several foodborne intoxications have resulted from confusion about the identity of food ingredients. Sodium nitrite can easily be confused with other salts, including sodium chloride, which is much less toxic. In one incident, a small grocery store was repackaging additives such as sodium chloride, sodium nitrite, and MSG from bulk containers. Somehow, sodium nitrite was labeled as MSG. The mislabeled product was used in hazardous amounts by consumers, resulting in acute methemoglobinemia and at least one death.

Sometimes it is less clear whether the mistake was entirely accidental. In 1983, an outbreak of niacin intoxication occurred among consumers of pumpernickel bagels in New York. Niacin intoxication is a rather mild, short-term illness involving rash, pruritis, and a sensation of warmth. The bagels had been excessively enriched with niacin, one of the B vitamins; they contained about 60 times the normal amount of niacin. This incident likely resulted from an accident, although intentional use of megadoses of vitamins is considered desirable by some consumers.

One very serious outbreak was clearly a result of ignorance on the part of the processor. A manufacturer of a soybean-based infant formula wished to decrease the sodium content of the formula despite a lack of evidence that such action would prevent the infants from developing hypertension later in life. The manufacturer removed the NaCl from the formula. Most infants receiving soybean formula survive on that formula alone because of allergies to formulae based on cows' milk. The removal of NaCl resulted in a deficiency of chloride in the formula. The result was a condition known as metabolic alkalosis, characterized by lethargy, poor appetite, failure to gain weight, vomiting, and diarrhea. Several infants died as a result.

More often, such errors of ignorance are perpetrated by consumers. A good example was a small outbreak of vitamin A intoxication. Twin infants were provided a diet consisting largely of pureed chicken livers, pureed carrots, milk, and vitamin supplements. The mother of these infants stated that she did not trust commercial baby foods. After several weeks on this diet, the infants began to vomit and developed a skin rash. The symptoms disappeared when a more normal diet was instituted.

The estimated intake of vitamin A and carotene was 44 000 IU/day compared to the RDA of 1500–4500 IU/day for infants.

VII. Summary

Man-made chemicals of many types occur in foods. However, only 5–7% of all acute foodborne disease outbreaks reported to the CDC in the US result from chemical intoxications. The proper use of man-made chemicals will prevent most intoxications arising from these agents.

Bibliography

Chin, H. B. (1992). Evaluating pesticide residues and food safety. *In* 'Food Safety Assessment' (J. W. Finley, S. F. Robinson and D. J. Armstrong, eds), pp. 48–58. American Chemical Society, Washington, DC.

Hotchkiss, J. H. (1989). Relative exposure to nitrite, nitrate, and N-nitroso compounds from endogenous and exogenous sources. *In* 'Food Toxicology – A Perspective on the Relative Risks' (S. L. Taylor and R. A. Scanlan, eds), pp. 57–100. Marcel Dekker, New York.

Irving, G. W., Jr (1982). Determination of the GRAS status of food ingredients. *In* 'Nutritional Toxicology' (J. N. Hathcock, ed.), Vol. 1, pp. 435–450. Academic Press, New York.

Linshaw, M. A., Harrison, H. L., Gruskin, A. B., Prebis, J., Harris, J., Stein, R., *et al.* (1980). Hypochloremic alkalosis in infants associated with soy protein formula. *J. Pediatr.* **96**, 635–640.

Maga, J. A. and Tu, A. T. (eds) (1994). 'Food Additive Toxicology'. Marcel Dekker, New York.

Munro, I. C. (1989). A case study: the safety evaluation of artificial sweeteners. *In* 'Food Toxicology – A Perspective on the Relative Risks' (S. L. Taylor and R. A. Scanlan, eds), pp. 151–167. Marcel Dekker, New York.

Munro, I. C. and Charbonneau, S. M. (1981). Environmental contaminants. *In* 'Food Safety' (H. R. Roberts, ed.), pp. 141–180. Wiley, New York.

Taylor, S. L. and Bush, R. K. (1986). Sulfites as food ingredients. *Food Technol.* **40(6)**, 47–52.

Taylor, S. L. and Byron, B. (1984). Probable case of sorbitol-induced diarrhea. *J. Food Prot.* **47**, 249.

Taylor, S. L., Bush, R. K. and Nordlee, J. A. (1996). Sulfites. *In* 'Food Allergies – Adverse Reactions to Foods and Food Additives', 2nd ed. (D. D. Metcalfe, H. A. Sampson and R. A. Simon, eds), pp. 339–357. Blackwell Scientific Publications, Boston.

Diet and Cancer

21

Elisabeth Garcia and Carl K. Winter

I. Introduction

It is widely recognized that one's risk of developing many types of cancers can be influenced by diet. Estimates suggest that roughly one-third of all human cancers can be attributed to diet, and dietary changes could significantly reduce the rates of certain cancers.

Although genetic factors are known to influence risks for developing cancer, heredity alone cannot explain cancer susceptibility as the extent of hereditary contribution to cancer appears to be limited. Some foods or food components may be associated with increases in dietary cancer risks while others may reduce dietary risks. Overall, cancer rates increase exponentially with age but, for some cancers, age-standardized rates are very different when rates are compared among countries. Some of these differences may be due to genetic and environmental factors, and lifestyle choices, such as smoking and physical activity levels, but significant evidence associates dietary factors with cancer incidence.

Many health experts suggest that dietary changes can bring a significant decrease in the incidence of cancer. In fact, changing dietary patterns and some aspects of lifestyle have been attributed with the potential of reducing the incidence of cancer in middle-aged and elderly populations by as much as 80–90%. While new research is still necessary to allow a greater understanding of the relationships between many individual dietary components and health, several recommendations have been developed from existing data that serve as guidelines for a healthy diet.

Foodborne Diseases 2nd Edn
ISBN: 0-12-176559-8

Table 21.1 Distribution of cancer deaths, according to age and sex, in relation to all deaths reported in the US in 1995

Age group (years)	Deaths due to cancer (% of total deaths)	
	Female	Male
1–19	0.08	0.11
20–39	0.57	0.48
40–59	3.95	3.93
60–79	11.73	14.05
80+	6.22	5.43

Data source: American Cancer Society (1999).

II. Incidence of Major Cancers

Throughout the world the incidence of cancer is increasing, particularly for some sites in the body such as lung, breast, prostate, colon, and rectum. There is a great variation in the incidence rates of cancers among countries. Among many socioeconomic factors, the per capita income of a country shows a strong correlation with cancer death rates.

High rates of stomach cancer are found in Japan and other parts of Asia. Stomach cancer has been associated with intake of large amounts of foods preserved by salt that presumably contain nitrosamine precursors. Primary liver cancer is common in some regions of Africa and Asia, and has been associated with early-life infection with hepatitis B virus and ingestion of foods containing aflatoxins. The main cause of lung cancer is cigarette smoking, which has also been established as a major risk factor for pancreatic cancer. The incidence rate of breast cancer in the US and Western countries is 5–6 times greater than in Asian countries; this variation is primarily attributed to dietary rather than genetic factors.

Another important consideration is the strong relationship between advancing age and cancer. Sixty percent of all cancers in the US are detected in persons aged 65 years and above. This segment of the population has expanded significantly throughout the century; while individuals 65 years and above represented only 4% of the population in 1900, their representation grew to nearly 13% in the early 1990s. It is expected that by 2030, one in five Americans will be 65 years or above. The distribution of cancer deaths in the US in 1995 relative to age and gender is summarized in Table 21.1; the percentage of cancer deaths clearly rises as age increases.

In the US, cancer is the second leading cause of death and is responsible for approximately 23% of all deaths. Excluding smoking, diet-related factors are believed to account for over 50% of all remaining cancer deaths in the US, or 150 000 deaths per year.

Statistics demonstrate that cancer mortality in the US has been declining. With the exception of lung cancer due to smoking, age-adjusted cancer mortality rates have decreased 16% since 1950. Figure 21.1 presents the variation in the occurrence of some major cancers among men and women in the US since 1930. The four leading cancers among males and females are reported in Table 21.2. Cancer mortality rates

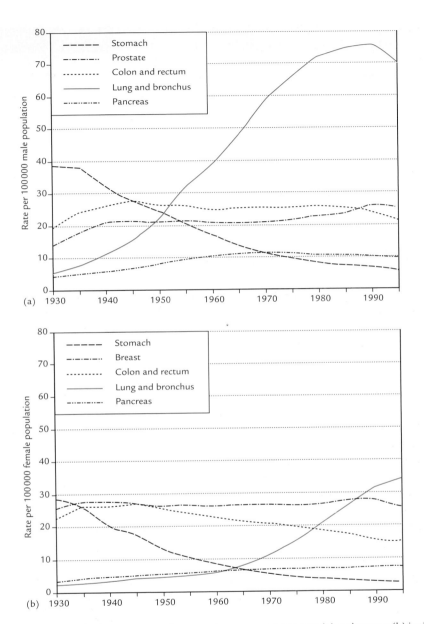

Figure 21.1 Variation in the occurrence of some major cancers among men (a) and women (b) in the US since 1930.

for women are typically lower than those for men when considering all racial and ethnic population groups. Cancer incidence has been higher among African-Americans than in any other population group; Native American women have the lowest incidence of breast cancer, while Caucasian women have the highest.

Studies of migrant populations entering new countries demonstrate that the incidence and types of cancer experienced by the migrant groups quickly approach those

Table 21.2 Leading cancer sites among males and females in the US

Male	Female
Lung and bronchus (32.7%)	Lung and bronchus (22.7%)
Prostate (12.4%)	Breast (17.2%)
Colon and rectum (10.2%)	Colon and rectum (11.4%)
Pancreas (4.6%)	Pancreas (5.5%)

Figures in parentheses represent percentage of cancer of all sites. Data are presented for all age groups; statistics from 1994. Modified from American Cancer Society (1999).

of their host populations. Examples include increases in breast cancer rates among Japanese migrant women in Hawaii and increases in colorectal cancer rates among migrants from eastern Europe and Asia moving to the US. There is plenty of epidemiological evidence linking a high consumption of vegetables with reduced rates of cancer in different regions of the world. Studies of Seventh Day Adventists, who eat fruits, vegetables, whole grains, and nuts in abundance, revealed that the cancer mortality rate among them is between one-half to two-thirds the rates of the general population.

III. Carcinogens in Food

Hundreds of carcinogenic (cancer-causing) chemicals have been shown to exist in food. Food carcinogens include chemicals produced naturally by plants or fungi as well as synthetic chemicals, added either intentionally (food additives) or unintentionally (pesticides, environmental contaminants). The label 'carcinogen' typically results from experimental studies in which laboratory animals exposed continuously to large levels of a chemical exhibit a statistically significant increase in cancers (or benign tumors) relative to those exposed to lower levels of the chemical. Human epidemiologic evidence relating exposure to food carcinogens and the incidence of cancer is severely limited. As a result, most estimates of human cancer risks from food carcinogens rely upon extrapolative measures that consider the likelihood of development of cancer in laboratory animals at high levels of exposure to predict possible effects on human populations exposed to low levels. These risk assessment measures include a myriad of conservative (risk-exaggerating) assumptions, including the assumed lack of a threshold dose for carcinogens, suggesting that any level of exposure to a carcinogen may pose some finite risk for the development of cancer. Mathematical models incorporating this nonthreshold assumption have been developed to 'quantify' predicted cancer risks; such models also routinely include statistical corrections that frequently present risks as the statistical upper bound. Such risk estimates may exceed the mean predicted risk at a given exposure level by several orders of magnitude.

In a few instances, enough epidemiologic evidence has been developed to link consumption of specific food carcinogens with increases in human cancer. The strongest links involve mycotoxins, which are naturally occurring toxins produced by fungi colonizing food crops (see Chapter 19). Elevated exposure to aflatoxins, mycotoxins produced by *Aspergillus flavus* and *Aspergillus parasiticus* that are frequently associated with nuts and grains such as peanuts and corn, appears to be a major contributor to human liver cancer. It is also likely that synergistic effects between aflatoxin exposure and both hepatitis B virus and chronic alcohol consumption have contributed to human liver cancer risk. Epidemiologic correlations have also been drawn between exposure to the fumonisins, mycotoxins produced by the widely distributed fungus *Fusarium moniliforme* (which commonly contaminates corn) and human esophageal cancer. Ochratoxin A, a mycotoxin produced by *Aspergillus ochraceus* and *Penicillium verrucosum*, has been implicated as a human urinary tract carcinogen. The ochratoxin A-producing fungi have been shown to occur in a number of human foods including coffee beans and grains.

While considerable public concern and debate centers on the possible cancer risks from synthetic chemicals in the diet, a comprehensive report of the US National Research Council concluded that the natural components of the diet may be of greater concern than synthetic compounds. This conclusion was based upon evidence suggesting that the human body did not make distinctions between natural and synthetic chemicals, and that both the number of naturally occurring chemicals and their levels are typically far greater than those for synthetic chemicals. Naturally occurring chemicals in the diet are also much more difficult to regulate compared with synthetic chemicals and much less is known about their toxicology relative to their synthetic counterparts because they have not been subject to the same requirements for toxicological testing. It should be emphasized that the same report also concluded that most naturally occurring and synthetic chemicals in the diet are present at levels below which significant biological effects are anticipated and therefore are unlikely to pose an appreciable human cancer risk. Macronutrients and excess calories were identified as the major dietary factors contributing to human cancer.

Epidemiologic studies have suggested that less than 2% of human cancers may be attributed to pesticides and the greatest pesticide risks may result from occupational, rather than from dietary exposures. Some studies have shown a link between occupational exposure to phenoxy herbicides such as 2,4-D and soft tissue sarcomas while others have not indicated any link; the evidence is suggestive enough that further study is anticipated. Epidemiologic studies concerning chlorinated hydrocarbon insecticides such as DDT have shown conflicting results. In one case-control study matching 58 breast cancer patients with 171 matched controls, a statistically significant increase of DDE (a DDT breakdown product and metabolite) levels in those women with breast cancer was noted relative to controls. In a subsequent and more comprehensive study, a similar association was not observed. In reviewing available information concerning the relationship between public exposure to pesticides and cancer, a panel of the National Cancer Institute of Canada concluded that it was unaware of definitive evidence suggesting that synthetic pesticides contribute significantly to overall cancer mortality. The Canadian council, in agreement with the US National Research Council, did not believe that any increase in exposure to pesticide residues resulting from increased consumption of fruits and vegetables posed a significant increased risk of cancer and that any cancer risks, however small, were out-weighed by the health benefits of consuming fruits and vegetables.

IV. Nutritional Factors and Cancer

A. Caloric Intake

The results of some studies examining the effects of total caloric intake on cancer require careful analysis, since, in some cases, the results cannot be attributed to caloric intake alone, but to the consumption of total fat and saturated fat.

In animal studies, significant reductions of mammary gland and colon tumors were observed when rats were fed diets restricted in calories, even in cases where the diet was high in fat. Studies of mice fed diets severely restricted in calories showed dras-tic reductions in the occurrence of spontaneous tumors. In rodents, calorie restriction leads to lower mitotic rates in several tissues, which may result in decreased tumor incidence. A higher human caloric intake than is physiologically required has been linked epidemiologically to an increase in the incidence of some cancers and repre-sents one of the major determinants of dietary-related cancer.

Additionally, obesity is closely associated with cancers of the gall bladder and endometrium, and possibly with postmenopausal breast cancer. This is important, as statistics indicate that more than three in ten adult Americans are considered over-weight as they exceed their ideal weight by at least 20%.

B. Fat

Lipids appear to be more highly associated with cancer than any other dietary com-ponents. Convincing evidence relating the incidence of certain cancers and fat intake has been provided by epidemiologic and laboratory animal studies. Diets in Western countries are frequently high in fat (contributing 38–45% of total calories) as com-pared to some Asian diets from which 5–25% of total calories come from fat. Total dietary fat correlates positively with cancer incidence and mortality data in epidemi-ologic studies of different countries. Strong associations were observed between dietary fat and cancer of the breast, colorectum, endometrium, and prostate, while kidney and testicle cancer rates had positive, yet lesser, associations with dietary fat. Laboratory studies indicate that cancers of the mammary gland, pancreas, and intes-tinal tract develop more promptly in animals receiving a high-fat diet than in animals fed a low-fat diet. Nevertheless, the effect of dietary fat can be partly attributed to an increase in total energy intake. No clear relationship has been identified between can-cer and dietary cholesterol.

There is strong evidence from comparisons of cancer rates among populations of different countries that consumption of red meat and fat of animal origin leads to increased numbers of colon and breast cancers. It is believed, however, that the dietary fat alone does not explain the potential cancer risk. Nevertheless, other studies reported no effect or only weak association between breast cancer incidence and consumption of fat. Some epidemiological studies associate breast cancer with energy intake from both saturated and polyunsaturated fats but not from monounsaturated fat. In regions such as Southern Europe, where olive oil is a staple and the consumption of saturated fats is low, colon cancer rates are also low. Some studies suggest that increased consumption of monounsaturated fatty acids (and reduced intake of total and saturated fat) is related to a decreased incidence of colorectal cancer, although other studies have not indicated such a relationship. According to other studies, a critical determinant of the effect of fat on cancer is the content of linoleic acid in the diet.

Some experimental studies using isolated fats and oils found that omega-6 polyunsaturated fatty acids were more efficient in promoting carcinogenesis than saturated fatty acids, while carcinogenesis was inhibited by omega-3 polyunsaturated fatty acids (found in a number of food forms such as marine fish and mammals, canola oil, wheat germ oil, and linseed and walnut oils).

C. Protein

It is difficult to evaluate the results of epidemiological studies on the association of protein consumption and risk of cancer because the consumption of protein of animal origin is highly correlated with the intake of saturated fats. When both dietary factors are evaluated, fats tend to present the strongest evidence of association. In animal studies when there is amino acid balance, high intakes of protein have been associated with increased carcinogenicity.

D. Dietary Fiber

Negative correlations have been observed between dietary fiber and colon cancer risk. Nonstarch polysaccharides and lignin are components of the total dietary fiber. Owing to some analytical methods used for quantification, dietary fiber is frequently separated into soluble (pectins, gums, mucilages, some hemicelluloses) and insoluble (cellulose, most hemicelluloses, and lignin) fractions for classification. Colonic function and intestinal microflora activities are influenced by dietary fiber. Fecal excretion of bile acids is stimulated by some soluble dietary fibers, such as pectin and oat bran. Decreases in intestinal transit time and increases of fecal bulk are caused by the insoluble dietary fiber fraction.

The protection by dietary fiber could be accomplished through reduced exposure to a potential carcinogen, either by acting as a diluent or by decreasing gastrointestinal transit time. The protective effect of dietary fiber on colon cancer risk has been confirmed by several laboratory animal studies, and the most consistent inhibitory effect is conferred by wheat bran.

E. Micronutrients

Vitamins and minerals may significantly influence cancer rates and development (Table 21.3). In the human body, as a consequence of normal cellular metabolism,

Table 21.3 Relationships between some micronutrients and cancer

Vitamin C

Most of the evidence from epidemiologic studies relating vitamin C to cancer is indirect, because it is based on the intake of foods containing vitamin C rather than on quantified intake of ascorbic acid. Vitamin C has antioxidant properties, and prevents the formation of carcinogens from some precursor compounds. Citrus fruits and vegetables rich in vitamin C offer protection against stomach cancer, as observed in epidemiologic studies.

Vitamin E

Low serum levels of vitamin E coupled with low serum levels of selenium have been related to increased risk of some cancers. The possible anticancer action of vitamin E has been attributed to its activity as a lipid antioxidant and as a free-radical scavenger. In experimental studies, vitamin E has been effective in blocking the formation of nitrosamines. Intake of vitamin E *supplements* has not been proven to reduce cancer risk.

Vitamin A and carotenoids

Vitamin A regulates cell differentiation and is important for maintaining healthy tissues. Preformed vitamin A is obtained only through animal sources; carotenoids can be obtained through plant and animal foods; only a fraction of carotenoids has provitamin A activity. In early studies on vitamin A and cancer, no distinction was made between preformed vitamin A and β-carotene. In laboratory animal studies, deficiency of vitamin A has shown to increase in chemically induced carcinogenesis, while retinoids can have a protective effect; tumor incidence has been lowered by vitamin A. Results from epidemiologic studies report negative correlations between the risk of lung cancer and consumption of carotenoid-rich foods. Carotenoids with no provitamin A activity, such as lycopene (found in tomatoes) have shown protective effects against cancer; it has been suggested that this effect is related to protecting cells against oxidative DNA-damage. Some authors believe that the chemopreventive action of β-carotene is related to its antioxidant activity, thus protecting against DNA oxidative damage. However, in a recent Finnish study, β-carotene supplementation has been related to a slight increase in the risk of lung cancer in cigarette smokers.

Folic acid (folate, folacin)

It has long been recognized that folic acid inhibits tumor growth. The incidence of several cancers, such as cancer of the colon and adenomas, has been related to folate deficiency, which is reputed to cause chromosome breaks in humans and animals. Low intakes of folate are not uncommon in American diets (see also vitamin B_{12}).

Vitamin B_{12}

Lipotropes (folic acid, choline, methionine and vitamin B_{12}) have been related to cancers of the bone marrow, cervix, esophagus, gastrointestinal tract, liver and respiratory tract. Lipotropes are involved in the metabolism of methyl groups, which have been suggested to be linked with cancer, owing to their effects on DNA methylation and gene expression. Lipotrope-deficient diets may affect many cellular activities such as xenobiotic metabolism and chromosome anomalies. In laboratory animals, cancer of the liver can be induced by lipotrope-deficient diets.

Vitamin B_6

It has been suggested that deficiency in vitamin B_6 could result in initial cell mutations that develop into a tumor, and that supplementation with vitamin B_6 could help in preventing some cancers.

Riboflavin

Apparently, deficiency of riboflavin increases the susceptibility to abnormal processing of potential carcinogens and affects liver detoxification mechanisms. Possible precancerous lesions in the esophageal epithelium of humans have been attributed to riboflavin deficiency. Diets poor in riboflavin, nicotinic acid, magnesium, and zinc have been correlated with increases in esophageal cancer.

Niacin

Deficiency of niacin has not been considered a major contributor to cancer development. Conflicting results suggest that niacin may be a carcinogen, a cocarcinogen, or an anticarcinogen.

Selenium

Selenium is an essential trace element; it is part of the enzyme glutathione peroxidase, an essential enzyme in the antioxidant defense system. In geographic regions where the soil contains more selenium, cancer mortality has been reported as lower than in areas poor in selenium. It should be noted that selenium is toxic, and the margin between toxic and safe doses is narrow.

Calcium

Some experimental studies suggest that the risk of colon cancer can be lowered by increasing dietary calcium to counteract the adverse effects on proliferation induced by high fat intake. Nevertheless, the epidemiologic and animal evidence associating colorectal cancer and calcium is not conclusive.

reactive species of oxygen or other radicals are formed. Exposure to such products of cellular metabolism can cause damage to DNA, protein, and lipids. While the body possesses many antioxidant defense mechanisms that utilize chemicals such as vitamin C, vitamin E, carotenoids, and others, some cellular components may still be oxidized. DNA repair pathways often effectively eliminate many of the damaged DNA lesions that may have been oxidized, but some DNA lesions remain and may accumulate with time.

F. Alcohol

Consumption of alcohol has been associated with cirrhosis and liver cancer, but there is consistent evidence that alcohol intake increases the risk of cancer of the oral cavity, pharynx, esophagus, and larynx, where alcohol interacts synergistically with tobacco. In the case of breast cancer, the consumption of alcohol is apparently related to increased estrogen levels leading to the disease. It is also suspected that the high incidence of esophageal cancer in parts of Asia is caused by alcohol in association with a nutritional factor.

G. Coffee

The results of epidemiologic studies regarding the risk of bladder cancer and coffee drinking are inconsistent.

V. Dietary Protective Factors

Humans have used plants as food, as well as sources of medicines and poisons, throughout time. The number of identified constituents of unprocessed plant foods has been estimated to be about 12 000 and these constituents represent many different chemical categories. Most plants seem to contain one or a few minor components of toxicological or pharmacological relevance, known as phytochemicals. The content of these constituents in plant tissues depends on many factors such as plant variety, growing conditions, stage of plant development, and length and conditions of storage. Their levels in the tissues typically represent a small percentage of the total composition of the plant, but occasionally may constitute several percent. While some of these chemicals are characteristic of specific fruits and vegetables, others are widely distributed.

In recent years, growing evidence from experimental studies has suggested that minor dietary components present in foods of plant origin may be capable of protecting against the development of cancer. These protective properties are conferred by a number of vitamins and minerals (Table 21.3) and nonnutrient components (Table 21.4). A high consumption of vegetables, fruits, whole grains, and legumes has been strongly associated epidemiologically with reductions in the risks for development

Table 21.4 Some dietary constituents that have been attributed protective activity against cancer

Source	Compounds
Cruciferous vegetables (broccoli, Brussels sprouts, cabbage, kale, cauliflower, mustard greens, etc.)	Sulforaphane Phenethyl isothiocyanate Benzyl isothiocyanate Glucosinolates (glucobrassicin, etc.) Indoles Phenols

Allium genus (onion, garlic, leek, shallots)	Organosulfur compounds:
	  Diallyl sulfide
	  Diallyl disulfide
	  Allyl mercaptan
	  Allyl methyl disulfide

Citrus fruit oils; other fruits and vegetables	D-Limonene
Caraway seed oil	D-Carvone
Soybeans	Isoflavones (genistein, daidzein) Protease inhibitors Inositol hexaphosphate Saponins
Green tea	Epigallocatechin gallate

Widely distributed	Flavonoids, such as:
	  Quercetin
	  Kaempferol
	  Apigenin
	  Myricetin
	  Rutin

Strawberry, raspberry, blackberry, walnut, pecan, chestnut, etc.	Ellagic acid
Oilseed, legumes, cereal seeds	Phytic acid, saponins, flavonoids, phytoestrogens
Ginger	Gingerol
Turmeric	Curcumin
Grapes, wine and other foods	Resveratrol, ellagic acid, flavenols
Whole grain products, various seeds, fruits, berries	Lignans
Ubiquitous in plants; abundant in cereals and legumes	Inositol hexaphosphate (phytic acid)
Ubiquitous in green plants	Chlorophyll
Dairy and animal fat	Conjugated dienoic linoleic acids

of several types of human cancer, owing to these protective effects and increased nutrient consumption.

A variety of mechanisms of action have been proposed to explain the activity of some nonnutrient protective factors. They include:

- antioxidant effects,
- preventing carcinogen activation,
- preventing initial genetic damage,
- increasing carcinogen detoxification and elimination,
- suppressing the development of cancer in cells already exposed to carcinogens,
- inhibiting cellular proliferation,
- inducing differentiation,

- inhibiting interaction of growth factors with their receptors,
- modulating hormonal effects, and
- increasing DNA repair.

Some compounds (Table 21.4) found in cruciferous vegetables (cabbage family) are believed to prevent the reaction of carcinogens with cellular macromolecules, induce the activity of detoxifying enzymes (such as glutathione-S-transferase), or suppress the development of cancer in previously initiated cells. Experimental studies reveal that compounds found in cruciferous and *Allium* plants have been effective against cancers of the esophagus, colon, lung, breast, and skin. Other compounds widespread in plants are flavonoids (Table 21.4), which appear to inhibit cancers of the skin, colon, and lung. Although some promising results have been seen in studies with some of the compounds listed in Table 21.4, it is important to realize that, in many cases, their beneficial effects in reducing human cancer risk have not been firmly established.

Several compounds have been reported as having both mutagenic and antimutagenic activities. Animal studies have shown that some of these compounds present in fruits and vegetables such as indole-3-carbinol, quercetin, and caffeic acid may actually be carcinogenic in their own right or may promote the carcinogenic effects of other chemicals. Indeed, this observation reinforces the general recommendation for consumption of a varied diet, with plenty of whole grains, legumes, and vegetables instead of consuming supplements of isolated phytochemicals or overindulging in specific food components.

VI. Conclusion

It is estimated that approximately one-third of human cancers may be attributed to diet. Within the human diet, a number of foods and food components that may influence human cancer risks have been identified. Consumption of significant amounts of dietary fat and specific types of carcinogens have been linked to increases in human cancer risks, while consumption of other foods and food components may have nutritive or other protective effects against the development of cancer. Health experts recommend liberal consumption of fruits, vegetables, whole grain cereals, and legumes owing to epidemiological evidence indicating correlations between consumption of such foods and decreased risks of various types of cancers.

Our knowledge about diet and nutritional factors and their relation to cancer is far from complete. There is strong evidence that diet influences most major types of cancers such as cancers of the esophagus, stomach, large bowel, breast, lung, and prostate. Nevertheless, there are many gaps in our knowledge that remain to be filled. Much evidence results from epidemiologic and animal studies, which must be carefully analyzed to avoid overinterpreting the findings because of limitations in experimental design and the sensitivities of the analytical tools used. Interpretation of such results frequently present challenges owing to factors such as confounding variables (such as high fat/high calorie diets), inaccuracy of data from human nutritional studies due to recall errors, and the difficulty in extrapolating results of animal experiments to humans.

There is, however, a general consensus among experts about the beneficial effects of diets high in consumption of plant foods such as fruits, vegetables, whole grain cereals, and legumes, owing to their low-fat and high-fiber content as well as vitamins, minerals, and nonnutrient protective factors. It is well documented that relatively large intakes of vegetables and fruits are associated with decreased risks of developing many cancers.

Bibliography

Alfin-Slater, R. B. and Kritchevsky, D. (1991). 'Cancer and Nutrition. Human Nutrition: A Comprehensive Treatise'. Plenum, New York.

American Cancer Society (http://www.cancer.org)

Ames, B. N. (1998). Micronutrients prevent cancer and delay aging. *Toxicol. Lett.* 102–103, 5–18.

Ames, B. N., Gold, L. S. and Willett, W. C. (1995). The causes and prevention of cancer. Review. *Proc. Natl. Acad. Sci.* **92**, 5258–5265.

Doll, R. (1990). An overview of the epidemiologic evidence linking diet and cancer. (Symposium on Diet and Cancer). *Proc. Nutr. Soc.* **49**, 119–131.

Dragsted, L. O., Strube, M. and Larsen, J. C. (1993). Cancer protective factors in fruits and vegetables: biochemical and biological background. *Pharmacol. Toxicol.* **72**(suppl. 1), 116–135.

Hill, M. J., Giacosa, A. and Caygill, C. P. J. (1994). 'Epidemiology of Diet and Cancer'. Ellis Horwood, New York.

International Agency for Research on Cancer (http://www.iarc.fr/)

National Research Council (NRC) (1989). 'Diet and Health. Implications for Reducing Chronic Disease Risk'. National Research Council, National Academy Press, Washington, DC.

NRC (1996). 'Carcinogens and Anticarcinogens in the Human Diet'. National Research Council, National Academy Press, Washington, DC.

Ritter, L. (1997). Report of a panel on the relationship between public exposure to pesticides and cancer. *Cancer* **80**, 2019–2033.

Steinmetz, K. A. and Potter, J. D. (1991). Vegetables, fruit, and cancer. II. Mechanisms. *Cancer Causes Control* **2**, 427–442.

Steinmetz, K. A. and Potter, J. D. (1996). Vegetables, fruit, and cancer prevention: A review. *J. Am. Diet. Assoc.* **96**, 1027–1039.

Waldron, K. W., Johnson, I. T. and Fenwick. G. R. (1993). 'Food and Cancer Prevention: Chemical and Biological Aspects'. Royal Society of Chemistry, Cambridge.

Wattenberg, L. W. (1992). Inhibition of carcinogenesis by minor dietary constituents. *Cancer Res.* (suppl.) **52**, 2085s–2091s.

Winter, C. K. and Francis, F. J. (1997). Assessing, managing, and communicating chemical food risks. *Food Technol.* **51**, 85–92.

Yancik, R. (1997). Cancer burden in the aged. An epidemiological and demographic overview. *Cancer* **80**, 1273–1283.

Prevention

Microbiology of Food Preservation and Sanitation

22

Maha N. Hajmeer and Dean O. Cliver

I. Introduction

In the hunter–gatherer stage of human social evolution, people were probably hungry much of the time. When food was found, whether a ripe fruit or a dying animal, it was likely to be eaten on the spot, thus obviating preservation. However, at least some of the time, even at that early stage, there were surely food surpluses (e.g. a whole tree, or several, loaded with ripe fruit, or a dying mammoth). Then the problems became: how long dare we stay here, how much food can we carry, and how long will it stay fit to eat? Surely some attempts were made to *preserve* food.

Ever since agriculture was invented thousands of years ago, there has been a race to feed the earth's growing human population. Even in normal times, production of most foods is seasonal, so that preservation is needed to keep us alive from one harvest until the next. Sometimes production failed, owing to drought, war, or other disaster, and many people starved. At other times there was surplus ('bumper crop', 'seven fat years'), and the challenge was to keep that surplus against the inevitable future shortages.

Domestic food animals offered an attractive alternative, in that they could be fed with materials that were not nutritious for humans (e.g. grass), might offer sustained

Foodborne Diseases 2nd Edn
ISBN: 0-12-176559-8

food production (eggs, milk, blood), could travel with the family if relocation was needed (perhaps even carrying household goods), and could be killed and eaten for any reason, at any season of the year. Food of animal origin is generally more highly valued than vegetable matter, so preservation of animal foods has been attempted, with varying success, from those times until today.

Most available preservation methods were devised long before microorganisms were discovered, so the techniques were devised on a 'trial and error' basis. It was clear that stored food needed to be protected from competing species and perhaps from other, competing humans. When microorganisms were discovered to be agents of food spoilage and foodborne disease, food preservation added a sanitation dimension. Now that our level of microbiological knowledge is relatively high, we still must confront the other competitors, in addition to microbes.

II. Food Preservation – Why and How?

The food supply is never constant: seasonal variations happen every year, and year-to-year variations are sometimes large. Humans have been challenged throughout time with the need to preserve food from harvest to harvest, and an accumulated food surplus was a reason not to migrate with the seasons, as many animals do.

The value of food depends on its availability at the time and place that the consumer wants it. All food enterprise is driven by the need to add value, the storage and distribution of food add value, and someone typically is paid for performing the necessary services.

Preservation is keeping food for human consumption by denying it to competitors such as:

- Bacteria and fungi that may grow in the food and alter it so that it is no longer edible or is dangerous to health.
- Insects, rodents, and birds that attack stored food and eat it before the people who gathered or grew it are able to use it.
- Animals larger or more dangerous than humans, which may, in certain circumstances, help themselves to stored food.
- Other people who are either more powerful or hungrier than those whose food it is.
- The internal combustion engine, when using ethanol produced from corn or other edibles.

Preservation also entails delaying or preventing intrinsic deterioration processes in food, such as over-ripening or undesired dehydration.

The focus of this section will be microbiological preservation, from the standpoints of both quality and safety. Most preservation technology has been developed originally to sustain quality, but the importance of the safety impact is increasingly recognized. Preservation methods include physical, chemical, and biological measures.

A. Physical Processes

1. Heating

Cooking is generally a final preparation step, but may be done in processing 'ready-to-eat' foods. Cooking processes often involve boiling water or direct application of flame. Cooking with oil or fat allows application of temperatures above 100°C. Although most thermal processes apply heat to the outside of the food, microwave ovens induce heat inside the food.

Baking is usually done with dry heat at oven temperatures well above the boiling point of water (e.g. 350°F ≈ 175°C). It is important, in assessing the antimicrobial effects of baking, to remember that foods are typically high in moisture, so that evaporative cooling keeps the internal temperature of food <100°C, regardless of the oven temperature.

Processes at temperatures below the boiling point of water are usually used commercially, rather than in the home. *Blanching* prepares food for further processing by inactivating enzymes in foods that are to be frozen or by releasing gases from foods that are to be packed into rigid containers for retort processing. *Pasteurization* is specifically applied to kill microorganisms, including pathogens; spore-forming bacteria generally survive. Pasteurization of milk was originally intended to kill any foreseeable level of *Mycobacterium bovis*, the agent of bovine tuberculosis. Although the low-temperature, long-time pasteurization specifications are still based on *M. bovis*, the high-temperature, short-time pasteurization temperature was raised because *Coxiella burnetii* (the agent of Q-fever, sometimes present in milk) was found to be slightly more heat resistant.

Thermal processes are generally designed on the basis of '*D*-values'. For a given pathogen, in a given food at a selected temperature, the rate of death is generally linear when the logarithm of the concentration of viable organisms is plotted against the time of exposure. The time required to kill 90% of the organisms is the *D*-value (for decimal reduction). If *D*-values for this pathogen are determined in the same food at various temperatures, it is possible to compute a '*z*-value', which is the change in temperature required to change the *D*-value by a factor of 10 (see chapter 17).

Various deviations from linear relationships of both *D*- and *z*-values are likely to occur in practice, so it is important that inoculated pack studies be conducted. The pathogen of concern is inoculated, at reasonable levels, into the food product and heated for the projected kill time at the planned temperature, to establish that killing really occurs as predicted. It is important in such studies to ensure that the laboratory-scale trial accurately represents what will be done commercially; additionally, it is common practice to use a 'cocktail' of five different strains of the pathogen as inoculum, so that results are representative of the pathogen in general, rather than just one laboratory strain.

Bacterial spores, particularly those of *Clostridium botulinum*, must be heated to temperatures above the boiling point of water if total destruction is to be achieved without greatly damaging the food. For foods that have a pH ≥4.6 ('low-acid' foods), a thermal process sufficient to kill 10^{12} spores of *C. botulinum* is required in the US. A *retort* is a device by which pressure is applied so that the boiling point of water is raised, commonly to 121°C (250°F). A properly conducted process at this temperature results in 'commercial sterility', which is the absence of any viable organisms that would express themselves under the conditions of storage and distribution of that product. The commercially sterile product is then generally stored and distributed in a 'hermetically sealed' container (if it was not already in such a container at the time of thermal processing), so as to prevent recontamination. A hermetically sealed container (can, jar, flexible pouch) is one that prevents any interaction between the product inside and the outside environment, except for heat exchange.

2. Cooling

Low temperatures retard or stop physical, chemical, and biological processes. *Chilling* or *refrigeration* reduces product temperatures to delay spoilage and prevent the growth of pathogens. Refrigeration temperatures are generally in the 0–5°C range, but some products cannot be kept this cold. Some spoilage organisms, both bacteria and molds, can grow slowly at refrigeration temperatures, so spoilage is delayed but not prevented. There are also some pathogens, of which *Listeria monocytogenes* and *Yersinia enterocolitica* are the most infamous, that can grow at refrigeration temperatures; the optimal growth temperatures of these bacteria are much higher, so cooling delays their growth, though it cannot prevent growth entirely.

Refrigeration was originally done with ice or snow, but mechanical refrigeration is now almost universally used. Ice continues to be a contact refrigerant for beverages and a few other foods (e.g. fish) but may present important sanitation problems.

Freezing converts water to the solid state, which halts biological processes. However, some preformed enzymes in foods may continue to function and cause textural and other deterioration, which is why blanching is often used to destroy these enzymes. Some microorganisms are killed by the freezing process, but those that survive may be preserved indefinitely. These organisms are quiescent in the frozen state. Protozoa and parasites are killed, if the temperature is low enough for a long enough time. The very low temperatures (−20°C or less) at which most foods are supposed to be stored are to preserve quality, rather than for safety.

3. Drying

Ripe grains are generally dry enough to store without processing. However, it can be very difficult to keep stored grains adequately dry in some climates. Other food products are often dried for preservation, by any of several processes. Some foods (e.g. raisins) are simply spread in the sun; others are heated or sprayed into hot air to remove water. Some of the most elegant drying processes involve the sublimation of water from the frozen state directly to the gaseous state. Forerunners of freeze-drying processes were used to dry meat in the Alps (viande séché) and to dry potatoes in the Andes (chuño). An alternative to removal of water for preservation is to add solute (e.g. salt cod) to bind the water so that microorganisms cannot use it. Honey is, in a sense, flower nectar from which water has been removed by bees, to the degree that the concentration of sugar prevents microbial spoilage. The stability of products, as influenced by low water content or high levels of solute, is predicted on the basis of water activity (a_w), which is defined as the water vapor pressure of a food relative to the vapor pressure of pure water, with a range of 0 to 1.

4. Irradiation

Irradiation processes apply electromagnetic energy to goods. Three common types will be discussed.

As mentioned above, *microwave* ovens provide a means of inducing heating in food, both for commercial processing and in home food preparation. The process is apparently quite safe, except that heating is sometimes uneven, whereby some portions of the food mass may not be adequately cooked and may still contain viable pathogens. For some reason, the microwave process was never subjected to the intensive safety scrutiny that has been applied to ionizing radiation.

Ultraviolet radiation has become an important means of decontaminating food surfaces and of disinfecting water. Natural surface disinfection of foods by sunlight has been demonstrated, but is not reliable enough to be practical unless an effort is made to expose all surfaces of the food to the sun and the intensity of the sunlight is high. Only the shorter wavelengths of ultraviolet light are effective against microorganisms; these lack penetrating power, so the microorganisms must be quite superficially located on a solid food. In the case of water, multiple ultraviolet sources are used, so that the path of the light through the water is short; the water must also be very low in turbidity for the process to be effective.

Ionizing radiation is able to penetrate into foods, although not all forms of radiation do so equally well. Permitted sources for food irradiation include cobalt-60, cesium-137, electron accelerators, and the last directed against a target that converts the β-particle (electron) bombardment into x-rays. In every case, the selection is based on low enough energy levels that the food does not become radioactive. Low-dose (<1 kGy) applications include sprouting control in onions, garlic, and potatoes; killing or sterilization of insects; and killing of *Trichinella spiralis* larvae. Protozoa are probably also killed, including tissue cysts of *Toxoplasma gondii*. Medium doses ($1–10$ kGy) perform the functions of pasteurization described above, without heating the food; most vegetative bacterial pathogens are killed, although bacterial endospores and viruses require higher doses than these to be inactivated. High doses (>10 kGy) are required to achieve commercial sterilization, as has been done with food for the US astronauts throughout the space program. The treatments must inactivate any bacterial spores or viruses that may be present. Only in the high dose range is destruction of nutrients a concern: conservation of nutrients and of flavor is accomplished by performing the irradiation at $\leqslant-30°C$, in the absence of oxygen.

B. Chemical Treatments

1. Acidification

Foods tend to have a neutral to acid pH, and bacterial action often lowers the pH even further. Many fruits are intrinsically acidic (e.g. apples contain malic acid and oranges contain citric acid). Other organic acids (e.g. acetic, lactic, propionic) are usually products of bacterial metabolism and, in turn, often exert strong antibacterial effects. Possibly the human development of a taste preference for acidity, which might be imparted by condiments such as vinegar or wine, was a step in the direction of safer food. The usual measure of acidity is pH, but organic acids are much more effective against most bacteria than are mineral acids such as hydrochloric, sulfuric, and phosphoric. This is related to the limited ionization of organic acids, whereby the acid molecules are better able to pass the plasma membranes of bacteria and take their effect internally. However, the molar level of an organic acid required to lower pH to some level (say, pH 4) is much greater than that of a mineral acid, so there is typically much more of the organic acid present at a given measured acidity in a food. A further factor is that many molds can use organic acids as substrate; as the mold grows, the level of acid is reduced, and the pH rises. Mineral acids, as far as is known, are quite capable of imposing a pH below 4.6 to prevent the growth of *Clostridium botulinum* in a food.

2. Enzyme treatments

Specific enzymes may be used to modify properties of food raw materials. A crude enzyme preparation, rennet (from calves' stomachs until a biosynthetic version was developed), has long been used in milk for making many types of cheese. Enzyme treatments may also affect the suitability of the food as substrate for microbial growth. For example, enzymatic digestion of starch may afford a source of oligosaccharides that bacteria can use as carbon sources. Enzymes that produce effects that discourage microbial growth are probably less common.

3. Antimicrobial additives

As contrasted to salt, which principally lowers water activity to inhibit microbial growth, some additives exert specific antimicrobial effects. Sulfites have apparently been used to preserve wine, and perhaps other fruit products, since antiquity. Curing meats by addition of nitrate (now nitrite is used directly, rather than leaving the nitrate-to-nitrite reduction to the resident microflora) has been found highly effective in preventing growth of *Clostridium botulinum*, which otherwise had such a bad reputation in the sausage context that the name *botulinum* derives from the Latin for sausage, *botulus*. Antimicrobial agents produced by microorganisms include bacteriocins, which are produced by Gram-positive bacteria and act against other, closely related species. Nisin is the only bacteriocin presently approved for addition to food in the US.

C. Biological Processes

Biological preservation of food has been practiced since prehistory. Some of these processes require a degree of control to succeed: sauerkraut production requires the addition of salt to the shredded cabbage leaves and the imposition of anaerobic conditions during the lactic acid fermentation. Most fermented milk products probably resulted first from spontaneous, uncontrolled microbial action, but specific conditions that promoted the desired result were gradually defined.

A process is often more predictable when a defined microbial inoculum is used to begin the biological process. For example, highly selected starter cultures are used in producing many cheeses, fermented sausages, beer, and bread. An old alternative is inoculation of a new batch with material saved from the preceding batch, a process sometimes called 'back-slopping'. This is simple, but there are sometimes failures due to bacteriophage action or overgrowth of undesirable organisms, and some of these failures have led to foodborne illnesses. In addition to primary inoculation in making cheese, some cheeses are later inoculated on their surfaces or more deeply with molds that further determine the character of the final product. These domesticated molds are probably a significant deterrent to invasion by other molds. The specific fermentation that produces Swiss (Emmenthaler) cheese includes production of propionic acid, which also combats mold spoilage.

D. Summary

Preservation, which includes many means to keep food safe and fit to eat, has been practiced for a very long time. Physical processes, especially drying and heating, are probably the oldest of these methods. Precooking, chemical preservation, and irradiation are some of the preservation procedures of the future. Controlled and defined biological processes are an important part of the art of food science, but may also present some safety hazards if a science base is not established.

III. Sources of Foodborne Disease Agents

Eating seldom leads to disease. Even under hygienic conditions that are far from ideal, most food is harmless. On the other hand, some foods are intrinsically hazardous and are not safe to eat until they have been processed, as in the case of the cassava varieties that must be leached efficiently and cooked in order not to cause cyanide poisoning on ingestion. During the hunter–gatherer phase of human social evolution, it seems certain that people ate the meat of dead or dying animals, as well as partially decomposed vegetable matter, because that was the alternative to starvation on occasion. Very likely, some illnesses and deaths occurred in such circumstances.

Most materials that are potentially food are not intrinsically hazardous, so one must suppose that disease agents have been introduced in some way. Whether the food *vehicle* by which an illness is transmitted is animal or vegetable, the pathogen may be of human or animal origin, or may be from the environment. Agents surely of animal origin are called *zoonoses*; there is no comparable name for agents that are surely of human origin, and many specific pathogens may have been produced in the body of a human or other animal, depending on the specific occasion. What will be called 'environmental pathogens' here are those whose reservoir is not humans or other animals. The basis of classification, then, is where the pathogen was produced, rather than the proximate source of food contamination.

A. Human-specific Agents

Viruses are said to be the leading cause of foodborne disease in the US, and they are certainly transmitted via food and water throughout the world. Almost without exception, viruses transmissible in this way infect only humans, so every outbreak represents contamination of food with material, usually fecal, of human origin. Other human-specific pathogens that are sometimes transmitted via food are bacteria (*Shigella* spp., probably *Vibrio cholerae*) and the protozoon, *Entamoeba histolytica*.

B. Zoonoses

Animals are known reservoirs of many foodborne pathogens. An extreme case is the beef tapeworm, *Taenia saginata*, which is transmissible only back-and-forth between cattle (and related species) and humans. The tapeworm stage occurs in the intestines of humans, who are called the definitive host because this is the sexual reproduction stage of the parasite. Eggs produced by the tapeworm in the human intestines are shed in feces. If the feces are ingested by cattle, they progress through stages that culminate in the production of larval inclusions in the beef muscle. If a human eats this infested beef without proper cooking or freezing, a tapeworm is likely to result, assuming that the new host does not already have a tapeworm. *Trichinella spiralis* infects any animal that eats raw meat, including bears, cougars, and rats. Swine have traditionally been the most important vehicle of human trichinellosis, but infections from swine are now rare in the US. Humans could also be infected by ingestion of raw flesh of other, infected humans, but this evidently does not happen.

Many other pathogens transmitted via foods are essentially always of animal origin and not transmitted from person to person, via food or even via direct contact. Historically, *Brucella* spp. and *Mycobacterium bovis* were important agents of serious disease in humans, transmitted via raw milk and occasionally other animal products. Other, more recently recognized bacterial species transmitted to humans via food are also usually produced in infected animals: *Campylobacter* spp., enterohemorrhagic *Escherichia coli*, *Salmonella enterica*, *Yersinia enterocolitica*, etc.

C. Mixed Animal Sources

Some of the bacteria named above (e.g. enterohemorrhagic *E. coli* and *S. enterica*) can be efficiently transmitted from person to person, whereas others (e.g. *Campylobacter* spp.) seldom are. Although secondary infections from foodborne enterohemorrhagic *E. coli* are well known, people infected with this agent are seldom a source of food contamination in outbreaks. Some *Salmonella* serovars are known to contaminate food from the human intestines – the difference seems to lie in how the agents establish themselves in the human digestive tract. A pathogenic bacterium that causes foodborne intoxications, rather than infections, is *Staphylococcus aureus*. It may come from cattle, in which it causes mastitis and is shed in milk, but also from humans who harbor it in skin lesions or in their nasal passages. In addition to bacteria that may be shed either in human or animal feces that contaminate food and water, the protozoa *Cryptosporidium parvum* and *Giardia lamblia* (also called *G. duodenalis* and *G. intestinalis*) may come both from humans or other animals. A strain of *C. parvum* that is human-specific has been identified; it is not known how prevalent this is as a cause of human disease, as compared to the strains that can alternate host species.

A special case is the pork tapeworm, *Taenia solium*. Humans are the definitive host, in which the tapeworm develops as a result of ingesting pork that contains cysticerci. The eggs shed in human feces may infect swine, but they are also infectious if ingested by humans. Autoinfection, from ingestion of the tapeworm host's feces on his own unwashed hands, occurs. People carrying the tapeworm are also capable of applying their feces, via unwashed hands, to foods eaten by others; this makes cysticercosis a foodborne disease. Human cysticercosis often involves the brain and is a much more severe condition than harboring a tapeworm.

D. Environmental Pathogens

Important foodborne disease agents may derive from the environment, having no association with fecal contamination. For example, proteolytic strains of *Clostridium botulinum* are common in soil, whereas nonproteolytic strains are often found in aqueous environments. Other spore-forming bacterial pathogens, including *Bacillus cereus* and *Clostridium perfringens*, often occur in soil. *Vibrio parahaemolyticus* and *V. vulnificus* are common in marine habitats (warm waters, or bottom sediments under cooler waters), and *V. cholerae* may also colonize water. Other pathogens that may occur in the environment include *Listeria monocytogenes*, *Staphylococcus aureus*, and various toxigenic molds and algae. It is important to recognize that these agents occur naturally wherever they happen to be, and are generally not present as a result of human-source contamination. Some pathogens (e.g. enterohemorrhagic *E. coli* and *Salmonella enterica*) are said to colonize farm environments, and others (e.g. *L. monocytogenes* and *S. enterica*) can colonize food-processing establishments and contaminate product as it passes through the facility. These last, at least, would not be regarded as *natural* contamination, but that does not mean they are easy to combat.

IV. Microbial Ecology of Foods

To have effective control measures over spoilage and pathogenic agents (bacteria or molds), it is important that one understands the ecology of these microorganisms in the system being evaluated. All living things, including spoilage and pathogenic microorganisms, interact with the environment they inhabit and are influenced by their surroundings. In order for these agents to cause food spoilage or foodborne illness a number of factors have to occur at the same time. For example, the microorganism has to be in or on the food, and the food as well as the environment must support growth and multiplication of the microbe. The food matrix, nature, and relative amounts of its components (e.g. water, carbohydrates, proteins, lipids, vitamins, enzymes, minerals, and additives) have an effect on the food's stability during processing, storage or preparation, and eventually on the fate of microbes in the system.

Spoilage and pathogenic microorganisms (usually bacteria, but also molds) can grow in or on a wide spectrum of foods, raw and cooked, such as meat, poultry, seafood, milk and milk products, eggs, and rice. Bacteria can also grow on food contact surfaces. Under favorable conditions, the microbes multiply, leading to food spoilage, owing to the production of undesirable metabolites. The food also may become health threatening if disease-causing microorganisms reach a dangerous level or if toxins are produced by these microorganisms. Many pathogenic bacteria, in particular, have to multiply and reach a high enough number in or on food before they can pose an infectious threat or cause illness. In other cases (e.g. growth of *Staphylococcus aureus*), high numbers of the pathogen are needed before toxins are produced. Presence of toxin in a food need not be accompanied by signs of spoilage. Under unfavorable conditions or in a stressful ecosystem, bacteria might be dying, or injured and adapting to the environment.

As mentioned earlier, many factors affect microbial growth in foods. The most important ones can be classified into two categories: intrinsic and extrinsic. Intrinsic factors are properties inherent to the food, including pH, moisture content, nutrient content, antimicrobial agents, biological structures, and redox (oxidation–reduction) potential. Extrinsic factors are properties or conditions pertaining to the environment in which the food and the microorganism occur. Unlike the intrinsic factors, extrinsic variables derive from the environment and can be manipulated. Such variables include time, temperature, relative humidity, presence of gases, and physical stress. Intrinsic and extrinsic factors in a system interact to provide environments for microbial growth, survival, or death. The following is a brief description of some intrinsic and extrinsic factors.

pH Microorganisms grow at a wide range of pH, from 1.5 to 11.0. The presence of microorganisms and the extent of their growth is affected by the pH of the system. Foods tend to have a pH ≤ 7, except egg whites, which are mildly alkaline. Bacteria grow in the pH range of 4.4–9.0; the majority grow well in basic conditions. Yeasts and molds typically grow in acidic conditions (pH ranges from 1.5 to 8.5 for yeasts and 1.5 to 11.0 for molds). Therefore, yeasts and molds will probably spoil strongly acidic foods (e.g. tomatoes and citrus fruits), and bacteria will spoil more nearly neutral foods (e.g. meats and vegetables). Some foods have a buffering capacity or ability to resist change in pH. Such buffering capabilities assist in minimizing food deterioration. Meats are a good example of foods with a buffering capacity, which resides largely in their proteins. An important pH value in food safety is 4.6, the lower limit for growth of *Clostridium botulinum*.

Moisture content/water activity (a_w) Water in a food system can be in two forms: free and bound. The free form (in a food system) is available for microbial metabolic activity. Microorganisms cannot use bound water because it is tied up with solutes in the system and unavailable. Moisture content and a_w are two parameters that indicate availability of free water in the system with higher values of both indicating higher availability of water. Moisture content reflects the free water in the system and ranges from 0% to 100%. The idea behind a_w is similar, but it is defined as the water vapor pressure of a food relative to the vapor pressure of pure water, and it ranges from 0 to 1. Any change in the water activity of a food system is an indication of changes occurring in a product. Determination of a_w can indicate how suitable conditions are for microbial growth and the kind of microbes expected to grow.

Nutrient content   Bacteria and molds require certain nutrients to support their growth. There must at least be a source of carbon and a source of nitrogen. Nutrient content of the food may support or hinder growth of microorganisms, depending on what the microbes require and what is available. Certain solutes (e.g. sugars or salts) might bind water in the system and render it unavailable for microbial metabolic activities.

Antimicrobial agents   Certain foods naturally contain antimicrobial agents. Examples include lactenin in milk, which inhibits growth of Gram-positive bacteria, and benzoic acid in cranberries, which prevents mold growth. These natural antimicrobials are useful, but they cannot preserve food indefinitely, especially if the food has been abused.

Physical/biological structures   These are structures inherent to the food itself that maintain the physical integrity of the food. Examples of such structures include nut shells and animal hide. These structures (as long as they are intact) render conditions unfavorable for microorganisms to invade and colonize the food.

Redox or oxidation/reduction potential   Redox or E_h potential presents a basis for classification of microorganisms based on their affinity for oxygen or lack of oxygen. Molds and some bacteria require oxygen, whereas other bacteria require the absence of oxygen. Some bacteria and yeasts can grow either under aerobic (oxygen available) or anaerobic (oxygen not available) conditions and thus are facultative anaerobes; others are microaerophiles, which prefer low oxygen levels. An $E_h > 0\,mV$ (i.e. positive) reflects aerobic conditions for growth.

Temperature   Most bacteria and fungi grow well at ambient or room temperature ($\sim 25°C$), but they all have temperature preferences. Microbes are classified in terms of their need and tolerance of environmental temperatures into four categories: thermophiles, mesophiles, psychrophiles, and psychrotrophs. Table 22.1 summarizes the temperature ranges defining these categories and gives some examples. Thermophiles live at high temperatures, and mesophiles live at warm temperatures, close to the temperature of the human body. Psychrophiles must grow in cold temperatures, whereas psychrotrophs grow at refrigerator temperatures ($\leqslant 10°C$), but at a slower rate than at their optimum. Pathogenic bacteria often fall into the mesophile or psychrotroph categories. Although thermophiles are usually members of the genera *Bacillus* and *Clostridium*, pathogens in these genera are mesophiles or psychrotrophs. To prevent spoilage, and growth of most bacterial pathogens, food needs to be kept out of the 'danger zone' (i.e. temperatures from 4°C to 60°C), a range that allows microbial proliferation.

Table 22.1 Classification of microorganisms based on the temperatures at which they will grow

Category	Temperature range (°C)	Optimum temperature (°C)	Microbial example
Thermophiles	50–75	65	Bacillus spp., Clostridium spp.
Mesophiles	10–45	37	Escherichia coli, yeast
Psychrophiles	0–15	8	Pseudomonas spp.
Psychrotrophs	0–35	21	Listeria spp., Yersinia spp., Enterococcus, yeast

V. HACCP

HACCP is an acronym for the *hazard analysis* and *critical control point* system. It is a system to prevent food contamination and assure safe production and processing of foods. HACCP is based on science and common sense; its objective is provision of safe food to consumers. A proper HACCP system at a food-processing or preparation facility is intended to prevent, eliminate, or reduce the likelihood of a biological, chemical, or physical hazard occurring in the food, which might eventually threaten the health of consumers. HACCP has economic advantages, including reducing costs from production of unsafe food by decreasing potential shutdowns, recalls, medical claims, and lawsuits. It is important to emphasize that HACCP concerns food safety, not quality; HACCP is often discussed as a means of quality enhancement, which is simply incorrect. There are several programs or systems specifically designed to address quality issues, including quality assurance programs and the International Standards Organization (ISO) 9000 series system. The ISO 9000 system is intended to provide common standards of quality during production or manufacturing of products, to assure that two or more trading partners (nationally or internationally) agree on the quality of the product.

The HACCP concept originated in 1959 as a plan developed for the US National Aeronautics and Space Administration (NASA) by the Pillsbury Company. The objective then was to assure safe foods for the US space program. Later, from the 1960s–1980s, HACCP was voluntarily adopted by a number of food companies in the US as a measure to assure safer food. After the 1993 outbreak of hemorrhagic colitis from *E. coli* O157:H7 in fast food, US government agencies proposed stricter food safety measures. As a result, a 1996 rule from the US Department of Agriculture (USDA) required HACCP in all meat and poultry plants to follow the recommendation of the National Advisory Committee on Microbiological Criteria for Foods. Information on this ruling can be found in the Pathogen Reduction Final Rule (also known as the MegaReg) published in the US Federal Register. For meat and poultry plants, the rule set deadlines for implementation of HACCP: January 26, 1998 for large ($\geq$500 employees) meat and poultry plants; January 25, 1999 for smaller ($\geq$10 and $\leq$500 employees) plants; and January 25, 2000 very small ($<$10 employees or with annual sales $\leq$\$2.5 million) plants. HACCP has also been mandated by the US Food and Drug Administration (FDA) for seafoods and juices, which are in its domain.

The HACCP system is based on control of the process, rather than of the final food product. It begins by defining the product and its intended consumers, then gathering information about food processing or preparation, and conducting a hazard analysis (HA). The HA identifies what may go wrong in a process, and how, and where, and what needs to be done. HACCP is designed to identify the hazards (biological, chemical, or physical) associated with the food during production or preparation, and also to identify the critical control point(s) along the line. Examples of biological hazards include bacteria, viruses, molds, protozoa, and parasites. Sanitizers, antibiotics, flavoring, paints, and solvents are some examples of chemical hazards; and metal, glass, and bone are physical hazards. When devising a HACCP plan, there are seven principles that need to be applied. This is, mainly, the responsibility of the HACCP team (i.e. lead individuals in developing the HACCP plan at a facility). The HACCP team should comprise people who work in the food plant in a variety of roles. The seven principles are summarized below.

Principle #1: Hazard analysis   The hazard analysis accomplishes three purposes: (1) the hazards of significance are identified; (2) the likely hazards are selected; and (3) the identified hazards can be used for developing preventive measures. As mentioned earlier, hazards can be biological, chemical, or physical in nature; the potential risk of each hazard is assessed based on its likelihood of occurrence and its severity. Hazard assessment is based on a combination of experience, epidemiological data, and information in the technical literature.

Principle #2: Identify the critical control points   A critical control point (CCP) is a point, step, or procedure at which control can be applied and a food safety hazard can be prevented, eliminated, or reduced to acceptable levels. It is important to identify potential CCP(s) in food preparation. CCPs can be cooking, chilling, sanitation procedures, product formulation control (pH, salt, water activity), and prevention of cross-contamination. Different facilities preparing the same food may differ in the risk of hazards depending on the operation. A CCP *decision tree* (Fig. 22.1) is helpful in assigning CCPs.

Principle #3: Establish critical limits for preventive measures associated with each CCP   Critical limits are the boundaries for safety at each CCP and may be limits with respect to temperature, time, meat patty thickness, water activity, pH, available chlorine, etc. Critical limits may be derived from regulatory standards or guidelines, literature, experiments, and expert opinion.

Principle #4: Establish procedures to monitor CCPs   Monitoring is a planned sequence of observations and measurements to assess whether a CCP is under control and to produce an accurate record. This record can be used in case of complaints about the product, and is also used in the verification of HACCP. The measurements for monitoring are visual observations, temperature, time, pH, water activity, etc. The measurements must be done 'on-line'; there is no time to wait for lengthy laboratory tests. There must be written specification of who has the responsibility for monitoring, and each observation must be recorded in writing.

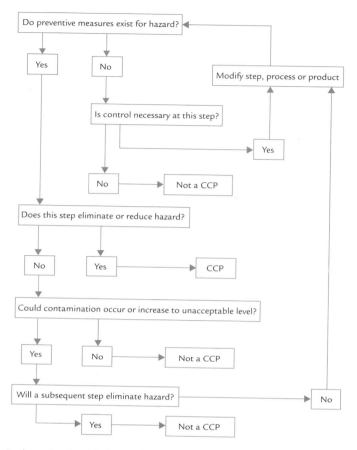

Figure 22.1 A schematic of a critical control point (CCP) decision tree.

Principle #5: Establish corrective actions Corrective actions are taken to gain control of the process when monitoring shows that a deviation has occurred and a critical limit has been exceeded. There must be written instructions for actions to be taken to correct the process and what to do with the product (reprocess, condemn, etc.) when critical limits have been exceeded. Additionally, the instructions should indicate who has the authority for the action. Sometimes regulatory agencies must be consulted. Actions taken must be recorded in writing.

Principle #6: Establish record keeping system This system is established to document the HACCP plan, including monitoring of CCPs and recording corrective actions taken when needed. This is necessary for internal audits and for verification of the HACCP system, sometimes by third parties. It is also important in case of a consumer complaint.

Principle #7: Establish verification procedures Verification procedures indicate whether the HACCP system in place is working properly. Verification is based on the HACCP documentation, and may include internal audits and/or verification done by a third party (e.g. outside consultant). Additionally, verification may include validation studies (e.g. laboratory testing of samples of food and/or the environment).

For a HACCP system to be implemented successfully in food processing, all of the seven principles must be followed and tailored for the specific process; the HACCP plan should be reassessed at least annually. Additionally, HACCP is not a stand-alone system. Other systems or procedures, such as standard operating procedures (SOPs), sanitation standard operating procedures (SSOPs), good manufacturing practices (GMPs), good agricultural practices (GAPs), recall procedures, and information on the establishment's prior recalls and customer complaints that are related to food safety are prerequisites of HACCP. These systems assist in provision of basic operating conditions that are necessary for the preparation of safe food. Application of HACCP in phases of the food system other than processing (e.g. farming, food service, retailing) can prove particularly challenging.

VI. Predictive Microbiology

As we strive to control microbial growth to our advantage, predictive modeling and predictive microbiology offer objective means to foresee the effects of actions on microbial growth, shelf-life, and safety of food products. Predictive microbiology can reduce the amount of laboratory testing required for these purposes. Also, predictive microbiology can aid HACCP implementation by assisting in identifying and evaluating microbiological hazards, identifying critical control points, establishing critical limits and monitoring methods, and defining corrective actions. Predictive modeling allows the identified hazard to be assessed, as affected by a variety of situations. Microbial predictive modeling uses mathematical equations to foretell the fate of microorganisms, particularly bacteria, under certain environmental conditions. These models or mathematical equations are developed after extensive study of the behavior of the microorganism(s) of concern.

These models not only help predict microbial growth in foods; they may also help assess the sources and likelihood of contamination. Predictions can help pinpoint conditions that should be avoided. Such information can later be applied to monitoring systems that give processors advance warning of problems. Predictive models can be applied in foreseeing effects of possible processing and handling measures, product formulation, optimizing conditions for growth of desirable microbes (e.g. starter cultures), process optimization and control, designing shelf-life studies, validation of regulations, and planning of further laboratory trials. However, it should be noted that predictive models have limitations and cannot serve as the entire solution of a specific food safety problem.

A. Growth Curves and Models

As discussed earlier, the growth of specific bacteria in foods is affected by factors such as temperature, pH, water activity, and oxygenation. Understanding the effects

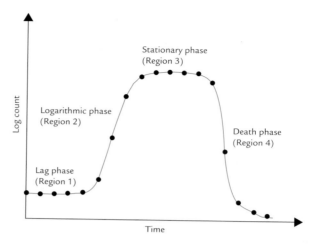

Figure 22.2 Generalized schematic of a typical growth curve under constant, initially favorable conditions, showing the four phases.

and interactions of such variables allows prediction of microbial growth and survival, identification of key control factors, and minimizing food spoilage and potential health risks. Under favorable conditions bacteria will multiply; plotting their numbers graphically over time yields a 'growth curve'. Changes in bacterial numbers with time can be compared to show the influence of varying environmental conditions, so growth curves can be used to predict the microbial safety or shelf-life of food products. As shown in Fig. 22.2, a typical bacterial growth curve comprises four phases: lag, log (exponential), stationary, and death (decay). A number of analytical growth models are found in the literature such as Gompertz, Richards, logistic, etc. Normally, one of the well-known analytical models is fitted to growth data and the specific growth parameters relevant to that model are calculated. Data do not always agree well with the fitted model, leading to misinterpretation. The microbiologist can only choose the most useful model by trial and error because different models use different parameters and so cannot be compared directly. The modified Gompertz equation (equation 22.1) is regarded as one of the best analytical tools.

$$\text{Log } N = A + D \exp\{-\exp[-B(t - M)]\} \tag{22.1}$$

where N is the number of bacteria; A, B, D, and M are empirical constants; and t is time (see Fig. 22.3). A is the value of the lower asymptote or initial number [i.e. log $N_{(-\infty)}$], D is the difference in value between the lower and upper asymptotes [i.e. log $N_{(\infty)}$ − log $N_{(-\infty)}$], M is the time at which exponential growth rate is maximum, and B is related to the slope of the curve at M.

One can see in Fig. 22.4 whether bacterial numbers are increasing or decreasing. Such curves depict growth and survival, aiding control of desirable and undesirable bacteria in food systems. Growth curves can help interpret trends observed in processing operations and assist in improving overall process effectiveness and in risk assessment.

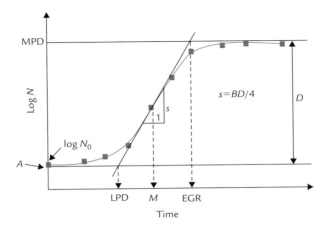

Figure 22.3 Schematic showing empirical constants of Gompertz equation, and growth kinetic parameters.

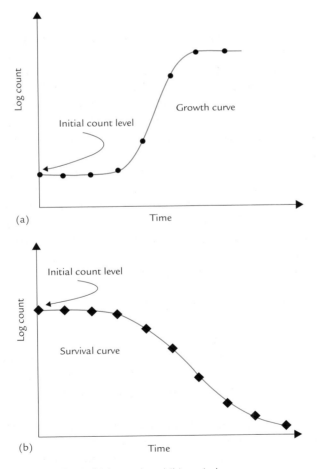

Figure 22.4 Schematic of typical (a) growth and (b) survival curves.

A characteristic curve (Fig. 22.4a) results when the environment favors bacterial growth; it comprises the four phases named previously and shown in Fig. 22.2:

1. The lag phase (region 1), in which the cells are adjusting their physiology and biochemistry to exploit the environment in which they find themselves.
2. The exponential (log) phase (region 2), in which the cells grow in their environment as rapidly as possible and at a relatively constant rate.
3. The stationary phase (region 3), in which the reproduction rate is equal to the death rate. The accumulation of waste metabolites and depletion of nutrients lead to some reduction in the growth rate of the microorganisms.
4. The death phase (region 4), in which a further increase in the accumulation of toxins increases bacterial lysis, and the rate of cell death exceeds the rate at which the cells divide.

Figure 22.4b shows a survival curve under conditions unfavorable for microbial proliferation. Only a death phase occurs in such a case.

B. Predictive Models and Growth Kinetic Parameters

By studying the growth kinetics (parameters characterizing growth or survival curves) and response of bacteria, one can obtain information for use in predicting the microbial safety or shelf-life of food products. Growth curve experiments are usually done with bacteria in laboratory growth medium because food systems are more complex and costly. Broth systems are more homogeneous and more readily sampled than solid matrices.

Simple parameters derived from the curves can illustrate and quantify differences between curves and enable simple comparisons of outcomes under various conditions. Predictive models developed to fit experimental growth curves usually yield a number of useful kinetic parameters (see Fig. 22.3):

- Lag phase duration (LPD) – the amount of time needed by the microorganisms to adjust to the growth environment.

$$\text{LPD} = M - B^{-1}\{1 - \exp[1 - \exp(BM)]\} \tag{22.2}$$

- Exponential growth rate (EGR) – the speed at which the population doubles within the exponential phase.

$$\text{EGR} = \frac{BD}{e} \tag{22.3}$$

- Generation time (GT) – the time taken for the population within the exponential growth phase to double, also called doubling time. GT can be calculated from the maximum slope within the exponential phase (GT $= \log_{10}2/\text{slope} = 0.301/\text{slope}$).

$$\text{GT} = \frac{0.301e}{BD} \tag{22.4}$$

- Maximum population density (MPD) – the highest microbial count attained.

$$MPD = D + A \qquad\qquad (22.5)$$

VII. Processing for Safety

There is a trend in the US, and perhaps other developed countries, to prefer foods that have not been processed – often not even cooked. This seems to be based on the mistaken idea that raw foods are more nutritious than those that have been processed or cooked. In fact, cooked foods can be as nutritious as raw foods (or more), and are certainly less hazardous. The recent trend to more and more outbreaks in which fruits and vegetables are vehicles is surely due to the increase in eating these foods raw. In that most fruits and vegetables are grown in open fields where they are subject to contamination from many sources, it is almost impossible to guarantee that no pathogens are present at harvest.

Processing for safety was once a commonly recognized principle. With the advent of HACCP, processing outcomes are even more predictable, and the safety of processed food is greater than ever. Several processes (e.g. heating, freezing, irradiation) described above can contribute significantly to the safety of food by either killing pathogens or preventing their growth. These are processes that can properly be designated critical control points, with attendant critical limits and means of monitoring. However, this does not apply to such processes as grinding (comminution) of meat, which are at best neutral from a safety standpoint. It seems most unlikely that currently proposed good agricultural practices can yield food whose safety is equal to that attained by processing.

VIII. Distribution and Preparation

Food is produced on farms and ranches, far from most consumers. Whether or not the food is processed, it must be distributed to retail outlets where consumers can buy it. Often, storage for various periods is involved, so that both time value and place value are added by the distributor. These value-added services cost money, but few consumers wish to save these costs by going to the various farms and ranches from which their food comes. Further, consumers do not have the space or facilities to store several months' supply of food in their homes. The services the distributor provides are largely based on convenience: the food is no more nutritious or safe at sale than it was when the distributor received it. On the other hand, there is a great potential to reduce nutritional value or to add a hazard. Storage and distribution of food do not lend themselves to the HACCP approach, as there are seldom decisive critical control points available. All the same, those who perform these services must be highly dedicated to the maintenance of food quality and safety. Government inspection can play an important role in seeing that the food is treated properly.

Preparation of food is often done by someone other than the person who will eat it. Even within a family, cooking tends to be part of the division of labor, whereby one person does more of it than others. However, cooking skills are less prevalent in the

US, and probably other developed countries, than in earlier generations. This means that foods are more often eaten away from home than formerly, and that foods eaten at home have often been prepared, partially or entirely, elsewhere. This is not necessarily bad from either the nutritional or safety standpoint, but it does require a high level of trust in the preparer.

A significant advantage of delegating food preparation is variety. One can eat what one wants at almost any time. Most restaurants and other food-service establishments have very good safety records, and convenience foods are often even better in this respect. An important consideration is that final preparation (especially cooking) has some possibility to undo any previous contamination of the food, but final serving has the potential to add contaminants just before the food is eaten. If final preparation and service occur away from home, they may be done by highly skilled professionals – or not. The quest for the cheapest possible food leads some restaurants to hire young or unskilled workers, and offer no sick leave, so that ill workers will continue to work with food, unless discovered and sent home, because they need the money. It is hard to put a price on food safety, but it is essential to recognize that the measures described in this chapter come at a cost. Consumers who are unwilling to pay the costs of food safety should expect to experience foodborne illness occasionally. On the other hand, the drive for a zero-risk food supply often leads to imposing measures that accomplish little to improve safety, but add significantly to the price of the food. Inasmuch as the US, at least, has a large class of people who are undernourished because of food prices, it is important that costs not be added unless a scientifically proven gain in safety will result.

IX. Inspection and Testing

Modern-day food regulations tend to set great store by inspection of food and the facilities in which it is produced, handled, and stored. Inspection surely serves a useful purpose up to a point, but a great deal of inspection is based upon perceptions of *wholesomeness* (concepts of fitness-to-eat) rather than on the safety of the consumer. Greater uniformity is also desirable. The FDA Food Code is advisory, rather than required, in the US, so states and counties often have regulations that conflict with neighboring government units. One might suppose that enforcement of food storage temperature requirements, at least, would be simple. However, it is noteworthy that the FDA says eggs must be brought to temperatures $\leq 45°F$ ($7°C$) as quickly as possible and held cold, whereas the European Union, which is in many respects more fastidious than the US, has no refrigeration. Where sanitation standard operating procedures are in use, it may well be that the best role of inspectors is to ensure that the SSOP is followed.

There is also a great push to increase testing of food. Methods for bacteriological testing of food are one of the fastest growing areas of science, perhaps because consumer activists believe that food safety is enhanced by testing. On the contrary, HACCP was developed to avoid foodborne disease in astronauts because testing was already recognized as inadequate for this purpose more than 40 years ago. The superimposition of testing on the HACCP program required by the USDA indicates that this is a lesson still to be learned. The essential problem with testing for safety is that it is almost impossible to obtain a sample that represents an entire lot of food. Pathogens are hazardous at very low levels and are not uniformly distributed in food; the positive tests for *E. coli* O157:H7 and *Listeria monocytogenes* that have led to destruction of huge quantities of food in the US in recent years may or may not have prevented a significant number of human illnesses. It is clear that foodborne disease continues to occur, and that much of the food that was destroyed probably did not contain the pathogen that was detected in the laboratory sample. There is also the problem that many foods have short shelf-lives, so the time loss if the food must be held until testing is completed leads to further waste. This is the reason that monitoring of critical control points in real HACCP systems is supposed to occur 'on-line', rather than waiting for laboratory results. The delay problem has been mitigated by development of more rapid test methods, but not eliminated.

Much of the bacteriological testing required by the USDA Pathogen Reduction Program is not based on holding the product during testing. Instead, the agency has established 'performance standards' for the levels of 'generic' *E. coli* and the prevalence of *Salmonella* spp. on carcass surfaces, with results being compiled cumulatively over days or weeks. This avoids arbitrary condemnation of food, but the performance standards are based on levels observed in surveys, rather than on risk assessments. This results in allowing much more *E. coli* on swine carcasses than on beef carcasses, just because there *is* more *E. coli* on the former than on the latter. Performance standards are supposed to upgrade in-plant sanitation, on the assumption that better sanitation will result in safer food. Imposing a critical control point such as irradiation would accomplish much more, but is not yet politically acceptable. Arguably, the best general purpose for food testing is to detect pathogens in foods suspected of having been the vehicle in an outbreak. In such instances, precise identification of the pathogen is as important as the initial detection.

Identification of foodborne pathogens once served simply to incriminate suspected vehicles in outbreaks. Newer methods can detect outbreaks on the basis that people are infected by the same agent, even when they are not ill at the same place or time. There is a whole spectrum of identification methods available for this purpose, from biochemical tests of isolated agents to serological, antibiotic, and bacteriophage typing of bacteria. Methods based on specific portions of the pathogen's genome (e.g. probing, polymerase chain reaction, sequencing, pulsed-field gel electrophoresis, random amplification of polymorphic DNA, and ribotyping) have added powerful tools. These also permit association of a specific food with an outbreak, assuming that the representative food samples are available by the time the outbreak is recognized.

X. Summary

This chapter has addressed preservation and sanitation of food, principally from a microbiological standpoint. Broad principles of control of food spoilage and safety have been discussed. One finds that there are a great variety of microbes that can occur in foods and cause human disease, and a great variety of sources from which these microbes may come. Bacteria and molds may multiply in foods, whereas viruses and parasites cannot. The HACCP system of food safety improvement has been described briefly. Applications of microbial ecology and predictive microbiology, particularly via HACCP in food processing, have been discussed in general terms. If more details are desired, these can be found in the pathogen-specific chapters and in the more extensive discussions listed in the Bibliography.

Bibliography

Bauman, H. E. (1990). HACCP: concept, development, and application. *Food Technol.* **44**, 156–158.

Bernard, D. (2001). Hazard analysis and critical control point systems: use in controlling microbiological hazards. *In* 'Food Microbiology: Fundamentals and Frontiers', 2nd ed. (M. P. Doyle, L. R. Beuchat and T. J. Montville, eds), pp. 833–846. American Society for Microbiology Press, Washington, DC.

Davidson, P. M. (2001). Chemical preservatives and natural antimicrobial compounds. *In* 'Food Microbiology: Fundamentals and Frontiers', 2nd ed. (M. P. Doyle, L. R. Beuchat and T. J. Montville, eds), pp. 593–627. American Society for Microbiology Press, Washington, DC.

Farkas, J. (2001). Physical methods of food preservation. *In* 'Food Microbiology: Fundamentals and Frontiers', 2nd ed. (M. P. Doyle, L. R. Beuchat and T. J. Montville, eds), pp. 567–591. American Society for Microbiology Press, Washington, DC.

Food Marketing Institute (1998). 'Trends in the United States: Consumers, Attitudes, and Supermarkets'. Food Marketing Institute, Washington, DC.

Jay, J. M. (1992). 'Modern Food Microbiology', 4th ed. Van Nostrand Reinhold, New York.

Kiple, K. F. and Ornelas, K. C. (2000). 'The Cambridge World History of Food'. Cambridge University Press, Cambridge, UK.

McMeekin, T. A., Olley, J. N., Ross, T. and Ratkowsky, D. A. (1993). 'Predictive Microbiology: Theory and Application'. RSP, Taunton, UK.

Montville, T. J. and Matthews, K. R. (2001). Principles which influence microbial growth, survival and death in foods. *In* 'Food Microbiology: Fundamentals and Frontiers', 2nd ed. (M. P. Doyle, L. R. Beuchat, and T. J. Montville, eds), pp. 13–32. American Society for Microbiology Press, Washington, DC.

Montville, T. J., Winkowski, K. and Chikindas, M. L. (2001). *In* 'Food Microbiology: Fundamentals and Frontiers', 2nd ed. (M. P. Doyle, L. R. Beuchat and T. J. Montville, eds), pp. 629–647. American Society for Microbiology Press, Washington, DC.

Sperber, W. (1991). The modern HACCP system. *Food Technol.* **45**, 116–118, 120.

Whiting, R. C. and Buchanan, R. W. (2001). Predictive modeling and risk assessment. *In* 'Food Microbiology: Fundamentals and Frontiers', 2nd ed. (M. P. Doyle, L. R. Beuchat and T. J. Montville, eds), pp. 813–831. American Society for Microbiology Press, Washington, DC.

Organizing a Safe Food Supply System

23

Rhona S. Applebaum, Dane T. Bernard, and Virginia N. Scott

I. What is 'Safe'?

As a society, we are concerned about risks, including the risk of becoming ill from something we eat. There is no such thing as a risk-free society. For example, staying at home for a 70-year life span holds 7700 chances in a million of incurring a fatal accident. We must learn to accept the fact that we cannot make food absolutely safe any more than we can make life risk-free. There are costs involved in reducing any risk, such as making our food safer. In many instances, moderate costs can result in a significant increase in safety, but there is a point of diminishing returns beyond which the costs greatly exceed the benefits. Furthermore, the concept of 'safe' is a relative one. The US Food, Drug and Cosmetic Act (FDCA) does not define 'safe' *per se*. Safety under FDCA has been interpreted to mean 'reasonable certainty of no harm'. Considering 'absolute safety' is unachievable, this interpretation of 'safe' is appropriate for purposes of this discussion.

A. Risks in the US

Several common causes of death, or 'risks to life', in the US are shown in Table 23.1. The US Centers for Disease Control and Prevention (CDC) have estimated there are 76 million cases of foodborne illness, with 5000 deaths, in the US annually.

Table 23.1 Selected risks to life in the US[a]

Cause of death	Deaths per year	Deaths per year per 100 000 people[b]
Heart disease	725 200	266
Cancer	549 800	202
Stroke	167 400	61
Automobile accidents	42 400	15.5
Suicide	29 200	10.7
Drug use	19 100	7.0
Homicide	16 900	6.2
HIV infection	14 800	5.4
Falls	13 200	4.8
Foodborne illness (estimated)	5 000	1.8
Malnutrition	4 000	1.5
Drowning	3 500	1.3
Fires	3 300	1.2
Bicycle accidents	950	0.35
Accidental shootings	824	0.30
Pregnancy	406	0.15
Lightning	93	0.034
Bees, wasps	64	0.023
Salmonella infections	38	0.014

[a] Based on 1999 data from the National Center for Health Statistics, on information from the Centers for Disease Control and Prevention, and from *Consumer Reports* (December, 1996).
[b] Based on a population of 272.7 million.

Actual documented deaths from foodborne disease, based on statistics for foodborne disease outbreaks for the years 1993–1997, ranged from 2 to 11 per year. While it is recognized that these data are significantly under-reported (see Chapter 2) and they do not take into account sporadic cases not linked to outbreaks, the CDC estimate of 5000 deaths per year is believed by some to be high. Estimated deaths from known pathogens totaled 1809; the remaining deaths are estimated from gastrointestinal illnesses from unknown causes. For persons aged 25–44, cancer and accidents are the most prevalent causes of death, causing over 20 000 and 27 000 deaths per year, respectively. Thus, although any death associated with food cannot or should not be trivialized, relative to other risks to life, food is safe. The US food supply is one of the safest in the world. The goal of this chapter is to show what is done to ensure that the food supply presents few risks for consumers.

B. The Human Food Network

Figure 23.1 is a simplified diagram of the human food network. Food comes from growers or producers (farmers, fishermen, ranchers). Most of this food goes to some sort of processor, who transforms the food in some manner. This may be as simple as cleaning and gutting fresh fish for the wholesaler, or as complicated as an operation that takes ingredients such as meat and vegetables and prepares a frozen meal. There may be many intermediary processors; for example, the processor preparing the frozen

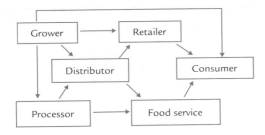

Figure 23.1 The human food network.

meal may purchase chicken that has been skinned, boned, and cooked by another processor, and combine that with frozen peas and carrots from another supplier. That supplier may have received peas that were shelled and frozen, and carrots that were diced and frozen by other processors, and so on. Some foods, such as fresh fruits and vegetables, may go directly to distribution without any processing, other than cleaning and boxing, or may even go directly to the consumer via farmers' markets and road-side stands. Distributors move foods from place to place and store foods until they are needed. They supply the retailer (e.g. grocery stores) and food service establishments (such as restaurants). Ultimately, food reaches the consumer.

C. Responsibility for Food Safety

The food network described above is greatly simplified. The production and marketing of food comprises a complex set of interactions. Included in these interactions are steps or interventions taken to ensure that the food to be eaten will be safe. Each of the above groups, from the grower to the consumer, has an important role in ensuring the safety of the food supply.

II. Producers' Role in Food Safety

When pathogenic organisms are present in foods, consumers may become sick. How do these pathogens get there? Pathogens may become associated with food through many routes, but primarily through contact of the food with soil, through animal feces, and through infected food handlers. Foods of animal origin are particularly susceptible to contamination with microbial pathogens, as in many cases the animal serves as the reservoir for the pathogen. We are beginning to develop the scientific knowledge needed to devise successful control strategies to reduce the risk of contamination of foods at the production level. Vaccines, competitive exclusion (feeding an animal microorganisms that can prevent pathogens from colonizing the gastrointestinal tract of the animal), control of pathogens in feed and water, feed supplements, and environmental controls are all being investigated for their potential in reducing pathogens in animals used for food. 'Best production practices' are being developed for production of animals, seafood produced by aquaculture, fruits, vegetables, and grains. New control strategies are also being implemented in the harvesting of fish and shellfish. Although steps can be taken to reduce pathogens in raw foods, it is unlikely that raw foods will ever be pathogen-free; we must assume that pathogens will be present with some frequency and treat food accordingly.

III. Industry's Role in Food Safety

The food industry is a diverse group, consisting of processors, transporters, distributors, retailers, food-service establishments, and their trade associations. The food industry is in business to provide a useful service and in return to be paid for that service. To be profitable, the industry must provide foods that meet the expectations of their consumers. One of these expectations is that food will be safe. In the interest of consumer good will, customer satisfaction, maintaining corporate and brand names, and avoiding a 'regulatory hazard' (see Section IV), companies strive to produce safe products to protect their assets. A lot of resources (time, equipment, money, and personnel) go into programs designed to produce safe, wholesome, and nutritious foods.

A. Programs for Food Safety

The food industry undertakes a number of programs to ensure that foods are safe and meet quality expectations. These include good manufacturing practices (GMPs), hazard analysis and critical control point (HACCP) systems, total quality management (TQM) systems, and International Organization for Standardization (ISO) 9000 quality management systems. These systems may be used individually or together by a company to meet the company's needs.

Procedures used to control conditions in a facility that contribute to the overall safety of a product are sometimes referred to as 'prerequisite programs' because they form the foundation for a HACCP program. Prerequisite programs may include pest control, cleaning and sanitation, equipment maintenance and calibration, programs for record keeping, recall procedures, and training programs. Many of these programs fall under GMPs. In the US, GMPs have a regulatory connotation, as both the US Food and Drug Administration (FDA) and the US Department of Agriculture (USDA) have established regulations that define the criteria that determine whether a food is considered adulterated under the meaning of the Federal Food Drug and Cosmetic Act, the Federal Meat Inspection Act, or the Poultry Products Inspection Act. These regulations generally have provisions regarding personnel, equipment, sanitary operations, and process controls. There are specific GMP regulations for low-acid canned foods and for acidified canned foods, in addition to general industry ('umbrella') GMPs that apply to all foods.

The HACCP system is a preventive system for assuring the safe production of food products. The focus of HACCP is the detection and prevention of hazards during production rather than relying on inspection of finished products to find and correct problems. There are seven HACCP principles:

1. Conduct a hazard analysis.
2. Determine the critical control points (CCPs).
3. Establish critical limits.
4. Establish a system to monitor control of the CCP.
5. Establish the corrective action to be taken when monitoring indicates that a particular CCP is not under control.
6. Establish procedures for verification to confirm that the HACCP system is working effectively.
7. Establish documentation concerning all procedures and records appropriate to these principles and their application.

The principles of HACCP are applicable to all segments of food production, including growing and harvesting, transportation, food processing, food preparation and handling, food service, and consumer handling and use (see Chapter 22). How HACCP is applied will differ depending on which segment of the food chain is being addressed.

TQM and ISO 9000 are quality management systems used by a number of food processors worldwide. In the past, many aspects of food safety have been handled as part of quality management systems. However, HACCP, a *safety* management system, has become the preferred system internationally for control of product safety, regardless of what 'umbrella' quality management system a company chooses to use.

B. How Processors Control Food Hazards

As noted above, food processors are turning increasingly to HACCP to control hazards (biological, chemical, or physical agents that are reasonably likely to cause an illness or injury in the absence of their control) in foods. HACCP is not a stand-alone program, as it cannot be effective without crucial prerequisite programs, including a solid foundation of GMPs, preventive maintenance programs, employee training programs, pest control programs, and the like. HACCP plans must be designed for specific products and process lines. To accomplish this, manufacturers generally assemble a multidisciplinary HACCP team, with members from production, quality control, engineering, etc., to develop the HACCP plan. The team describes the product and how it will be used, constructs a flow diagram for how the product is manufactured, and identifies all the hazards that may reasonably be expected to occur at each step from primary production, processing, and distribution to the point of consumption. The HACCP team then evaluates the identified hazards to determine which hazards are of such a nature that their elimination or reduction to acceptable levels is essential to produce a safe food. It is this evaluation step that separates hazards that warrant HACCP control from those which are of low risk and adequately controlled by prerequisite programs. The team determines what control measures must be applied at what points (CCPs) and what critical limits must be applied to ensure control. A monitoring system is established, along with specific corrective actions that must occur if critical limits are not met. Procedures are also established to verify that the HACCP plan is working correctly (record reviews, audits, sample analysis, etc.). The team will evaluate any changes in the processing of the product to determine if changes in the HACCP plan are required.

C. Distributors' Role in Food Safety

Many large processing companies operate their own distribution systems and can assure that their products are transported under conditions that prevent product contamination and ensure appropriate temperature control. Other companies must rely on outside companies that are in the business of transporting goods, which may not be limited to foods. Such companies may not be very knowledgeable in food safety. The shipper will generally specify conditions under which shipments must occur. While these conditions will include the use of clean conveyances and will specify temperature control where necessary, unless there is a means of monitoring these conditions, errors may occur. A major outbreak of salmonellosis from ice cream in 1994 was attributed to the transportation of pasteurized ice cream premix in a tanker that had been inadequately cleaned after carrying raw liquid eggs. In the 1960s, botulism from smoked fish was attributed to inadequate refrigeration during shipping. There have been few instances of illness from temperature abuse during transportation of perishable products, primarily because spoilage generally occurs and the products are not consumed; however, these losses cost the food industry a great deal of money every year. The food industry is beginning to look at ways to apply HACCP and other controls to the distribution segment to ensure the food reaching the retail and food service segments of the industry is safe and of high (maximum) quality.

D. Ensuring Safe Food at Retail

There are many opportunities for abuse of foods at the retail level. Refrigerated and frozen foods may be unloaded on the store's dock and stay there for hours before being placed at a proper temperature. Refrigerated display cases may be overloaded, so that some of the product is outside the cold zone. Cans and other containers may be handled so roughly that the integrity of their seams and/or seals is compromised. This is not to suggest that retail personnel deliberately abuse food. Although the retailers' knowledge of food safety may be limited, it is becoming more common for larger retailers to employ individuals with food safety knowledge. Today, many large retailers have food processing operations (e.g. dairies and bakeries) of their own and are increasingly preparing hot and cold foods on site (or in a central kitchen) for the convenience of consumers (note the variety of salads and prepared foods at the delicatessen counter, the roast chickens sold hot, the extensive salad bar, the prepared sandwiches). Although the implementation of HACCP in the retail sector is increasing, processors must continue to work with the retail sector to help ensure products are stored and displayed in a manner that ensures safety and quality are maintained, because the manufacturer of a product has ultimate responsibility for that product.

E. How Food Service Establishments Ensure Safe Food

Data on foodborne disease outbreaks indicate that the foods involved are primarily eaten at homes or at delicatessens, cafeterias, and restaurants, and that approximately twice as many outbreaks occur from eating out as eating at home. Food service establishments have been implicated in a number of very serious outbreaks such as the one in the Northwest in 1993 from hamburgers eaten at a fast-food chain. Food service operations in particular may suffer from the lack of food safety knowledge as a result of the low-wage, high-turnover nature of the business. In part because of this history of problems, the food service sector is second only to food processors in embracing HACCP for the control of food hazards. The Educational Foundation of the National Restaurant Association has developed its 'ServSafe' program to teach food service establishments how to apply HACCP in their operations. Moreover, the FDA (see Section IV.A.1) has developed a HACCP-based model *Food Code*, upon which state and local laws and regulations are based, for retail and food service operations. Compliance with the *Food Code* should minimize the potential for a foodborne disease outbreak.

F. The Role of Trade Associations and Professional Societies in Food Safety

Industry associations have made substantial contributions to food safety. Many of these trade associations have scientists and technical specialists on their staff to provide assistance to their members, and to interact with government agencies and legislators on food safety issues. Many trade associations are also actively involved in research on industry-wide problems. The National Food Processors Association (NFPA, formerly the National Canners Association) established its own laboratories in 1913 and developed industry-wide standards for the safe heat treatment of canned foods. NFPA's three laboratory centers continue their efforts to ensure the safe production of processed and packaged foods, in part through a research program conducted in the laboratories and in part through partnerships with university laboratories.

The food industry funds research on food safety at company laboratories and through its associations, including, but not limited to, NFPA and the NFPA Research Foundation, the American Meat Institute Foundation, National Cattlemen's Beef Association, National Fisheries Institute, and the International Life Sciences Institute. Such efforts are not unique to the US. Campden & Chorleywood Food Research Association (CCFRA) a large, independent food and drink research organization in the UK, serves as a major source of research and technical information on issues related to food and drink safety.

The food industry also works to enhance food safety through training courses in HACCP and other food-safety areas that are conducted by organizations such as NFPA's Food Processors Institute, the American Meat Institute, and the National Restaurant Association's Educational Foundation, as well as by industry alliances with academic institutions (the International HACCP Alliance – formerly the International Meat and Poultry HACCP Alliance – the Seafood HACCP Alliance, and the Juice

HACCP Alliance). CCFRA in Europe has an extensive training program and offers similar food safety courses.

Professional societies such as the Institute of Food Technologists (IFT) and the International Association for Food Protection (IAFP) conduct national and regional meetings, and publish journals that present current information on food safety. Their local affiliates or sections also hold food safety meetings. A comparable organization in the UK is the Institute of Food Science and Technology (IFST). Other groups, such as the American Public Health Association and the Association of Official Analytical Chemists (now AOAC International), publish standard methods for the microbiological examination of foods that are used worldwide by microbiologists in sampling and analyzing foods to verify that specifications have been met.

IV. Government's Role in Food Safety

One of the most important functions of government is to ensure public safety. By analogy to the police that protect the public against other kinds of hazards, there are agencies at each level of government that are responsible for various aspects of food safety. The discussion that follows will emphasize government activities in the US because these are most familiar to the authors. Clearly, there are many other ways to organize government activities in food safety to accomplish the task.

A. US Federal Agencies

1. Food and Drug Administration

The FDA is a unit of the Public Health Service of the Department of Health and Human Services and is directed by a Commissioner. Within FDA, the Center for Food Safety and Applied Nutrition (CFSAN) is responsible for ensuring that foods are safe, wholesome, and accurately labeled. The Agency has jurisdiction for all foods entering interstate commerce (including imports) except meat, poultry, and egg products (although FDA has responsibility for shell eggs). It operates under the authority of the Federal Food, Drug, and Cosmetic Act as amended, which prohibits the sale of adulterated or misbranded food. For purposes of the law, food is adulterated if it contains any poisonous or deleterious substance that could be injurious to health, contains any filthy or putrid substance or is otherwise unfit to eat, or has been prepared under unsanitary conditions whereby it might have been contaminated. Enforcement is conducted by a field force that is organized into regions and districts. FDA's 700 inspectors make unannounced inspections of food-processing facilities on an irregular basis (there are about 55 000 plants under FDA jurisdiction). They make notes of their observations and may collect samples for analysis by FDA laboratories. If they detect problems, they can get a court injunction, have the product seized, or 'suggest' that the product be recalled from distribution. The FDA regulatory approach is based partly on good manufacturing practice regulations, and more recently on regulations requiring HACCP for seafood and juice products. In addition to food processing and distribution, the FDA regulates eating establishments on common carriers, federal property, and interstate highways. The Agency conducts research in its own laboratories, funds some research at universities and other organizations, and drafts model ordinances and codes (e.g. the *Pasteurized Milk Ordinance* for grade A milk products and the *Food Code* for food service, retail food stores, and food-vending operations) for possible adoption by state and local governments.

2. US Department of Agriculture

USDA has jurisdiction over meat and poultry in interstate commerce and in intrastate commerce in about half the states, and over all plants producing egg products. Its authority derives from the Federal Meat Inspection Act, the Poultry Products Inspection Act, and the Egg Products Inspection Act. It also has quarantine responsibility for meats, vegetables, and fruits being imported into the US. Inspection activity is based in the Food Safety and Inspection Service (FSIS), whereas international quarantine is conducted by the Animal and Plant Health Inspection Service. FSIS conducts continuous, in-plant inspection of 6500 meat and poultry slaughtering and processing operations and egg-product processing plants. About 7600 inspectors are engaged in this work, and one or more is generally expected to be present during operating hours in larger meat and poultry slaughter and processing plants. FSIS also conducts examinations of warehouses and imported foods at point of entry into the US. In 1996, FSIS established new requirements for meat and poultry plants to operate under HACCP and to establish sanitation standard operating practices (SSOPs). In addition, slaughter plants must conduct testing for *Escherichia coli* to verify process control. FSIS conducts *Salmonella* testing in slaughter and raw meat grinding operations to assure that plants meet performance standards for this organism. As these requirements took effect between January 1997 and January 2000 (depending on the size of the establishment), FSIS began making fundamental changes in how it operates. The Agency is being reorganized to modernize its inspection service and is revising its existing regulations to be consistent with HACCP principles. Industry has assumed full responsibility for ensuring sanitary operations and safe food products, and FSIS performs a verification role, reviewing records and operations and taking samples for analysis as needed. This role is similar to that of an FDA inspector, except FSIS maintains continuous inspection. Countries exporting meat and poultry to the US are required to provide an 'equivalent' system. Failure to do so will lead to a ban on importation of meat and/or poultry products from individual countries. FSIS has announced plans to propose HACCP regulations for egg products.

The Animal and Plant Health Inspection Service (APHIS) is charged with responding to issues involving animal and plant health. Its mission is to protect America's animal and plant resources by safeguarding them from exotic invasive pests and diseases, monitoring and managing agricultural pests and diseases existing in the US, resolving and managing trade issues related to animal or plant health, and ensuring the humane care and treatment of animals. The APHIS mission is an integral part of USDA's efforts to provide the nation with safe and affordable food. Without APHIS protecting America's animal and plant resources from agricultural pests and diseases, threats to our food supply would be quite significant. APHIS inspectors operate at points of entry into the US (ports, airports, etc.) to prevent introduction of exotic insect pests and agents of plant and animal disease. APHIS also is involved in allowing for field testing of new plant varieties produced by genetic engineering.

In addition to the regulatory agencies within USDA, the Department has an in-house research arm, the Agricultural Research Service (ARS). ARS conducts research in support of the Department's regulatory activities and provides technical expertise to meet national food, food safety, and environmental emergencies.

3. Other US Federal Food Safety-related Activities

a. Centers for Disease Control and Prevention

The CDC, like FDA, is part of the Public Health Service of the Department of Health and Human Services. CDC supports surveillance, research, prevention efforts, and training in the area of infectious diseases through its National Center for Infectious Diseases. Within this Center is the Division of Bacterial and Mycotic Diseases, which contains the Foodborne and Diarrheal Diseases Branch, which focuses on the surveillance, outbreak investigation, and control and prevention of foodborne and diarrheal diseases. This Branch works closely with state health departments and serves as a consultative resource for federal agencies such as FDA and FSIS. It also works with the Pan American Health Organization and foreign governments to control and prevent diarrheal diseases. Laboratory sections provide reference, surveillance, and educational materials for cholera, shigellosis, and infections caused by *Salmonella*, *Listeria monocytogenes*, *Campylobacter*, and *E. coli*. The laboratories also evaluate methods for rapid diagnosis of enteric diseases, characterization of epidemic strains, and identification of virulence factors and toxins. The Biostatistics and Information Management Branch collects and processes surveillance data, and provides summaries. CDC periodically issues separate summaries of waterborne and foodborne disease outbreaks; the most recent report summarizes foodborne disease outbreaks for the years 1993–1997.

In 1995, CDC began a joint project with FSIS and FDA known as the Foodborne Diseases Active Surveillance Network, or FoodNet. This collaborative effort with state health departments and local investigators at a number of US locations (California, Oregon, Minnesota, Connecticut, and Georgia, with Maryland and New York added in 1998, Tennessee in 2000 and Colorado in 2001) is investigating diarrheal diseases to identify the incidence of foodborne illness and the risk factors for these illnesses more accurately. The project takes a proactive approach to gathering data on diarrheal illness through population-based surveys, laboratory surveillance, physician surveys, and case-control studies.

b. Environmental Protection Agency

The Environmental Protection Agency (EPA) sets safety goals and standards in the form of maximum contaminant limits for microbial and chemical contaminants in drinking water in the US. Because many water uses in food plants require the use of potable water and EPA's drinking water standards are the measure of potable water, EPA indirectly regulates food plants.

EPA also regulates pesticides and pesticide uses, including disinfectants and sanitizers used in food plants, which are defined by law as pesticides. All materials defined as pesticides used in and around a food plant must be registered with EPA. In addition, pesticides that may come in contact with food during agricultural production or post-harvest, as well as food contact-surface sanitizers, must have an EPA established tolerance (or exemption from a tolerance). Tolerances for pesticides are enforced by FDA.

c. National Marine Fisheries Service

The National Marine Fisheries Service (NMFS) of the National Oceanic and Atmospheric Administration, US Department of Commerce, manages and promotes the US's living marine resources. It operates a voluntary service to inspect, grade and certify domestic fishery products. In addition, NMFS has developed a HACCP-based inspection program that deals with safety, wholesomeness, and economic fraud (species substitution, overbreeding, etc.).

d. Bureau of Alcohol, Tobacco, and Firearms

The Bureau of Alcohol, Tobacco, and Firearms (ATF), part of the US Department of the Treasury, regulates the production and distribution of beverage alcohol in the US.

e. Department of Defense

The US Department of Defense (DoD) is responsible for the adequacy, quality, cost and availability of food, containers, and food service for the US armed forces. DoD contracts with the food industry to provide rations to feed US combat troops, as well as to supply base commissaries and food service operations. DoD supports a research program to apply new technologies to develop foods in packages to meet military requirements. Much of this research is carried out at the US Army Natick Research, Development and Engineering Center (now called the US Soldier System Command), in cooperation with industry.

B. State Agencies

State agencies have jurisdiction over food produced and distributed within their own boundaries. State departments of health and departments of agriculture generally carry out food-related responsibilities at the state level in accordance with laws passed by state legislatures. These laws may, in some cases, be more stringent than those of the federal government and may apply to foods traveling in interstate commerce when they are within that state.

1. Departments of health

Departments of health often are responsible for inspection of restaurants and milk-processing establishments, except where such activities are conducted by city or county agencies.

2. Departments of agriculture

Departments of agriculture may carry, at the state level, responsibilities comparable to those of the FDA and USDA combined. They can, if their state chooses to have them do so, inspect processing and distribution of all foods within the state, including those that will be sold outside the state. If the state does not undertake inspection of meat and poultry with a rigor comparable to that of the USDA, USDA is required by federal law to take over inspection even of meat and poultry destined for intrastate consumption.

3. Cooperation with federal agencies

Cooperation is established, where possible, between state and federal agencies to avoid duplication of efforts and to maximize efficient use of limited resources. Cooperative agreements may be established for analysis of food samples, to handle consumer complaints, or to conduct inspections in specific types of establishments within the state. Some agreements have been developed to allow a state regulatory agency to assume primary responsibility for all food manufacturing and storage industries within a state. (See Section IV.D below for other cooperative food safety efforts.)

C. Local Agencies

These include agencies of county and municipal governments, which function largely within their own political jurisdictions. Cities or counties may choose to supervise food service establishments and retail stores within their political boundaries. Many counties and larger cities have their own inspectors and laboratories to conduct inspection and testing, and perhaps to investigate outbreaks of foodborne illness. Interactions with state agencies depend on the local political climate. Relationships range from cordial and cooperative to strained.

D. Cooperative Food Safety Programs

One notable example of federal/state cooperation is the *National Conference on Food Protection*. In 1971, the American Public Health Association (APHA), through a grant from FDA, organized a conference with the objective of improving control of microbial contamination of food. The Second National Conference for Food Protection took place in 1984, and it began to evolve into the structure of councils and committees it uses today. Regulatory officials at the federal, state, and local level, with participation from industry, consumers, and academia, meet biennially to identify and address emerging problems of food safety and to formulate recommendations. Although the Conference has no formal regulatory authority, it is a powerful organization that influences model laws and regulations, and assures uniform interpretations and implementation of these among all government agencies. A major focus of the Conference on Food Protection is to update the *Food Code* to reflect the most current science and the best strategies to ensure safe food at retail and in institutions.

Another cooperative group, the *Association of Food and Drug Officials* (AFDO), established in 1896, fosters uniformity in the adoption and enforcement of food laws and regulations. The group works closely with industry to foster understanding and cooperation and to provide public health protection.

The *National Shellfish Sanitation Program* (NSSP) was established in 1925 in response to a typhoid outbreak traced to sewage-polluted oysters. The NSSP is a voluntary cooperative program of state and federal regulatory agencies and the shellfish industry that provides recommendations, through an NSSP Manual of Operations, to enhance shellfish safety. FDA coordinates and administers the NSSP. The NSSP Manual of Operations also serves as the basis for FDA 'certifying' foreign shellfish sanitation programs. Foreign countries can only export molluscan shellfish to the US if they have signed a Memorandum of Understanding with FDA agreeing to abide by the NSSP.

Further efforts to strengthen shellfish safety resulted in the formation of the *Interstate Shellfish Sanitation Conference* (ISSC) in 1982. The ISSC is composed of federal, state, and local regulatory agencies responsible for shellfish sanitation, the shellfish industry, and the academic community. The Conference holds annual meetings to discuss issues related to shellfish sanitation and incorporates appropriate changes into the NSSP Manual, which contains the sanitary requirements for the growing, harvesting, handling, and distribution of shellfish throughout the US and abroad. The ISSC developed a Model Ordinance containing the compliance requirements of the NSSP Manual of Operations written in regulatory language that can be easily adopted by states. The Model Ordinance, which can be found on the FDA website, presents guidance for the safe and sanitary control of molluscan shellfish, and is periodically updated in a manner similar to that for the *Food Code*. Through their participation in NSSP and membership in ISSC, states have agreed to enforce the Model Ordinance for molluscan shellfish.

The ISSC was patterned after another program, the *National Conference on Interstate Milk Shipments* (NCIMS), which operates to assure a nationwide safe and wholesome milk supply. NCIMS is a voluntary organization controlled by the member states and governed by an executive board whose members include representatives from state departments of health and agriculture, FDA, USDA, and industry. NCIMS holds biennial conferences and recommends modifications to the *Grade 'A' Pasteurized Milk Ordinance* (PMO, the standard for Grade 'A' milk that has been adopted by all 50 states, the District of Columbia, and the US Trust Territories) and related documents that specify sanitary standards, requirements, and procedures to ensure the safety and wholesomeness of Grade A milk and milk products. Anyone who wishes to ship milk must be certified by the state that they meet the standards of the PMO, which contains requirements for both the farm and the plant.

Other cooperative programs that bear mentioning have arisen from regulatory requirements that seafood, meat, and poultry be processed under HACCP. The HACCP regulations specify that certain functions be carried out by individuals trained in HACCP principles and their application. The *Seafood HACCP Alliance* and the *International HACCP Alliance* have produced standardized educational programs that meet these regulatory requirements. Academia, industry and regulatory representatives serve on the Seafood HACCP Alliance. Although members of the International HACCP Alliance come primarily from industry associations, industry educational foundations, and consulting groups, this Alliance works closely with the regulatory agencies and academia. With the publication in 2001 of HACCP regulations for juices, a *Juice HACCP Alliance* has also been established, also with representatives from industry, academia, and regulatory agencies.

E. Regulatory Bodies in Other Countries

1. Canada

The Canadian Food Inspection Agency Act, introduced in 1996, created the Canadian Food Inspection Agency (CFIA), which consolidates food inspection and animal and plant health services of Agriculture and Agri-Food Canada, Health Canada, and Fisheries and Oceans Canada. The new Agency, directed by a President who reports to the Minister of Agriculture and Agri-Food, began operation in April of 1997. CFIA's role is to enforce the food safety and nutritional quality standards established by Health Canada and, for animal health and plant protection, to set standards and carry out enforcement and inspection. The Agency covers all inspection services related to food safety, consumer fraud, and animal and plant health programs. CFIA also sets labeling and advertising policy and commodity standards for processed products. Responsibility for food safety policy, standard setting (except as noted for CFIA), risk assessment, analytical testing research, and audit resides in Health Canada.

2. The European Union

In the European Union (EU), the decision-making process is shared by the Commission (EU Executive Body), the Council (representatives from 15 member states), and the Parliament (elected by EU citizens). The EU develops harmonized community laws in areas where the existence of national laws of member states could affect the functioning of the internal market. Uniform food safety legislation is a prime candidate for harmonization, in order for member states to have confidence in the safety of foods and ensure free movement of goods. Directives are the primary means employed by the EU; adoption of a Directive is very complex, and involves the Commission, the Council and the Parliament of the EU. Directives are binding on the member states as to the objectives to be achieved, but the form and methods by which these are attained are left to the discretion of the member states. However, many of the directives in the area of food safety are spelled out in such detail that there is little flexibility for member states.

Food safety directives for red meat, poultry meat, wild and farmed game, minced meat, meat products, fish and shellfish, fishery products, egg products, and milk and milk-based products exist. Council Directive 93/43/EEC (June 14, 1993) on the hygiene of foodstuffs covers the preparation, processing, manufacturing, packaging, storing, transportation, distribution, handling, and offering for sale of foods not covered by product-specific Directives. The Food Safety (General Food Hygiene) Regulations 1995 of the UK formalize Directive 93/43/EEC; they set out basic hygiene principles and focus on identifying and controlling food safety risks at each stage in the process of preparing and selling food in accordance with the Food Hygiene Directive. All of the EU hygiene directives specify the use of hazard analysis, risk assessment, and other safety management techniques to identify, control, and monitor critical points for the production of foods. The Official Control of Foodstuffs

Directive (89/397/EEC), supplemented by the Additional Food Control Measures Directive (93/99/EEC), sets out how compliance with food safety directives will be monitored. Enforcement authorities in member states must conduct food hygiene inspections in a specific manner and are required to submit information annually on the number, frequency, and type of inspections carried out, together with details of any infringements. Each of the member countries of the EU has agencies, such as the Food Standards Agency in the UK, that have responsibility for food safety and enforce, where applicable, the EU Directives, as well as any regulations applying specifically within that country. Directive 89/397/EEC and other EU Directives were implemented in the UK through the Food Safety Act of 1990 and through statutory Codes of Practice that guide UK enforcement authorities.

In January 2000, the EU released a white paper on food safety describing the establishment of a European Food Authority to provide scientific advice to the European Commission on food safety issues. In November 2000, the Commission released a proposal for a regulation laying down the general principles of food law and establishing the European Food Authority (2000/0286), which passed in January 2002 (Regulation EC178/2002).

3. Australia

In Australia, food is the constitutional responsibility of each of the six State and two Territory Governments. The National Food Authority was established as a federal agency by legislation in 1991 to provide a national focus on food and food safety. In 1996, as the result of a treaty between Australia and New Zealand to adopt uniform food standards progressively, the National Food Authority became part of the Australia New Zealand Food Authority (ANZFA), a partnership between the New Zealand and the Australian Commonwealth, State, and Territory governments (In mid-2002 ANZFA will become Standards Australia New Zealand – FSANZ). ANZFA administers the Australia New Zealand Food Standards Code, which is a collection of individual food standards. The Food Standards Code has three major chapters that apply as specified to one or both of the countries. The Australia New Zealand Food Standards Council is the decision-making body for food standards. Standards adopted by the Council are published and adopted by reference into the food laws of New Zealand and the Australian States and Territories. ANZFA is also responsible for coordinating national food surveillance and recall systems, conducting research and developing codes of practice with industry.

The Australian Quarantine and Inspection Service (AQIS) is a federal agency responsible for quarantine and for the inspection of all food and agricultural products exported from Australia covered by the Export Control Act of 1982. Products such as meat, fish, dairy, and fresh fruit and vegetables must be issued an export permit by AQIS, which implies that the product meets the minimum standards for export and any additional requirements of the importing country. ANZFA develops and sets policy for food imported into Australia, which must comply with the Food Standard Code. AQIS carries out import inspection in accordance with policy determined by ANZFA. Similar services are provided in New Zealand by the Ministry of Agriculture and Forestry (MAFF) Quarantine Service.

4. Japan

Responsibility for food in Japan resides within the Ministry of Agriculture, Forestry and Fishery (produce, fish, food labeling policy), and the Ministry of Health and Welfare (Meat and Dairy Section, Food Hygiene and Inspection Section, Import Inspection Section). The ministries prepare regulations that go through the Diet (their legislative body), as well as food-related directives. Inspections are conducted at the local level by health and welfare officers.

F. International Agencies

Several international organizations concern themselves with various aspects of food safety. Cooperative efforts by many countries attempt to bring order in the chaotic field of international trade in food. These groups develop *advisory* recommendations and guidelines to national governments.

1. United Nations agencies

Some United Nations agencies carry out food safety activities on behalf of their member nations, which are usually also members of the United Nations.

a. World Health Organization

The World Health Organization (WHO) conducts many activities in the area of food safety. In addition to staff functions, it convenes 'expert committees' to consider topics such as food microbiology and hygiene and food additives. In May 2000, WHO adopted a food safety resolution to foster the development and implementation of food safety programs among member countries and to support food safety efforts at WHO.

b. Food and Agricultural Organization

The Food and Agricultural Organization (FAO) is concerned with the production and preservation of food worldwide. It interacts with the WHO in the areas of wholesomeness and safety and is the sponsor of Codex Alimentarius activities. It also convenes 'expert consultations', often in conjunction with WHO. Recent consultations have dealt with HACCP and risk management in food safety. In addition FAO and WHO, at the request of the Codex Committee on Food Hygiene, have established a joint program to conduct risk assessments for various pathogen/commodity combinations (e.g. *Salmonella* spp. in broilers/eggs and *Listeria monocytogenes* in ready-to-eat foods).

c. Joint FAO/WHO Codex Alimentarius Commission

The FAO/WHO Codex Alimentarius Commission ('Codex') was established in 1962 to implement the Joint FAO/WHO Food Standards Program. Codex develops international food standards and guidelines with two major purposes in mind: (1) protecting the health and economic interests of consumers; and (2) facilitating international trade. Codex has been responsible for setting a variety of commodity standards, hygiene practices, and maximum residue limits for pesticides. Over 160 countries participate in Codex.

2. International Commission on Microbiological Specifications for Foods

The International Commission on Microbiological Specifications for Foods (ICMSF) is a nonprofit, scientific advisory body established in 1962 under the auspices of the International Union of Microbiological Societies (IUMS). Membership consists of 16–20 food microbiologists from more than ten countries. ICMSF's goal is to foster the movement of microbiologically safe foods in international commerce. ICMSF has published a number of books that deal with: microbial significance and enumeration; sampling for microbiological analysis; microbial ecology of foods; HACCP; characteristics of microbial pathogens; and food safety management. These books have become standard reference books for food microbiologists throughout the world.

V. The Role of Universities in Food Safety

Without the research conducted at universities throughout the world, it would be difficult to know how to address many of the food safety issues that arise. Whenever a new foodborne pathogen is identified, it is necessary to determine its characteristics, its ecology, and how to control it. While the food industry and regulatory agencies conduct some research (USDA has the Agricultural Research Service, and FDA maintains research laboratories), the majority of research on foodborne pathogens and their control is conducted by university research laboratories. Much of this research is funded by government grants. For example, USDA's Cooperative State Research Education and Extension Service funds research in agriculture, food, and the environment at US institutions.

Many universities also have extension programs that provide advice to consumers and to industry within the state. It is through such extension programs that many small food businesses obtain the advice needed to ensure that the foods they manufacture are safe.

VI. Consumers' Role in Food Safety

A. Basic Self-protection

The consumer must recognize that most foods, especially raw foods of animal origin and foods that are grown in the dirt, are contaminated with microorganisms, and that some of these are pathogenic to humans. There is no zero risk. Consumers must take steps to educate themselves as to the potential risks and the ways to control them. They must learn that raw red meat may carry *Salmonella* and *E. coli* O157:H7; that raw poultry will likely carry *Salmonella* and *Campylobacter*; that oysters may be contaminated with hepatitis A virus or with *Vibrio vulnificus*; that vegetables may carry spores of *Clostridium botulinum* or enteric pathogens such as *Salmonella*. Proper handling and cooking practices will minimize the risks from these and other hazards. Immunocompromised persons must recognize they are at greater risk than the population in general and take appropriate steps to protect themselves. Persons with food allergies must learn to read food labels carefully and question restaurant personnel about the presence of specific food allergens in prepared dishes. As noted earlier in this chapter, the responsibility for food safety rests with everyone along the food chain, from the grower to the consumer. However, the consumer is the final critical link in the prevention of foodborne illness.

B. Shopping

Consumers should make a habit of reading the labels on the foods they buy. In addition to the product name and brand, virtually all packaged foods will contain a list of ingredients in descending order of predominance by weight and a nutrition label. There may also be special handling statements about storing the food ('keep refrigerated', 'refrigerate after opening', 'keep frozen until use', etc.), a shelf-life date serving as a quality guideline ('use by ...', 'best if used by ...', 'use within 7 days after opening') and cooking instructions, where applicable. Raw and partially cooked meat and poultry products carry a label explaining safe cooking, storing, and handling practices. Reading and understanding these labels is essential for consumers to have the highest quality food, and may be important for consumers to protect themselves: improperly cooked foods and refrigerated foods left on the shelf have resulted in illness on numerous occasions. Persons with food allergies and sensitivities must learn to recognize in the ingredient statement those ingredients that may trigger an adverse reaction for them. Consumers must also learn to recognize food that has been mishandled or is in poor condition: refrigerated foods should be cold to the touch and frozen foods should be hard; do not buy canned foods that have dented seams, bulge at the ends, or show signs of leakage (stained labels or product on the exterior). Tamper-evident seals, if present, should be intact. Consumers should use common sense in shopping: buy cold food last and get it home fast.

Grocery stores in recent years have made an effort to provide consumers with a variety of information about the foods they purchase from the store, including information on food safety. Such displays of information are often found at the meat counter, outside the manager's office and at the store entrances and exits.

C. Food Storage, Preparation, and Handling at Home

Groceries should not be left in the trunk of the car or on the kitchen counter for an extended period of time: put refrigerated and frozen foods in the refrigerator or freezer as soon as possible. One should pay careful attention to labels indicating the need for refrigeration – several outbreaks of botulism have been caused by refrigerated foods that were inadvertently stored at room temperature.

Consumers should use appliance thermometers to check the temperature of the refrigerator (it should be about $40°F \approx 4°C$) and freezer (it should be close to $0°F \approx -18°C$). Fresh meat, poultry, and seafood should be frozen if it will not be used in 2–3 days. Juices from raw meats must not drip on other foods, especially ready-to-eat foods.

Food preparation practices in the home are often the key to safe eating. Foods must be properly cooked to destroy harmful bacteria. A general rule would be to cook red meats (beef, pork, veal and lamb) to 160°F (~71°C) and poultry to 180°F (~82°C). A meat thermometer will indicate that proper temperatures have been reached. If no thermometer is available, one should cook until the red is gone and juices run clear. (However, studies have shown that with hamburger this is an inadequate indicator that all pathogens have been killed; consumers should use a meat thermometer.) In poultry, one should check for the absence of pink color near the joints. In the US, many consumers prefer red meats to be cooked rare or medium rare. This may be acceptable for a steak, where contamination is primarily on the surface, but not for ground beef, where the bacteria are distributed throughout the product. Studies have shown that cooking to the medium-well to well-done stage is necessary to ensure destruction of *E. coli* O157:H7. Eggs must also be thoroughly cooked until the yolk and white are firm, not runny, to destroy *Salmonella*. Recipes in which eggs are not cooked or are only partially cooked should be avoided, or pasteurized, liquid egg products may be substituted.

Consumers' biggest mistakes occur through cross-contamination: a utensil or cutting board or plate that has been used for a highly contaminated item such as raw meat is subsequently used for a product that will be eaten without further heat treatment (as when the same cutting board is used for raw poultry and then, with little or no washing, for salad vegetables; or when the grilled chicken is put back into the dish in which the raw chicken was marinated). Cross-contamination also occurs when liquid from raw meat or poultry inadvertently contacts ready-to-eat items, such as when the cook puts raw meat in the skillet and then prepares a salad without washing his hands. Handling ready-to-eat foods without adequate hand-washing after changing the baby's diapers or petting the dog (or the family iguana) also can result in contamination with pathogens such as *Salmonella*.

Consumers need to assure proper care of leftovers. They should be refrigerated within 2 hours. Large amounts should be divided into small, shallow containers so they will cool quickly. Leftovers should be thoroughly reheated and should be used within a few days, or frozen. If they look or smell strange, one should not taste them. 'When in doubt, throw it out!' is a good adage to follow.

D. Amateurs in the Food Network

Many episodes of foodborne illness can be traced to improper food preparation practices by nonprofessional food handlers who have prepared foods for a group function such as a picnic, a party or a 'potluck' dinner. Foods for such occasions may be prepared by persons who lack the facilities, skills, and food safety knowledge necessary for large-scale food preparation. Large amounts of food may remain for extended times at temperatures that allow harmful bacteria to multiply. Food preparation and consumption may take place at sites that have inadequate facilities for sanitation and hot or cold holding of food. Consumers must be knowledgeable enough about practices that can result in foodborne illness to avoid certain foods at such gatherings ('cold' food that isn't, foods that have been insufficiently cooked, 'hot' foods that have been sitting for several hours at room temperature) and to make sure that they do not contribute to illnesses by providing a dish that has been improperly prepared or temperature abused.

E. Rights and Responsibilities

Consumers have the right to expect that the foods they purchase are safe, unadulterated, and properly labeled. However, as noted earlier in this chapter, absolute safety cannot be attained. Consumers must recognize this and learn what can and cannot be expected from the foods they purchase. Raw foods of animal origin should be expected to contain pathogens such as *Salmonella*; cooked foods of animal origin should not. However, the amount of contamination on raw foods such as meat and poultry has become an issue being argued in both the political and the scientific arenas. Outbreaks of foodborne illness on raw produce such as lettuce, raspberries, tomatoes, melons, alfalfa sprouts, etc. demonstrate that contamination problems are not limited to foods of animal origin. Controlling pathogens on such products will be much more difficult, and will entail a significant amount of research to establish the source of contamination and the appropriate controls. There is no easy solution at this time. To make all foods safe for all consumers would require sterilization of foods, a solution that is neither acceptable nor desirable. Furthermore, recontamination after sterilization could still result in foodborne illness.

Consumers have ways to participate in decisions concerning the safety of their food. They can voice complaints to the manufacturer of a product or to the manager of the store in which a product was purchased. On larger food safety issues, they can write to the FDA or USDA. If the concerns are related to food safety legislation, consumers can contact their elected officials. All of this has become much simpler thanks to the Internet. Consumers can even send e-mail to the President at the White House! There are also a number of groups that represent consumers in food safety-related matters. However, consumers should investigate carefully the goals and strategies of any consumer group they join to ensure that the views expressed accurately represent the consumer's position. Many of these groups are effective at getting their rhetoric in print and on the airwaves, but they may not necessarily serve the interests of those who wish to continue to have an abundant, varied, safe food supply at reasonable cost.

F. Information Needs and Sources

There is an incredible amount of food safety information available, almost all of it in the public domain. Unfortunately, much food safety research is published in scientific journals and may never be presented to the public in a useful way. The need for food safety information presented in terms the public can understand and apply is tremendous. We also lack effective means of disseminating the information that has been developed (primarily by government agencies such as USDA). In fact, both USDA and FDA have 'hotlines' that consumers can call for information. USDA's

Meat and Poultry Hotline offers one-on-one personal food safety advice, as well as a fax-on-demand service for food safety information. FDA's hotline provides access to recorded messages; public affairs specialists are available for consultation during specific hours.

Ready access to food safety information is undergoing a major transformation, thanks to the Internet. 'Surfing the net' will lead to a surprising amount of excellent information on food safety (see the Internet URLs), as well as much that is worthless. One must consider the source of the information to assess its credibility. Individuals offering food safety-related recommendations should be able to provide scientific back-up for these recommendations. Not everything can be taken on faith – information must be evaluated critically. One can look for connections to other sites that may be able to substantiate the information. As food safety comes to the forefront in the US and elsewhere, the amount of educational information on food safety will increase. An educated consumer has a critical role in advancing and enhancing the safety of our collective food supply.

VII. Summary

Safe food does not happen by accident, nor is the government responsible for making it safe. The producer, the processor, the distributor, the retailer, the consumer, all play a role in food safety. The US enjoys one of the most abundant, diverse, inexpensive, and safest food supplies in the world. However, there is room for improvement, especially in communicating to consumers ways in which they can protect themselves. If the food industry, the government and consumers all accept their roles in food safety, whether in the US or in other countries, we can expect a safe food supply. If any one group abdicates its responsibility, this expectation is unlikely to be met.

Bibliography

FDA Center for Food Safety and Applied Nutrition Outreach and Information Center: 888-SAFEFOOD (888-723-3366); recorded messages 24 hours a day; information specialists available 10 a.m. to 4 p.m. Eastern time, weekdays.

USDA Meat and Poultry Hotline: 800-535-4555 (202-720-3333 in the Washington, DC area); recorded information 24 hours a day; staff 10 a.m. to 4 p.m. Eastern time, weekdays.

Internet URLs

Australia New Zealand Food Authority: http://www.anzfa.gov.au
Canadian Food Inspection Agency: http://cfia.acia.agr.ca
Centers for Disease Control and Prevention: http://www.cdc.gov

Food Allergy & Anaphylaxis Network: http://www.foodallergy.org

Food and Drug Administration: http://www.fda.gov

FDA Consumer Advice: http://www.cfsan.fda.gov/~lrd/advice.html

Government Food Safety Information: http://www.foodsafety.gov

United States Department of Agriculture: http://www.usda.gov

USDA FSIS: http://www.usda.gov/fsis

USDA consumer publications: http://www.fsis.usda.gov/oa/pubs/consumerpubs.htm

    'Foodborne Illness: What Consumers Need to Know' (September 2000): http://www.fsis.usda.gov/oa/pubs/fact_fbi.htm

    'Basics for Handling Food Safely' (May 2001): http://www.fsis.usda.gov/oa/pubs/facts_basics.htm

    'Cooking for Groups: A Volunteer's Guide to Food Safety' (August 2001): http://www.fsis.usda.gov/oa/pubs/cfg/cfg.htm

    'Use a Food Thermometer' (April 2000): http://www.fsis.usda.gov/oa/thermy/bro_text.htm

USDA/FDA Foodborne Illness Education Information Center: http://www.nalusda.gov/foodborne/index.html

World Health Organization Food Safety Program: http://www.who.int./fsf

Index

Food Science and Technology
International Series

Maynard A. Amerine, Rose Marie Pangborn, and Edward B. Roessler, *Principles of Sensory Evaluation of Food*. 1965.

Martin Glicksman, *Gum Technology in the Food Industry*. 1970.

Maynard A. Joslyn, *Methods in Food Analysis*, second edition. 1970.

C. R. Stumbo, *Thermobacteriology in Food Processing*, second edition. 1973.

Aaron M. Altschul (ed.), *New Protein Foods:* Volume 1, *Technology, Part A*—1974. Volume 2, *Technology, Part B*—1976. Volume 3, *Animal Protein Supplies, Part A*—1978. Volume 4, *Animal Protein Supplies, Part B*—1981. Volume 5, *Seed Storage Proteins*—1985.

S. A. Goldblith, L. Rey, and W. W. Rothmayr, *Freeze Drying and Advanced Food Technology*. 1975.

R. B. Duckworth (ed.), *Water Relations of Food*. 1975.

John A. Troller and J. H. B. Christian, *Water Activity and Food*. 1978.

A. E. Bender, *Food Processing and Nutrition*. 1978.

D. R. Osborne and P. Voogt, *The Analysis of Nutrients in Foods*. 1978.

Marcel Loncin and R. L. Merson, *Food Engineering: Principles and Selected Applications*. 1979.

J. G. Vaughan (ed.), *Food Microscopy*. 1979.

J. R. A. Pollock (ed.), *Brewing Science*, Volume 1—1979. Volume 2—1980. Volume 3—1987.

J. Christopher Bauernfeind (ed.), *Carotenoids as Colorants and Vitamin A Precursors: Technological and Nutritional Applications*. 1981.

Pericles Markakis (ed.), *Anthocyanins as Food Colors*. 1982.

George F. Stewart and Maynard A. Amerine (eds.), *Introduction to Food Science and Technology*, second edition. 1982.

Malcolm C. Bourne, *Food Texture and Viscosity: Concept and Measurement*. 1982.

Hector A. Iglesias and Jorge Chirife, *Handbook of Food Isotherms: Water Sorption Parameters for Food and Food Components*. 1982.

Colin Dennis (ed.), *Post-Harvest Pathology of Fruits and Vegetables*. 1983.

P. J. Barnes (ed.), *Lipids in Cereal Technology*. 1983.

David Pimentel and Carl W. Hall (eds.), *Food and Energy Resources*. 1984.

Joe M. Regenstein and Carrie E. Regenstein, *Food Protein Chemistry: An Introduction for Food Scientists*. 1984.

Maximo C. Gacula, Jr., and Jagbir Singh, *Statistical Methods in Food and Consumer Research*. 1984.

Fergus M. Clydesdale and Kathryn L. Wiemer (eds.), *Iron Fortification of Foods*. 1985.

Robert V. Decareau, *Microwaves in the Food Processing Industry*. 1985.

S. M. Herschdoerfer (ed.), *Quality Control in the Food Industry*, second edition. Volume 1—1985. Volume 2—1985. Volume 3—1986. Volume 4—1987.

F. E. Cunningham and N. A. Cox (eds.), *Microbiology of Poultry Meat Products*. 1987.

Walter M. Urbain, *Food Irradiation*. 1986.

Peter J. Bechtel, *Muscle as Food*. 1986.

H. W.-S. Chan, *Autoxidation of Unsaturated Lipids*. 1986.

Chester O. McCorkle, Jr., *Economics of Food Processing in the United States*. 1987.

Jethro Japtiani, Harvey T. Chan, Jr., and William S. Sakai, *Tropical Fruit Processing*. 1987.

J. Solms, D. A. Booth, R. M. Dangborn, and O. Raunhardt, *Food Acceptance and Nutrition*. 1987.

R. Macrae, *HPLC in Food Analysis*, second edition. 1988.

A. M. Pearson and R. B. Young, *Muscle and Meat Biochemistry*. 1989.

Dean O. Cliver (ed.), *Foodborne Diseases*. 1990.

Marjorie P. Penfield and Ada Marie Campbell, *Experimental Food Science*, third edition 1990.

Leroy C. Blankenship, *Colonization Control of Human Bacterial Enteropathogens in Poultry*. 1991.

Yeshajahu Pomeranz, *Functional Properties of Food Components*, second edition. 1991.

Reginald H. Walter, *The Chemistry and Technology of Pectin*. 1991.

Herbert Stone and Joel L. Sidel, *Sensory Evaluation Practices*, second edition. 1993.

Robert L. Shewfelt and Stanley E. Prussia, *Postharvest Handling: A Systems Approach*. 1993.

R. Paul Singh and Dennis R. Heldman, *Introduction to Food Engineering*, second edition. 1993.

Tilak Nagodawithana and Gerald Reed, *Enzymes in Food Processing*, third edition. 1993.

Dallas G. Hoover and Larry R. Steenson, *Bacteriocins*. 1993.

Takayaki Shibamoto and Leonard Bjeldanes, *Introduction to Food Toxicology*. 1993.

John A. Troller, *Sanitation in Food Processing*, second edition. 1993.

Ronald S. Jackson, *Wine Science: Principles and Applications*. 1994.

Harold D. Hafs and Robert G. Zimbelman, *Low-fat Meats*. 1994.

Lance G. Phillips, Dana M. Whitehead, and John Kinsella, *Structure-Function Properties of Food Proteins*. 1994.

Robert G. Jensen, *Handbook of Milk Composition*. 1995.

Yrjö H. Roos, *Phase Transitions in Foods*. 1995.

Reginald H. Walter, *Polysaccharide Dispersions*. 1997.

Gustavo V. Barbosa-Cánovas, M. Marcela Góngora-Nieto, Usha R. Pothakamury, and Barry G. Swanson, *Preservation of Foods with Pulsed Electric Fields*. 1999.

Ronald S. Jackson, *Wine Science: Principles, Practice, Perception*, second edition. 2000.

R. Paul Singh and Dennis R. Heldman, *Introduction to Food Engineering*, third edition. 2001.

Ronald S. Jackson, *Wine Tasting: A Professional Handbook*. 2002.

Malcolm C. Bourne, *Food Texture and Viscosity: Concept and Measurement*, second edition. 2002.